POLLINATORS & POLLINATION

POLLINATORS & POLLINATION

Nature and Society

JEFF OLLERTON

PELAGIC PUBLISHING

Published by Pelagic Publishing
PO Box 874
Exeter
EX3 9BR
UK

www.pelagicpublishing.com

Pollinators & Pollination: Nature and Society

ISBN 978-1-78427-228-9 *Paperback*
ISBN 978-1-78427-229-6 *ePub*
ISBN 978-1-78427-230-2 *PDF*

A CIP record for this book is available from the British Library

Front cover: Buff-tailed bumblebee (*Bombus terrestris*) feeding on Clover (*Trifolium*) flowers in unmown lawn, Monmouthshire, Wales, UK. June. © Phil Savoie/naturepl.com

Rear cover: Left photo shows a female Dawson's burrowing bee (*Amegilla dawsoni*) pollinating camel bush (*Trichodesma zeylanicum*); on the right is a male great carpenter bee (*Xylocopa aruana*) pollinating snake vine (*Hibbertia scandens*). Both photos were taken in Australia, © Kit Prendergast. Centre photo shows a peacock butterfly (*Aglais io*) pollinating creeping thistle (*Cirsium arvense*) in Britain, © Mark Hindmarch.

This book is dedicated to my family: Karin, Ellen, Oli, Patrick and James,
whose love and support I cherish – thank you!

Contents

Preface

In writing *Pollinators & Pollination: Nature and Society* I have tried to provide a personal, state-of-the-art overview of what pollinators are, where they are to be found, how they contribute to the pollination of both wild and agricultural plants, supporting the wild ecosystems and agricultural yields on which our civilisations depend, and how individuals, governments and organisations of all kinds can and do play a role in their conservation. While the book is aimed at a very broad audience, and is intended to be comprehensible to anyone with an interest in science and the environment, and their intersection with human societies, I hope it will also be of interest to those dealing professionally with plants and pollinators. The subject is vast, and those working on bee or hoverfly biology, for example, or plant reproductive ecology, may learn something new about topics adjacent to their specialisms. I certainly learned a lot from writing the book.

The range of species and examples that I cover reflects my own interests and experiences. Inevitably such a personal assessment of the current state of such an extensive subject will miss out topics and studies that others will see as being crucial to the wider implications of how important pollinators are to the natural world and to human society. There is also, given my nationality and where I live and work, a bias towards Britain and the rest of Europe. To those who feel neglected, I apologise: there are only so many pages in a book, and only so many hours in a day.

Any errors of fact are my own responsibility; I would appreciate these being pointed out to me, and I will aim to provide regular updates on my blog. Any differences of opinion are an opportunity for discussion that will move the field on and increase our understanding of a subject that has fascinated me for over thirty years and in which there is still much to be discovered.

So welcome, reader, to my book. Are you sitting comfortably, perhaps with a cup of coffee in hand? Thank the pollinators, and their trillions of flower visits, for your beverage of choice. Are you wearing cotton clothes? Again, thank the pollinators for supporting a fabric industry worth many billions of dollars. Perhaps you have been given this book as a Christmas gift? Did you know that pollinators are responsible for adding aesthetic and financial value to the holly and mistletoe that decorate your house, and a high proportion of the food in your Christmas dinner is the result of insect visits to crop flowers? Have I begun to convince you of the importance of pollinators and the pollination services they provide? Then read on!

Acknowledgements

Academics at universities interact with a lot of people: their students, colleagues at their institution, and other academics further afield with whom they engage through email, conferences, social media, and via publications. If I had to thank, individually, everyone who has in some way made a contribution to this book and my career, it would probably exceed the word count. And I would no doubt miss out a lot of names. So please forgive these blanket thanks and acknowledgements, but you know who you are: my colleagues and students at the University of Northampton for their constant support and stimulation; my many collaborators, co-authors and correspondents over the years; everyone who has read and commented on my blog (jeffollerton.co.uk), much of which provided inspiration for this book; the attendees of the annual Scandinavian Association for Pollination Ecology meetings (once a SCAPEr, always a SCAPEr); the friends and colleagues I've made from serving on regional project boards and committees, especially the Northamptonshire Local Nature Partnership, the Nene Valley Nature Improvement Area project, and the Nenescape project; those colleagues who critically read draft chapters and stopped me making some embarrassing errors; everyone who gave me permission to use images, unpublished data and information; and the people at Pelagic Publishing, especially Nigel Massen for his patience with a book that long exceeded its contractual deadline, and Hugh Brazier for his astonishing skills as a copy-editor. Finally, thanks to the many, many organisations and institutions, large and small, who have funded my travel and research over the past thirty years, including the University of Northampton. Parts of this book were completed during a short sabbatical at the University of New South Wales, and I thank them for awarding me a Visiting Research Fellowship.

Chapter 1

The importance of pollinators and pollination

It's easy to underestimate, and impossible to exaggerate, the importance of pollinators and the pollination services they provide to plants.

As a scientist I'm expected to weigh evidence dispassionately and come to conclusions based on that evidence, without exaggeration or hyperbole. Statements that I make should be balanced, nuanced, carefully considered and derived from the facts. With that in mind, in my opinion, the opening sentence of this chapter is a careful judgement of the importance of pollinators and what they do. I strongly suspect that they are more important than we currently know, and we currently know that they are hugely important. That importance goes far beyond simply the production of seeds and fruit for human consumption, or indeed beyond sustaining plant populations. Most ecosystems of land plants, animals, fungi and microbes are ultimately reliant upon the flower-visiting activities of pollinators. They are essential both for the functioning of these ecosystems from year to year, and, in the long term, for the evolution of biological diversity. In this book I will explore these ideas from the perspectives of the ecology and evolution of the pollinators and the plants that they service. I will also go beyond the fundamental science to consider the role of pollinators in human society, and the threats to them and the pollination services they provide to agriculture. Along the way we will also consider the influence of their interactions with flowers on culture, and their political power. But before we get to all that, we need to begin with a consideration of what a pollinator, and pollination, actually are.

Pollination and the pollinators that perform it

Plants move, but they do so slowly, growing gradually upwards and outwards, or taking generations to disperse their seeds long distances. That is not a good strategy for finding a mate, especially if you wish to reproduce sexually every year, as most plants do. This is where pollination by an external vector comes in: rather than moving to find mates, plants entrust their pollen to a third party to ensure delivery to a suitable partner. As we'll see, in most cases this involves co-opting animals (vertebrates and invertebrates) to their service by attracting and rewarding them, what we term a 'mutualism' or mutualistic interaction. Some plants deceive their pollinators and give no reward. Other plants have dispensed with animals and use wind or water, of self-pollinate, to ensure reproduction.

The majority of flowering plants (with some exceptions) have both male and female reproductive organs within their flowers (Figure 1.1). The male organs (anthers) produce pollen, which contains the male sex cells (gametes) that are broadly equivalent to sperm in animals. When pollen arrives on the stigma, *this* is the act of pollination. At this point the pollen germinates and sex cells pass down the style to the ovary, where they fertilise the female sex cells contained within the ovules. Thus, plant sex and animal sex show some similarities, but also many differences. To begin with, many plants (because they are co-sexual) can self-reproduce; most animals cannot do this, though there are for example some hermaphroditic slugs and snails. This self-pollination, in which pollen from a flower lands on the stigma of the same flower, is fine over a few generations, or as a temporary way to ensure reproduction. Over the long term, however, it can cause inbreeding problems and loss of genetic diversity. Though there are examples of plants that seem to be perpetual inbreeders, suffering no obvious ill effects, pollen exchange between different individuals of the same species is much more common. That's why so many species have evolved strategies to achieve cross-pollination. I'll explore this in more detail in Chapter 3.

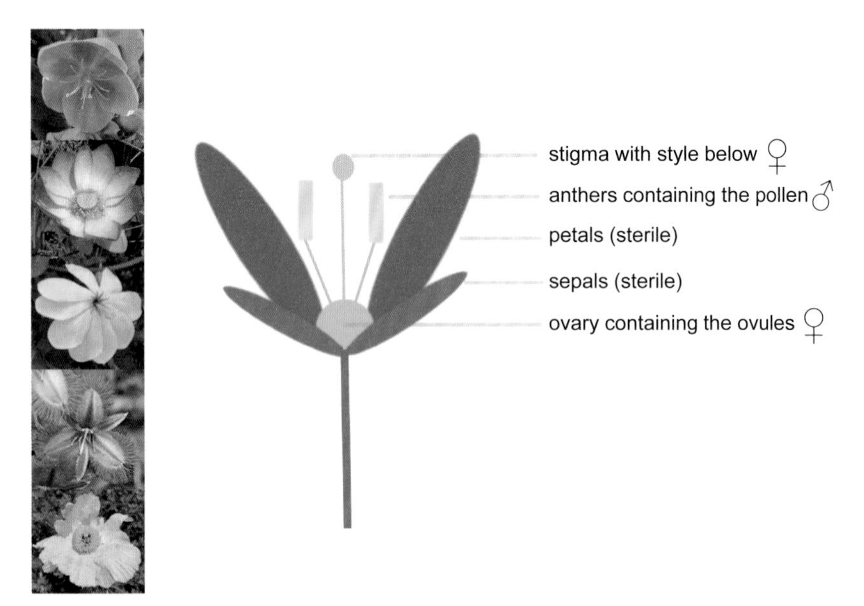

FIGURE 1.1 The basic structure of a typical flower showing the female, male and sterile parts. Also shown are five examples of flowers with open flowers and/or exposed sexual parts, demonstrating how variable the reproductive structures can be. From top to bottom they are: *Helleborus argutifolius, Nelumbo nucifera, Gardenia thunbergia, Thysanotus tuberosus, Romneya coulteri.*

FIGURE 1.2 The wind-pollinated flowers of a sedge (*Cyperus* sp.). Note the exposed, dangling stigmas and anthers, and the lack of a colourful perianth, all hallmarks of wind-pollinated flowers.

There are broadly three ways in which to move pollen from one plant to another: using air or water currents, or using animals. Water pollination is rare and restricted to a small number of aquatic plants. Wind pollination is more common, but still a minority strategy in terms of numbers of species, though not in total numbers of plants: anyone who suffers with hay fever knows that the grasses, trees and other plants that produce wind-blown pollen are far from rare (Figure 1.2). Pollination by animals, including vertebrates such as birds and bats, and invertebrate insects, is by far the most common strategy. It is this that we will focus on in the rest of the chapter. We will begin by examining the ecological and evolutionary importance of pollinators, and their value to agriculture (including food security and human health). These ideas will be explored in more detail in later chapters.

A diverse set of animal pollinators (see Chapter 2) is crucial for the reproduction of most plants. By our best current estimate, 87.5% of the *c.*352,000 living species of flowering plants have evolved to be partially or wholly dependent upon them (Ollerton *et al.* 2011), a statistic that I will examine in detail below. A smaller, but not dissimilar, proportion (about 75%) of the world's main crop species are animal-pollinated (Klein *et al.* 2007), but that in itself is not a surprise because the crop plants were originally derived from wild species. That domestication has fed human societies – which in turn, through art, literature, music and science,

FIGURE 1.3 Hummingbirds visiting a flower on a piece of pre-Columbian South American pottery.

have scrutinised the behaviour of pollinators, and the adaptations plants use to attract and reward them. Observations of pollinators and flowers have been appropriated, directly and indirectly, by human creativity into the cultures of societies that are supported by the ecological interaction of pollination. Thus, we can see a continuous chain of influence and importance from ecological functions that long pre-date human evolution, through to present-day agriculture and broader cultural influences. Animal-pollinated flowers have always played a huge role in human societies from the earliest times, including burial rites that go back far into prehistory. However, the cultural value of pollinators themselves, while implicit in the ecology and evolution of flowers, is also explicitly expressed in a variety of ways. These insects are part of our language ('a bee in my bonnet', 'busy as a bee', 'butterflies in my stomach'), our music (Rimsky-Korsakov's 'Flight of the Bumblebee', 'Hummingbirds', an instrumental track by American psychedelic band Love, and more recently my colleague Lars Chittka's efforts to raise money for the charity Buglife, as featured at killerbeequeens.bandcamp.com/releases), and our literature (such as the John Clare poem that opens Chapter 10).

Pollinators also feature in the visual arts, and have done for thousands of years, beginning with cave art of bees and honey hunters that may be as much as 10,000 years old (Dams and Dams 1977). Butterflies and bees are illustrated in ancient Egyptian paintings and sculpture, hummingbirds visiting flowers were often depicted on pre-Columbian pottery (Figure 1.3), and these sorts of motifs have persisted to the present day. And of course we appropriate the aromas of flowers for scents, both naturally by the distillation of odour compounds from petals, and artificially, in chemically synthesised air fresheners that loosely mimic natural fragrances. I'll come back to these cultural aspects in several subsequent chapters. However, I suspect that these ubiquitous, embedded cultural references to pollinators, and the flowers they pollinate, are one reason why most people do not explicitly appreciate the importance of pollinators.

What makes an animal a pollinator?

At first glance this may seem to be a trite question: clearly, visiting a flower is a prerequisite for an animal to be a pollinator. But that is not enough; just landing on a flower is insufficient to begin the chain reaction that finally results in the production of a seed. The evidence required to determine the effective pollen vector of a plant was formalised by Paul Alan Cox and Bruce Knox and is termed the 'Cox–Knox postulates', analogous to 'Koch's postulates' in medicine (Cox and Knox 1988). The Cox–Knox postulates state that, under field conditions (i.e. not in the laboratory or garden) the following must be observed:

1. transfer of pollen onto a vector
2. transport of pollen by that vector
3. transfer of pollen from the vector to the stigma of a flower
4. the deposited pollen effecting fertilisation of ovules

In relation to animals, as a starting point, it must come into contact with a flower in the correct way, orientated in such a fashion that it touches the reproductive parts of the flower (Figure 1.4). Assuming that the flower visitor is indeed such a potentially legitimate pollinator, as opposed to a visitor that just steals nectar or pollen but does not transfer pollen, visiting a single flower is also not enough. Pollination of that flower may occur if the insect is already carrying pollen from that species, but in order to further transfer pollen it must move to another of the same species. Some pollinators, including species of bees and butterflies, are very good at this as they develop behavioural routines in which they search for flowers of the same size, colour and smell. Other flower visitors, including many flies and beetles, are less fussy and move between any flowers in the vicinity, if they move at all. That said, there are exceptions to this: fussy flies that seek certain flowers, and bees that are less than loyal. One of the things that you will discover in this book is that, while there are many broad generalities that we can confidently state about pollinators and pollination, there are also usually exceptions to those statements. Something else that may surprise some readers is the sheer diversity of pollinators, not only in terms of the number of species, but also in the types of animals that act as pollen vectors. As I'll show in Chapter 2, bees are only part of

FIGURE 1.4 Examples of legitimate flower visits by pollinators in which some part of the insect touches the reproductive structures of the flower. Top left images by Judith Trunschke; bottom left image by David James; right image by Kit Prendergast.

this diversity, and not even the most abundant. Bees, some flies, birds, butterflies and moths you may be prepared for. But lizards? Cockroaches? Wasps? Marine worms? The list is surprisingly long and grows every year as we discover new types of interactions between flowers and animals.

For any of these pollinators to be effective, body size, shape and hairiness (or featheriness) can also be important. One often hears or reads about insects being 'adapted to be good pollinators' because they possess certain features. A good example of this are the bumblebees (*Bombus* spp.), which are large and hairy 'to make them effective pollinators'. Clearly this is not the case: they are large and hairy mainly because they are adapted to cold conditions where larger body size is an advantage, and because they can generate some heat internally, so the hairs provide insulation. Although the hairs do pick up a lot of pollen that can then be groomed into pollen baskets, this is not the main function of the hairs. For most insects there is no immediate advantage to them in pollinating a particular flower, and all they are concerned with is obtaining food such as nectar and pollen from those flowers. Active pollination, in which a pollinator deliberately places pollen onto the stigma of a flower, is extremely rare. In fact it is known only from seed parasite pollination systems such as those involving fig wasps and figs, and yucca moths and their yucca flowers (see Chapter 5). In these cases the insects lay their eggs within the ovules of a flower where their developing larvae can feed on some of the fertilised seeds as they grow. Thus the flower's reward for the insects is a food supply for their young, and it's in the female insect's interest to provision her offspring by pollinating the flower to produce seeds.

As we will see in Chapter 4, many of the animals that we see visiting flowers do not pollinate those flowers, or do so poorly, or are actually interested in eating the flowers or the developing seeds. Determining which animals are actually pollinators is a time-consuming task and requires careful observations and experiments, but can certainly be done by amateur naturalists with no specialist equipment (see Chapter 14). Anyone who spends time looking at flowers and their visitors will soon begin to appreciate how ecologically ubiquitous and important they are. However, they may be surprised to learn that, although the details of the process of pollination have been known for over two hundred years, hard, quotable evidence to back up its ecological importance has only recently been published, as I'll explain in the next section.

The ecological importance of pollinators

As I mentioned at the start of this chapter, scientists such as myself need to base our judgements on the best evidence available. But sometimes that evidence is not available, or at least not as firm as we might think. Data on the importance of pollinators is a good example. In 2010 I was writing a research paper on pollinators and found myself in a quandary. I wanted to provide a figure for the global ecological importance of pollination by animals, and specifically what proportion of the world's 352,000 or so flowering plant species was actually animal-pollinated. I had seen lots of proportions cited in the scientific literature, ranging from 67% to 96%, and lots of vague statements about 'most' and 'the majority' of plants needing pollinators. So I did what any good scientist should do when faced with a range of published answers: I followed the cited references back to their sources to assess how the figures had been arrived at. Frustratingly, it seemed that none of the numbers had any solid basis. The 67% estimate turned out to be a mis-citation of a different estimate, and other figures related just to surveys of trees or were based on what strategies were found in different plant families, often assuming that all families used a single strategy. That's bound to lead to inaccuracies because, for example, we know that even in archetypical wind-pollinated families, such as the sedges (Cyperaceae) and grasses (Poaceae), some species have evolved insect pollination. The reverse is true in families such as the poppies (Papaveraceae) and the daisies (Asteraceae), so clearly evolutionary flexibility abounds. More worryingly, some of the higher estimates disappeared like fast-flying bees in search of flowers. They were guestimates based purely on what the scientist in question had decided was 'about right', founded on personal observations. While valid as an opinion, this really needed to be backed up with some firm data.

As with many of the simple and obvious questions in ecology, the assumption was that we 'just know' the answer. Scientists sometimes have a tendency to rely on another scientist's account of what a source says, rather than checking it directly for themselves. That's bad practice, but all of us do it occasionally.

Now, one could argue that it doesn't matter, that it's a trivial question, not important enough to waste time researching. Of course, I disagree: speculative arm-waving about basic facts and figures is no basis for a deep understanding of the natural world, let alone scientifically informed conservation policy (see Chapters 10 and 12). More fundamentally, having reliable data is a vital starting point for any argument about the importance of conserving pollinators. Policy makers and non-governmental organisations (NGOs) like to be able to present accurate, evidence-based numbers and facts in their reports, and those numbers and facts are often scrutinised in detail by those opposed to a particular policy or campaign. Arguments about the importance and effects of climate change, for example, cannot be made without reference to past and future global temperatures, current trends, and so forth. Likewise, the loss of grassland and woodland to agriculture and development needs to be quantified if a case is to be made for their preservation. If we have no firm foundation from which to start a discussion, it makes the whole exercise of convincing businesses, politicians and the public all the more difficult – something I will come back to when we discuss the politics of pollination in Chapter 13.

Having convinced myself that there was a significant question to be asked, and after establishing that there was no reliable, quotable figure for the proportion of flowering plants that are animal-pollinated, I set out to do the calculation.

The starting point for this was the existing scientific literature on plant–pollinator communities that had been accumulating over the past century or more. Most of these descriptive, natural history studies had focused only on the animal pollinators and the flowers that they visited. They included no information about which plant species in the community were wind-pollinated (a vital statistic if you are going to calculate the proportion of animal-pollinated plants). But a reasonable number had included counts of the wind-pollinated grasses, trees and so forth. The earliest studies included work by botanists John Willis and Isaac Burkill of surveys of plants and pollinators in England and Scotland, published between 1895 and 1908 (Willis and Burkill 1895, 1903a, 1903b, 1908, Burkill 1897). There was also a short, privately published book from 1928 called *Flowers and Insects*, written by the American entomologist Charles Robertson. This summarised his field work over the previous thirty-odd years in Carlinville, Illinois, and is still (to date) the most intensive study of a plant–pollinator community in a single locality. We'll encounter Robertson again in Chapter 4; his observations were a great early contribution to pollination ecology, and whenever I think of him I can't help mis-quoting Simon and Garfunkel with a short burst of 'Here's to you, Mr Robertson ...'

The fact that so many of the most detailed studies are so old probably tells us something about the time needed to carry out such surveys, time that few modern naturalists or scientists have available. It's also time that the various funding agencies would not pay for, as most research grants are on the order of five years or less. It needs scientists to make time to accumulate their own observations over a long period. Fortunately, there are still some who take this long view of their science. It's an area where amateur natural historians can also make a contribution, as I will discuss in Chapter 14.

In addition to published studies I also had some unpublished data from work that I had done during field campaigns in parts of South America, Africa and the UK. In total I amassed a list of 42 plant communities that had been assessed for the number of plant species that were wind- versus animal-pollinated. Along the way I also picked up a couple of collaborators. Sam Tarrant was then working on his PhD on pollinator communities of restored landfill sites (see Chapter 12), and he helped with data collation. Rachael Winfree, an American scientist who had taken an interest in the work, helped with the writing and framing of the subsequent paper. Using these data we were able to calculate the average proportion of plants that were animal-pollinated across all of the communities.

But that figure was less than satisfactory, because we knew from our data that the proportion rose steadily from the temperate zone to the tropics. Walk around a patch of grassland or woodland in Europe or North America. Observe the plants that are in flower and whether or not they are visited by insects. Look at the flower morphology. Are they drab with exposed stigma and anthers, or brightly coloured, perhaps tubular (see Figures 1.1 and 1.2)? Do the flowers have a scent or bright petals to attract insects? When you shake the flowers, does the pollen billow out in a big cloud, to be caught on the wind? Or is it sticky, adhering to the anthers until it's caught on a foraging insect? Observations such as these tell us whether or not a plant is wind- or animal-pollinated (though there's a small proportion of species that are both – a condition termed ambophily). Tally up the plants in that community and you will typically find that somewhere between 70% and 80% of the species will be animal-pollinated. Try the same exercise in a patch of tropical dry scrub or rainforest, in Africa or South America, and the figure will usually be between 90% and 100%. There's a lot of variation around these averages (the Willis and Burkill study in Scotland recorded 95% of plant species as animal-pollinated, for instance), but these are exceptions and the overall trend is strong. Not only that, it is consistent in both the northern and the southern hemispheres of our planet, despite the different geological and evolutionary histories of those two regions (see Figure 1.5 – note that since the original work described here we have added many more community surveys to the database).

This increase in the proportion of animal-pollinated plants in the tropics is likely to be a result of a number of factors. These include greater seasonality in the temperate zone favouring wind pollination, and the higher humidity to be found in many tropical habitats, particularly rainforest (Rech *et al.* 2016). Pollen clumps together when it's wet, which is why grasses and other wind-pollinated plants typically release their pollen when the weather is warm and dry. The significance

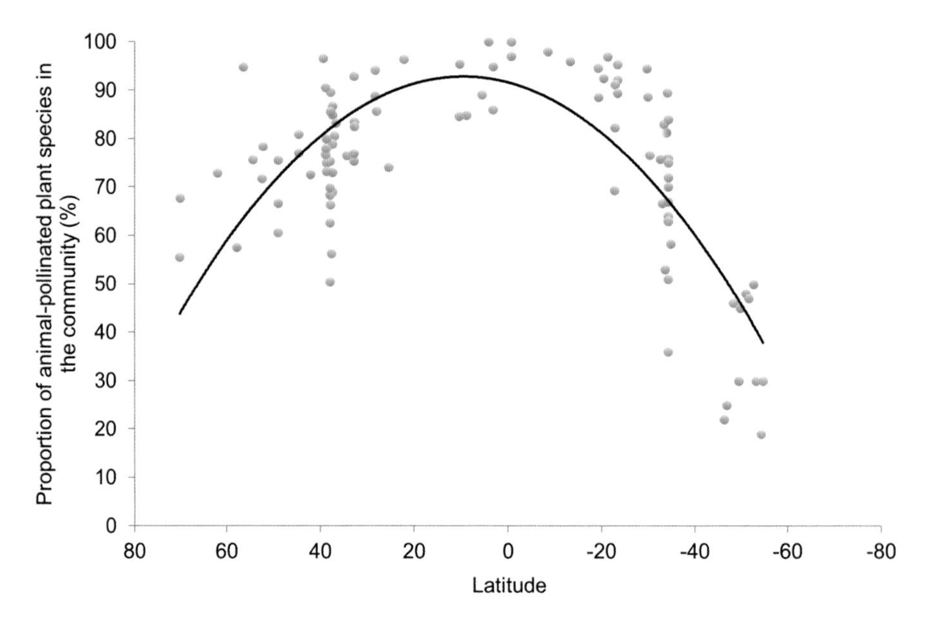

FIGURE 1.5 The proportion of animal-pollinated plant species found in 101 different plant communities across the world, from approximately 54 degrees south in the southern hemisphere to 70 degrees north in the northern hemisphere.

of this pattern for the calculation that we were trying to do lies in the fact that the tropics have many more plant species than regions further north or south. In order to take that into account I used a published source to calculate the proportion of the approximately 352,000 species of flowering plants that live in the tropics (about half of them, as it turned out), as well as how many were in the subtropical and temperate zones (almost a third and just over a fifth, respectively). Using these figures we refined our estimates and calculated that 308,006 species, or 87.5% of all flowering plants, are animal-pollinated.

As with all such calculations, this is a provisional figure open to future revision, not least because we know that there are several sources of inaccuracy in its calculation. To begin with, our estimate was based on a small sample of plant communities containing in total some 4,000 flowering plant species. This is just a little more than 1% of the estimated diversity of flowering plants. In addition, these plant communities were biased geographically: most were from Europe, and North and South America, with few from Asia or Africa, and none from Australia. Finally, we did not include water-pollinated species, or those that are obligately self-pollinating and therefore require neither wind nor an animal pollinator. However, these are a small fraction of the total number of plants compared to the wind- and animal-pollinated species.

Having acknowledged these sources of errors, Sam, Rachael and I wrote up a short manuscript with the matter-of-fact title 'How many flowering plants are pollinated by animals?' and submitted it to the ecology journal *Oikos*. Following peer review (a vital part of the scientific publishing process – I've never published anything that wasn't improved by being critiqued by other scientists, including this book), it appeared in February 2011. Almost immediately Ollerton *et al.* (2011) started to be referred to in the publications of other pollination ecologists who were keen to have a solid figure to cite. Indeed, it fast became the most highly cited paper in *Oikos* over that period. This is because pollination ecology has become a hugely important area of research over the last 25 years (Figure 1.6). Prior to that rather few scientists (especially in Britain) were interested in pollinators and pollination. I suspect that it was seen as old-fashioned, or as an area of research where all the interesting questions had been answered. After all, Charles Darwin wrote several books on the topic.

However, by the 1990s it had become clear to researchers in this field that there was still a lot that we did not understand even about some of the basic aspects of pollination ecology. In addition, concerns about the loss of pollinator diversity was fast moving up the agenda for conservation organisations (see Chapter 13, *The politics of pollination*). Lots of studies are now focusing on the ecology and conservation of plant–pollinator interactions, and our paper provided one argument for why it is important to study them: without these pollinators to maintain the reproductive outputs and genetic diversity of plant populations, we would ultimately lose a high proportion of botanical diversity. As well as being highly cited in the scientific literature (more than 1,900 citations at the end of 2020) the *Oikos* paper is also referred to in governmental policy documents and reports such as the IPBES *Assessment Report on Pollinators, Pollination and Food Production* (2016) and the *All-Ireland Pollinator Plan 2015–2020* (National Biodiversity Data Centre 2015 – see Chapter 13). So in one sense the question of how many flowering plants are animal-pollinated was basic and trivial, but in another it was an important calculation to make, not least because it reinforces the point about the importance of pollinators and the role they play in wild plant communities.

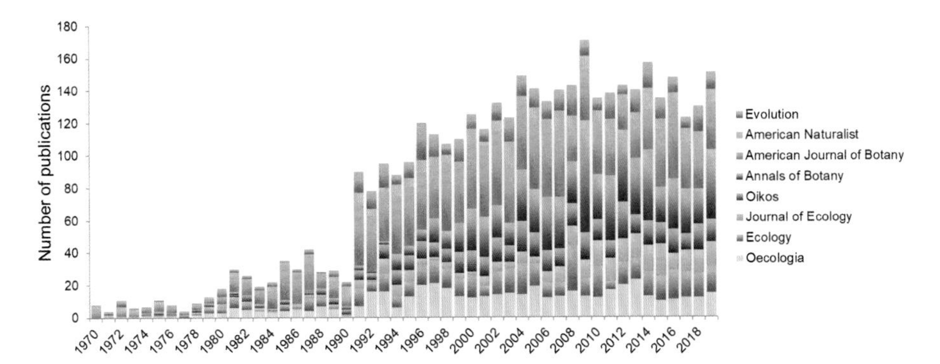

FIGURE 1.6 The number of scientific papers, commentaries and editorials being published each year that relate to pollination ecology in eight of the world's leading ecological, evolutionary and botanical research journals. Data are from ISI *Web of Science* from 1970 to 2019, using the search term pollinat*.

Pollinators as ecological connectors

Pollinators do not stand alone; they are embedded within food webs. One way to think of them is as an ecological bridge between the plants that they pollinate, the food those plants provide for herbivores, frugivores and seed parasites (see Chapter 4), and predators that in turn prey upon the pollinators. In this respect they provide not only the vital ecological function of pollination, but by feeding on nectar and pollen, either exclusively or as part of a broader diet, they themselves are a nutritious meal. Pollinating insects are a food source for rather a lot of vertebrate and invertebrate predators, including birds, bats, dragonflies and spiders, as well as parasites such as ichneumonid wasps and conopid flies. Some birds are particularly fond of bumblebees. Many years ago, I was approached by one of the gardeners at Oxford University Botanic Garden who was surprised, and perturbed, that she kept finding eviscerated bumblebees beneath one of the small trees in the garden. It was certainly a bird (perhaps a robin or great tit) that was catching these bumblebees as they foraged on flowers, and taking them back to the same branch to process. Despite the risks of being stung, these birds value social bees as a prey item as their honey stomach is an immediate, sugar-rich source of food.

Although we understand much about the role of pollinators in nature, there's still lots that we don't know about their ecological importance. This is especially true at an ecosystem level: for example, how much of the energy and nutrients within an ecosystem flows through the flower visitation pathway? That's a question that has hardly been addressed at all in the scientific literature, and yet it's an important one for understanding the long-term maintenance of terrestrial biodiversity. The Earth is not a botanic garden, and we cannot ensure that every corner is tended, maintained by human hands. For the most part it gets on in its own way, at its own pace, as it has for billions of years. The ecological importance of pollinators stretches back at least 200 million years, and conceivably longer, and has had a profound effect not only on the ecology of our planet, but also on the evolution of the dominant land plants.

The evolutionary importance of pollinators

Having established that pollinators play a hugely important ecological role on the planet at the moment, let's consider their evolutionary importance, starting with the *c.*352,000 species of flowering plants I mentioned above. That figure is another estimate, of course; it may be as low as 300,000 or as high as 400,000. The number of described species is over 295,000, and many newly discovered species are added to that total every year. But regardless of the actual figure, that order of magnitude is greater than the total number of land plants that have previously lived, and gone extinct, on our planet since the group evolved at least 450 million years ago. To put it in perspective, what we know of the current diversity of spore-producing plants (mosses, liverworts, ferns, horsetails and their

TABLE 1.1 The global diversity of described plants as currently understood, in descending order of species richness. After Christenhusz and Byng (2016).

Flowering plants	295,383 species
Ferns	10,560 species
Mosses	12,700 species
Liverworts	9,000 species
Lycopods	1,290 species
Gymnosperms	1,079 species
Hornworts	225 species

relatives) and seed-producing gymnosperms (conifers, cycads and their allies) and angiosperms (flowering plants) is shown in Table 1.1.

These figures are increasing year-on-year as botanists discover previously unknown species, and over the past two decades the number described each year has remained quite steady at about 2,000 species per annum. The reasons for such high diversity in the flowering plants are still actively debated by botanists and are thought to relate to a range of biological attributes of this group that are not shared with others. The reliance of most flowering plants on pollinators is probably a significant factor, though it is not unique to the angiosperms. In fact, as many as one-third of the living gymnosperms are insect-pollinated. Furthermore, the fossil record indicates that their ancestors, a much more diverse group of plants, were mainly adapted for insect pollination. The implications for this in determining how plants evolve will be explored further in Chapter 5, but for now it's worth considering just what it means for a flower to have evolved under natural selection from pollinators.

The role of pollinators in shaping flowers

Taxonomists have traditionally defined flowering plant species on the basis of their morphology and ecology (and more recently using genetic data). Plant leaves, stems and roots tend to be variable; they often don't make for good characters with which to define plant groups (though there are exceptions).

Take the spurges (the genus *Euphorbia* in the family Euphorbiaceae), for example. The spurges that a British naturalist might regularly encounter in the field, such as the sun spurge (*Euphorbia helioscopia*), tend to be rather short, herbaceous species. Pay a visit to a plant collection such as at Oxford Botanic Garden, however, and you will see another side to the spurges. The garden holds one of the National Collections of hardy spurges. In the beds devoted to their cultivation you will find Mediterranean species that are taller, more robust, largely evergreen and very different to our native spurges. Step inside the glasshouse that's home to the succulent collection and it's a different story again: here you will find a magnificent species at least five metres in height, with thick, water-storing stems on its upper sections and, lower down, a massive corky trunk. Growing close by is a true cactus, a species of *Cereus*, similarly massive, and similarly fleshy and corky. When I visit the gardens with my students I show them these two fine specimens that have been growing in this glasshouse for many years. I point out that these two species belong to very unrelated, and biogeographically separated, groups of plants, the highly succulent spurges (Euphorbiaceae) from Asia and Africa, and the cacti (Cactaceae) from the New World. They look similar because they have evolved to grow in similar conditions: semi-arid habitats with only occasional rainfall, where water must be stored in order to allow plants to survive until the next shower. It's an example of convergent evolution, in which species have evolved to look superficially similar as they adapt to similar climatic conditions or ecological lifestyles. The students often ask, 'so, how can we tell them apart?' and it's a fair question. The best way, I respond, is to look at the flowers, not the stems.

All spurges, whether in a British woodland or in a garden, or elsewhere in the world, share similar floral characteristics, and it is on these that we classify the genus. This is true of most groups of plants: they can be taxonomically defined first and foremost by the features of their flowers, though fruits are also important in this regard. This fact has been recognised since at least the work of British naturalist John Ray in the seventeenth century, and then later the Swede Carl Linnaeus in the eighteenth century. Floral characters tend to be more fixed than vegetative characters, at least at higher taxonomic levels such as genus or family. But that doesn't mean that flowers never vary in these groups, far from it, and it is this variation that helps us to define species. And this is where pollinators play a large role.

A good example can be found among British campion species. Red campion (*Silene dioica*), as the name helpfully suggests, has reddish flowers that are mainly pollinated by bumblebees and butterflies during the day. In contrast, white campion (*S. latifolia*) has white flowers that produce scent in the evening, and are pollinated by moths. However, there is some overlap in their pollinators: day-flying insects occasionally visit the white and night-active pollinators the red, and hybrids between these plants do occur (Goulson and Jerrim 1997, Goulson 2009 – these are not, but perhaps should be, called rosé campions). But overall there is a clear indication that the two species have evolved to exploit rather different pollination niches, and natural selection by nocturnal moths has driven the evolution of pale flowers that emit their scent at night in white campion.

Complex flower adaptations to particular pollinators are found throughout the flowering plants, and these will be discussed further in Chapter 5. However, in many cases it is really quite difficult to dissect how a plant's flowers have evolved to attract and reward particular pollinators, and to construct a story that makes sense. There's nothing simple about the ecology and evolution of flowers and their pollinators. Plant evolution is complex, and pollinators are only part of the story. For example, almost all of the grasses are wind-pollinated, though there are some tropical rainforest species that attract insects and at least one genus of bees – *Lipotriches* – that is a specialist collector of grass pollen. Flowers of the wind-pollinated species have not evolved in response to pollinators, but rather as a function of dispersing and catching wind-blown pollen in different ways, as well as protecting anthers and ovaries from herbivores. The evolution of new wind-pollinated plant species is therefore more likely to be a result of local adaptation to non-biological aspects of the environment, such as soil chemistry or moisture levels, rather than due to biotic processes via interactions with other species.

Pollinators in agriculture

Any farmer or gardener who grows carrots will tell you that pollinators are not required to obtain a good crop. This is true for most of the world's crop plants, both by number of species and by weight of crop. A drive through the British countryside will confirm this. In any one year most fields are devoted to wind-pollinated cereals such as wheat and barley, or root crops like potatoes, turnips and carrots which don't need pollinators to obtain a yield (though they may need them to produce seed for sowing the next year – see below). Crops that require insect pollination to produce the fruit or seeds that we directly consume, such as oilseed rape (canola), field beans and apples, are less common. Indeed, Department for Environment, Food and Rural Affairs (Defra) statistics for its annual review of agriculture in 2019 show that most of the crops grown in the UK (by area) were wind-pollinated (Table 1.2). The contribution of insect pollinators to this agricultural production varies enormously according to the type of crop. For example, some orchard fruit varieties are 100% reliant on insects for pollination, whereas different types of oilseed rape vary in their ability to be self- or wind-pollinated, as well as insect-pollinated.

It's clear that most of the UK's arable output relies on either wind pollination or on the production of edible roots and tubers. This is a pattern that is repeated elsewhere in the world where most crop production (at least by area or volume) involves plants that do not require animal pollination. That does not at all minimise the importance of pollinators to agriculture, however, for several reasons. First of all,

TABLE 1.2 Defra statistics for crops planted in the UK in 2019, showing the area planted and whether or not the crop requires animal pollination for its yield. Note that not all crops are represented. These are the most recent publicly available data from Defra (2020).

Crop	Animal-pollinated?	Thousands of hectares
Cereals	No (wind)	3,211
Potatoes	No (for the crop)	144
Sugar beet	No	108
Oilseed rape	Yes	530
Linseed and borage	Yes	17
Field beans and peas	Yes	178
Orchard and soft fruit	Yes	35

raw statistics such as these tell only part of the story. Some of those crops that don't need pollinators for their yield, such as carrots, are sown from seed each year; many varieties (particularly F1 hybrids) certainly do require pollinators to produce that seed (Wyns 2014). Commercial seed producers rent honey bee hives or use mass-produced blowflies for their pollination services. Second, there has been big change over time in the relative proportions of animal- versus non-animal-pollinated crops grown in Britain, with the area of plants pollinated by animals steadily increasing as the other crops decrease. Third, fruits and seeds from animal-pollinated crops provide many of the essential micronutrients in our diet, as well as increasing the diversity and richness of flavours in our food. Finally, a significant amount of insect-pollinated crops and non-food plants are either grown in home or community gardens, or harvested from the wild. We're going to look at all of this in more detail, and consider the pollinators involved in agricultural production, in Chapter 7.

A note on terminology

In this book I'm going to use the term 'pollination system', and I think it needs to be defined here as it is one of those expressions that seems to mean different things to different people. The way I use it, and I think the way that it was originally intended when the term was used in the ecology literature of the 1970s, is that the pollination system = flower characteristics + pollinators. That is to say, the colour, shape, size, odour, rewards and so forth produced by a flower (or an inflorescence functioning as a single reproductive unit – see Chapter 3) plus the animals that effectively transfer pollen. This is distinct from a 'pollination syndrome', which refers only to the flower traits themselves, or a 'pollinator guild' or 'pollinator functional group', which refers only to the flower visitors. I have seen some authors use 'pollination syndrome' as synonymous with 'pollination system', but to my mind they are distinct things. I have also seen other authors use 'pollination system' to mean the community of plants and pollinators in an area, or as analogous to the breeding system, but my use of the term in this book excludes those.

Conclusion

In providing this initial overview of the importance of pollinators and of pollination as an ecological function, hopefully encouraging you to read on further into the book, I have treated the three perspectives of ecology, evolution and agriculture in separate sections. However, they are not truly separate subjects. They are linked not only by the importance of pollinators to the wider environment and to agriculture, but by our cultural perception and exploitation of those pollinators and the flowers that they service. As scientists and citizens we see pollinators as important not only for the roles I have described above, but because we appreciate watching their behaviour and life histories. It's time now to meet some of those pollinators and learn about them in more detail.

Chapter 2

More than just bees: the diversity of pollinators

The diversity of animals that act as pollinators is truly astonishing, with an estimated 350,000 mainly terrestrial vertebrate and invertebrate species moving pollen between flowers (Ollerton 2017). And that's only the described species. Given that around 90% of insect species remain undescribed by taxonomists, the true diversity is much higher. Even the best-studied group of animals, the birds, continues to surprise us: as recently as 2007 researchers described a new species of hummingbird (Brewer 2018). Hummingbirds are, of course, important pollinators of many plants in the Americas. In my experience, though, if you ask people to name a pollinator most will say 'bee', usually with the honey bee in mind. When I give talks to non-specialist groups I enjoy the slow reveal of informing the audience that there are about 270 species of bee in the UK; that there are more than 20,000 species worldwide; and that this represents less than 10% of the total known diversity of pollinators. Blowing minds with facts is one of the perks of my job.

In this chapter I want to explore the global diversity of flower-visiting animals, introducing the major and minor groups that include known pollinators. In later chapters we will discuss what this taxonomic diversity means for the evolution of flowers, for their ecological and agricultural functions, and for future conservation. The sequence in which these groups is presented is both taxonomic (invertebrates first, followed by the animals with backbones), and, within these groups, in decreasing order of diversity. This is to give a sense of the evolutionary breadth of the animal pollinators, and the relative importance of different groups, at least as far as numbers of species is concerned (Table 2.1). However, bear in mind that this may not reflect their overall importance as pollinators, because size, behaviour and (especially) abundance within an ecological community (i.e. population size) are all important in this regard. For example there are currently only 24 species of bumblebee in Britain, but the high numbers of six or seven of these make them one of our most important groups of pollinators of both wild plants and crops.

For this first section I've drawn heavily on the study by Wardhaugh (2015) and my own recent assessment (Ollerton 2017).

TABLE 2.1 The estimated number of described species in the major pollinator groups.

Taxon	Estimated number of pollinating species in the major groups	Diversity of significant subgroups	Sources
Lepidoptera (butterflies and moths)	141,600		Wardhaugh (2015)
• Moths (Heterocera)		123,100	
• Butterflies (Rhopalocera)		18,500	
Coleoptera (beetles)	77,300		Wardhaugh (2015)
• Flower chafers (Cetoniinae)		4,000	Sakai & Nagai (1998)
Hymenoptera (bees, wasps, ants)	70,000		Wardhaugh (2015)
• Bees (Anthophila)		20,446	Ruggiero *et al.* (2020)
• Spider wasps (Pompilidae)		5,000	Pitts *et al.* 2005
• Social wasps (Vespidae)		5,000	
Diptera (flies)	55,000		Wardhaugh (2015)
• Hoverflies (Syrphidae)		6,000	
• Bee-flies (Bombyliidae)		4,500	
Thysanoptera (thrips)	1,500		Wardhaugh (2015)
Aves (birds)	1,089		Regan *et al.* (2015)
• Hummingbirds (Trochilidae)		365	
• Honeyeaters (Meliphagidae)		177	
• Sunbirds (Nectariniidae)		124	
• White eyes (Zosteropidae)		100	
• Parrots (Psittacidae)		93	
Hemiptera (bugs)	1,000		Wardhaugh (2015)
Collembola (springtails)	400		Wardhaugh (2015)
Blattodea (termites and cockroaches)	360		Wardhaugh (2015)
Mammalia (mammals)	344		Regan *et al.* (2015)
• Bats (Chiroptera)		236	
• Non-flying mammals (various)		108	
Neuroptera (lacewings)	293		Wardhaugh (2015)
Trichoptera (caddisflies)	144		Wardhaugh (2015)
Orthoptera (crickets)	100		Wardhaugh (2015)*
Mecoptera (scorpionflies)	76		Wardhaugh (2015)
Psocoptera (barkflies)	57		Wardhaugh (2015)
Plecoptera (stoneflies)	37		Wardhaugh (2015)
Lacertilia (lizards)	37		Olesen & Valido (2003)
Dermaptera (earwigs)	20		Wardhaugh (2015)
Crustacea (mainly Isopoda)	11		Ollerton (1999), van Tussenbroek *et al.* (2016)
Polychaeta (marine worms)	3		van Tussenbroek *et al.* (2016)
Total	**349,371**		

* Possibly an overestimate, as only one species confirmed as a pollinator (Micheneau *et al.* 2010).

Butterflies and moths (Lepidoptera)

Lepidoptera is the most species-rich pollinator group, by a big margin. Globally, over 140,000 species are thought to visit flowers, which is about the diversity of the beetles and bees and wasps combined (Table 2.1). Most of this diversity is within the moths, of which there are ten times as many species as butterflies. However, this assessment is based on the assumption that all butterflies and moths visit flowers, which we know to be false. In Britain, for instance, the purple emperor (*Apatura iris*) feeds mainly on sap from trees and honeydew from aphids, while in the tropics butterflies often feed on the juices from fruit. However, we don't know how common these food habits are worldwide, compared to flower feeding. In addition, we know that some moths do not have functional adult mouthparts. Strange as it may seem, these moths do not feed as adults, relying on the energy consumed as caterpillars to fuel their brief period of mating. While lepidopterists think that only a small proportion (perhaps 10%) opt for this gorge-then-starve strategy, in reality we don't know for sure: the life histories of most moth species are largely unknown. The nocturnal moths are among the least well-studied pollinators, because in general pollination biologists do not undertake surveys at night. When they do, however, many of the moths are found to carry pollen (Devoto *et al.* 2011). In fact adaptations of flowers to moth pollination are not at all uncommon,

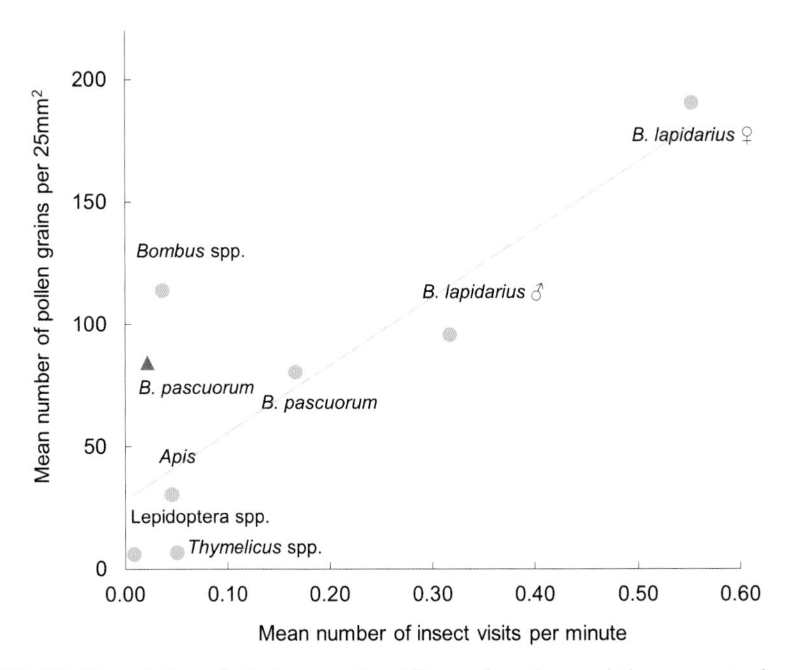

FIGURE 2.1 The relationship between rate of flower foraging and the amount of pollen carried on an insect's body, for visitors to greater knapweed (*Centaurea scabiosa* – green circles) and its parasitic plant knapweed broomrape (*Orobanche elatior* – orange triangle). Redrawn from data originally presented in Ollerton *et al.* (2007a).

FIGURE 2.2 Six-spot burnet moth (*Zygaena filipendulae*) feeding on the flower head of common teasel (*Dipsacus fullonum*).

especially in the tropics, where plants are frequently pollinated by hawkmoths (Sphingidae) (S.D. Johnson *et al.* 2017, Ollerton *et al.* 2019a).

More is known about butterflies as flower visitors. However, they are often rather dismissed as pollinators because they carry less pollen than large hairy bees, they visit flowers less frequently, and they forage only for themselves rather than for a nest (social or solitary) that they are provisioning. Indeed, we often find that insects which forage more frequently accumulate greater amounts of pollen (Figure 2.1). However, in some generalist plant species butterflies can be just as effective as bees at pollinating flowers, and may move greater distances between flowers, so reducing the chance of self-pollination by that plant and making the butterflies more effective as pollinators (Ollerton *et al.* unpublished data).

The distinction between nocturnal moths and diurnal butterflies is broadly true, but there are exceptions: some moths fly during the day and some butterflies fly at night. Among the most distinctive of the day-flying species are the cinnabar and tiger moths in the family Erebidae, and the burnet moths (Zygaenidae) (Figure 2.2). The stunning colours of these moths reflect the fact that their caterpillars accumulate toxins from the plants on which they feed. This has important implications for the conservation of some species, as we will see in Chapter 9 when we consider ragwort and cinnabar moths.

Worldwide there exist some very intimate and specialised interactions between moths and the plants that they pollinate, including yuccas (*Yucca* spp.) and yucca moths in the genera *Tegeticula* and *Parategeticula* (Powell 1992, Pellmyr *et al.* 1996, Yoder *et al.* 2010); senita cacti (*Pachycereus* (*Lophocereus*) *schottii*) and *Upiga virescens*, the senita moth (Fleming and Holland 1998, Holland and Fleming 1999); and leafflower trees (*Glochidion* and *Breynia* spp.) and *Epicephala* leafflower moths (Kato *et al.* 2003, Kawakita and Kato 2004, Finch *et al.* 2018). In some of these interactions the female moths actively pollinate the flowers.

Beetles (Coleoptera)

More than 77,000 species of beetles are thought to be active flower visitors, and many of these are pollinators. The old received wisdom was that most beetles did damage to flowers because of their chewing, rather than sucking, mouthparts. However, that's been overturned in recent years and we know that flowers in southern African habitats, for instance, are well adapted to beetle pollination. For example flower chafers of the scarab beetle subfamily Cetoniinae are important pollinators of a wide range of grassland plants including some proteas (Proteaceae) and asclepiads (Apocynaceae subfamily Asclepiadoideae – Ollerton *et al.* 2003, Shuttleworth and Johnson 2009a; see also Chapters 3, 4 and 5). The distantly related monkey beetles (Hopliini) pollinate other groups (Steiner 1998, Mayer *et al.* 2006). One of the reasons why pollination by beetles (and groups such as spider-hunting wasps – see below) is so common in southern Africa is that there

FIGURE 2.3 Common red soldier beetles (*Rhagonycha fulva*) on flowers of hogweed (*Heracleum sphondylium*).

are fewer bee species than we might expect based on landmass and climate; these other groups fill the niches that bees might otherwise have exploited (Ollerton *et al.* 2006).

The effectiveness of beetles as pollinators is still poorly understood in most cases, and there are few studies of their ability to pick up and transfer pollen. This is despite the fact that beetles can be extremely abundant on flowers and may use them as sites for mating. For example, in some years soldier beetles (Figure 2.3) have a population explosion and are commonly found on a wide variety of plants with dense flower heads such as members of the daisy (Asteraceae) and umbellifer (Apiaceae) families (Lamborn and Ollerton 2000).

Bees, wasps, ants and sawflies (Hymenoptera)

There are perhaps 70,000 pollinating species within this taxonomic order, of which the bees make up only around 20,400. The others comprise an evolutionarily mixed bag of families that collectively include the social and solitary wasps, sawflies, ants, and their relatives. What they lack in diversity compared to some other groups, the bees more than make up for in importance as pollinators of both wild and agricultural plants, and in their cultural significance. The general notion of what a bee is, and how it behaves, looks to the honey bee (*Apis mellifera*) as a model: social, with a hierarchy, a queen, and a large nest (termed a hive for colonies in captivity). In fact, this view of bee-ness, though long embedded within our psyche, is far removed from the biology of the average bee: most of them have no social structure at all, and a fair proportion of them are parasitic. In Britain and Ireland we have about 270 species of bees, give or take (Falk and Lewington 2015), though there have been a number of extinctions and additions to this fauna over the past century and a half (see Chapters 10 and 11). These species provide a reasonable sample of the different lifestyles adopted by bees globally. They can be divided into four broad groups.

Honey bees include several highly social species and subspecies of the genus *Apis*, of which the ubiquitous western honey bee (*A. mellifera*) is the most familiar. Most colonies are found in managed hives, though persistent feral colonies can be found in hollow trees, wall cavities and other suitable spaces. They are widely introduced into parts of the world where they are not native (e.g. the Americas, Australia, New Zealand) and there is some debate as to whether they are truly native to the British Isles and northern Europe, with supporting evidence and arguments on both sides. Colonies can be enormous and contain thousands of individuals, mostly female workers, with a single queen. Unmated queens and males (drones) are produced by the colony later in the season.

Bumblebees (belonging to the genus *Bombus*) are also mainly social, though their nests are much smaller (tens to hundreds of individuals). Depending on the species these nests can be in long grass, rodent holes, or cavities in buildings and trees. Twenty-seven of the more than 250 species have been recorded in Britain and Ireland, but six of these are not strictly social; they are parasitic and belong to the subgenus *Psithyrus*, which will be described below.

The so-called solitary bees are by far the largest group in the British Isles (about 170 species) and worldwide (more than 90% of all species). In Britain and Ireland they belong to fifteen genera, including *Andrena, Anthophora, Osmia, Megachile,* etc. The females of most of these bees, once they have mated, construct nests that they alone provision with pollen for their developing young. Nesting sites can be genus- or species-specific, and include soil, cavities in stone or wood, and snail shells. Some species are not strictly solitary at all and may produce colonies with varying levels of social structure, though without a queen or a strict caste system; we term them 'primitively eusocial'. In fact sociality has evolved and been lost numerous times in the bees and in the rest of the Hymenoptera (Danforth 2002, Hughes *et al.* 2008, Danforth *et al.* 2019). Even within closely related species it can vary; for example in the carpenter bee genus *Ceratina* (Apidae: Xylocopinae)

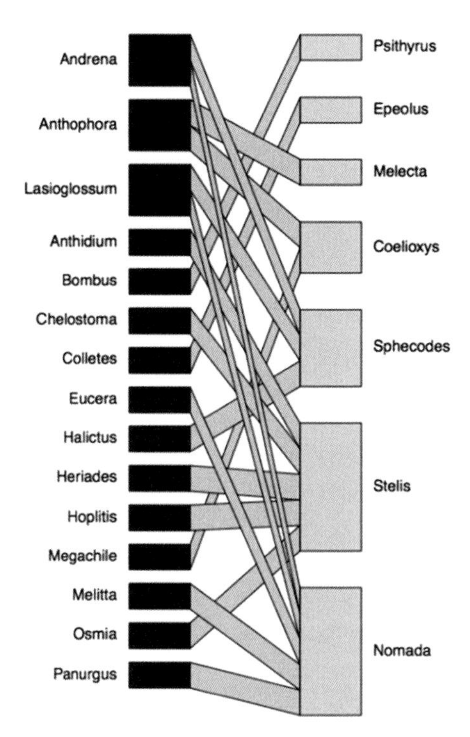

FIGURE 2.4 Relationships between British bee genera and their cuckoo bee parasites. This bipartite graph is structured such that the hosts (to the left, in black) are ranked top to bottom from most to least parasitised (in terms of number of cuckoo genera that interact with them). The cuckoo bees (on the right, in grey) go in the reverse order, from most specialised to least specialised. Note that this set of interactions only applies to Great Britain and Ireland; breadth of host–parasite interactions is wider in continental Europe and elsewhere in the world. Information on hosts and parasites from Falk and Lewington (2015).

tropical species are more often social than their temperate relatives (Groom and Rehan 2018).

The final group is termed the cuckoo bees. Like their avian namesakes, they parasitise the nests of both social and solitary bees (though never, interestingly, honey bees). There are about 70 species in seven genera, including the bumblebee subgenus, *Psithyrus*. Other genera include *Melecta*, *Nomada* and *Sphecodes*. In some cases the parasitic species have a close evolutionary relationship to their hosts and may resemble them, as seen in some *Psithyrus* species. In other cases they may be only distantly related and in fact look more like wasps – for example, *Nomada* species. Some genera of cuckoo bees are restricted to parasitising only a single genus of bees, while others are parasites of a range of genera (Figure 2.4).

Although we often think of bees, overall, as being the most important pollinators, in fact species vary hugely in their effectiveness at transferring pollen between flowers. Pollinating ability depends upon factors such as abundance, hairiness, behaviour, body size, and visitation rate to flowers (Figure 2.1). Size is especially important for three reasons. First of all, larger animals can pick up more pollen on their bodies, all other things being equal. Second, in order to bridge the gap between picking up pollen and depositing it, flower visitors must be at least as large as the distance between anthers and stigma, unless they visit the stigma for other reasons. Finally, larger bee species tend to forage over greater distances on average (Greenleaf *et al.* 2007), thus increasing the movement of pollen between plants. However, most of the world's bees are relatively small, as we can see from the analysis of British bees in Figure 2.5. Many species have a maximum forewing length of only 4 or 5 mm, and the majority of species are smaller than honey bees. Remember also that these are maximum sizes measured from a sample; individual bees can vary a lot within populations and even (in the case of *Bombus* spp.) within nests (Goulson *et al.* 2002). So the assumption that all bees are good pollinators needs to be tempered by an acknowledgement that some are much better than others.

The bees form a natural monophyletic group or 'clade', which is to say that they evolved from a single common ancestor. As far as we know, this ancestor lived in the Cretaceous period at least 100 million years ago. In contrast, the group that we call wasps comprise a number of only distantly related clades and is a bit of a dumping group as far as classifying pollinators is concerned. However, one thing that they have in common, and which distinguishes them from most bees, is that they are not so reliant on flowers: the majority of species are carnivorous or parasitic, and feed on nectar only for their own energy needs. In Britain and Ireland the most familiar pollinating wasps are the social Vespidae, of which there are 10 species. However, in these islands we also have solitary Vespidae (23 species) plus hundreds of species of solitary and parasitic wasps from families such as Ichneumonidae, Sphecidae, Pompilidae and so forth. In addition there are about 550 species of sawflies (suborder Symphyta), many of which visit flowers and can be abundant on plants such as wild carrot (*Daucus carota* – Lamborn and Ollerton

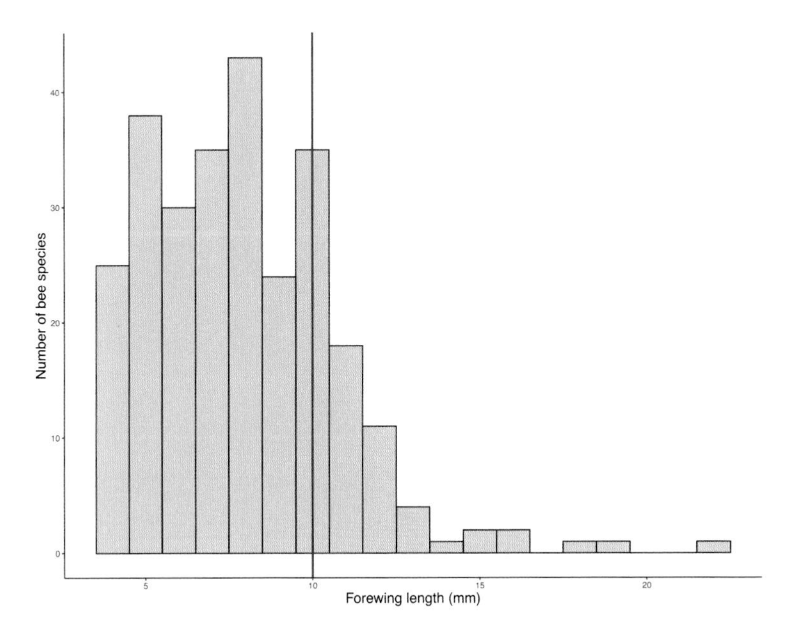

FIGURE 2.5 The sizes of British bees. Forewing length is a good measure of overall body size and the data are maximum lengths recorded for species, except for the social bumblebees and honey bee, where I have used maximum size of workers (queens are often much larger). The blue line indicates the honey bee (*Apis mellifera*). The biggest bee in this data set is the violet carpenter bee (*Xylocopa violacea*), which, while not generally considered a native species (yet), has bred in Britain in the past. Data from Falk and Lewington (2015).

2000). Almost all wasps are carnivorous, but a notable exception is the pollen wasps (Vespidae subfamily Masarinae) which, alone of all the wasps, provision their larvae with nectar and pollen, much like solitary bees. They don't occur in the British Isles, and are found mainly in the drier parts of southern Africa, as well as in North and South America (Gess 1996).

The ichneumonid wasps (Hymenoptera: Ichneumonidae) are a fantastically diverse group of insects that mostly share a similar parasitic life history: they lay their eggs in or on a host insect. Around 24,000 species have been described, and estimates for their full diversity range between 60,000 and 100,000 species. In Britain there are approximately 2,500 species, almost 10 times our bee diversity. Many species visit flowers, particularly umbellifers, and they can therefore be quite significant (though under-studied) pollinators of plants like hogweed (*Heracleum sphondylium*) and its relatives in the family Apiaceae.

Wasp-pollinated flowers include some that are specifically adapted to pollination by particular groups of these insects. For example, spider-hunting wasps (Pompilidae) act as specialised pollinators of orchids (*Eucomis* spp.) and asclepiads in southern Africa, with the flowers showing quite distinctive adaptations to pollination by these insects (see Chapter 3). In Britain, the vespid-pollinated members of the genus *Scrophularia* are another example. More commonly, however, flowers

FIGURE 2.6 Hornet (*Vespa crabro*) taking nectar from common ivy (*Hedera helix*). She seems to be a queen stocking up on energy prior to hibernating and is carrying a large amount of pollen. There's every chance that she's a very effective pollinator of ivy, which is a key nectar resource in autumn and early winter.

attract a wide range of both wasps and other types of pollinator, for example umbellifers (Apiaceae) and common ivy (*Hedera helix* – Figure 2.6). Wasps of the genus *Vespula* have been shown to be especially effective pollinators of ivy (Jacobs *et al.* 2010).

Finally for this group, it's not possible to discuss wasps without considering the figs (*Ficus* spp.) and the fig wasps, but I'm going to save that for Chapter 5 when we examine the evolution of specialised and generalised interactions between plants and pollinators.

Sawflies (Symphyta) often visit flowers such as umbellifers (see the example of wild carrot in Chapter 5) and are no doubt effective in such generalist pollination systems. I know of no examples of specialised sawfly pollination but they may well exist, waiting to be described.

Ants (Formicidae) are frequent flower visitors but relatively rare as pollinators. There's a long-standing idea in pollination ecology, which originated with Australian ecologist Andrew Beattie, that they are ineffective because of the production of an antibiotic, myrmicacin, by their metathoracic glands which inhibits the growth of bacterial and fungal spores, and, crucially, pollen (Beattie *et al.* 1984). Twenty years after this was proposed, Andrew reviewed the existing literature on specialised ant pollination systems and could document only a few cases (Beattie 2006). However, since then, more examples have come to light, though the ants are acting as pollinators of plants that attract other types of insect as well and may not be the most effective pollinators (Domingos *et al.* 2017, Xiong *et al.* in press). The contribution of ants to more generalist pollination systems might therefore be more widespread than we realise, and it deserves greater attention. However there's no doubt that pollination by ants remains rare in comparison to their diversity, which is on a par with that of the bees.

True flies (Diptera)

In many respects it is the flies that surprise us with their diversity as pollinators and their importance for agricultural and natural ecological systems (Orford *et al.* 2015, Rader *et al.* 2016). That's not because of the number of species that visit flowers; in fact, of the four main flower-visiting insect orders, Diptera has the fewest known pollinating species (Table 2.1). What's surprising is first of all the number of fly families involved, and secondly that most of these families do not routinely visit flowers. The most frequently encountered pollinating family is the hoverflies (Syrphidae), which are common flower visitors and often mimic bees and wasps in their colouration and hairiness, probably to protect them from predators. The bee-flies (Bombyliidae) are also strongly associated with flowers, as are other long-tongued flies belonging to families such as the tangle-veined flies (Nemestrinidae) and horse flies (Tabanidae) in southern Africa and elsewhere. But most other groups do not make a habit of visiting flowers. In a review of pollinating

FIGURE 2.7 A species of Anthomyiidae (Diptera) sunning itself within the bowl-shaped flower of a buttercup (*Ranunculus repens*).

Diptera diversity, Larson *et al.* (2001) listed 55 families that are known to contain species that pollinate flowers. To put this into perspective, there are reckoned to be between 160 and 190 families of flies, so potentially about one-third of these families includes pollinators.

Flies visit flowers for a wide range of reasons. Some require nectar to fuel their flight; others also need to consume the proteins in pollen in order to complete egg production in females. For predatory flies, flowers provide a lure that will bring prey into range, while in cold areas, especially the Arctic and high mountains, bowl-shaped flowers focus heat on an insect, allowing its body to warm faster (Figure 2.7). Finally, it's not unusual to find flies and other insects mating on the flat platforms produced by umbellifer flower heads – which function as a kind of insect pick-up joint.

However, a large proportion of the groups of flies that visit flowers do so because the flowers mimic their food or egg-laying sites. Such flowers are adept at manipulating hard-wired, inflexible fly behaviour – and we'll return to this idea in Chapter 3.

Miscellaneous invertebrate pollinators

During the time I've spent in the field, in sites across the world, I've seen all sorts of other invertebrates visiting flowers, including many of the different groups listed in Table 2.1. Some of them are specialist pollinators of just a few plants, others are occasional pollinators of generalist plants, and many may also be herbivores that visit flowers to eat them, for example crickets and grasshoppers (Orthoptera), earwigs (Dermaptera), true bugs (Hemiptera), and slugs and snails (Mollusca). They can cause considerable damage to these flowers, and in extreme cases prevent seed set entirely. The specialist exceptions to this really push at the boundaries of our understanding of what pollinators are, however, as some examples demonstrate. Cricket pollination has been recorded in at least one tropical orchid species (Micheneau *et al.* 2010), and hemipteran bugs are known to be pollinators of some southern African plants (Ollerton *et al.* 2003). Perhaps even more surprisingly, cockroaches (Blattodea) pollinate at least twelve species in a diverse range of families (Nagamitsu and Inoue 1997, Mertens *et al.* 2017, Vlasáková *et al.* 2019, Xiong *et al.* in press). Some recently discovered fossil cockroaches from amber deposits in Myanmar suggest that cockroach pollination of both gymnosperms and angiosperms may be ancient, dating at least to the Cretaceous period (Hinkelman 2020, Hinkelman and Vršanská 2020).

The idea that slugs and snails might act as pollinators seems more preposterous, but has been part of the mythos of pollination biology for many years. Several purported examples have been published, but all of them fail one or more fundamental tests of either pollinator effectiveness or adaptation to pulmonate molluscs as agents of pollen transfer. Perhaps the most interesting and widely cited study is by Sarma *et al.* (2007), who provided evidence from India that the graceful awlsnail (*Allopeas gracile*) pollinates a species in the morning glory family (Convolvulaceae) called *Evolvulus nummularius*.[1] However, there are issues with that study. The plant is not native to India (it originated in the Americas) and the flowers show no specific adaptations to snail pollination. It's also very likely that the snail is not a native either; it may be from South America, though it is now very widespread in warmer parts of the world (Capinera 2017). Thus there is no way that these two species could have evolved a close relationship. Slugs and snails often visit flowers and may well pick up pollen as they move around, but in my opinion malacophily isn't yet proven, as no plant has been conclusively shown to be adapted to molluscs for pollination. Such species may well exist, however, waiting to be discovered. The wet tropics would be a good place to look.

Spiders (Araneae) are almost exclusively carnivorous, though some will feed on pollen that is trapped in their webs and on extra-floral nectar produced by stems and leaves (Eggs and Sanders 2013, Nahas *et al.* 2017). Those which are found in flowers, however, are usually predators waiting for a legitimate pollinator that they can hijack. Crab spiders (Thomisidae) are the best-known example, some of which can change colour to match that of the flower in which they are sitting. They may indeed occasionally self-pollinate flowers, but the extent to which they move around between flowers is limited. Once again, no specifically spider-adapted flowers have been found.

Thrips (Thysanoptera) are generally tiny insects that individually carry relatively little pollen. But they can be hugely abundant in flowers, and a number of tropical rainforest trees are thought to be specialised for thrips pollination. In addition, they have been shown to contribute to pollen movement in species that are pollinated by other vectors, including wind (Varatharajan *et al.* 2016, Scott-Brown *et al.* 2019).

Finally, life on Earth never ceases to surprise us when it comes to interactions between species. In the last few years some marine seagrasses have been shown to be pollinated by flower-visiting crustacean shrimps and polychaete worms (van Tussenbroek *et al.* 2016). It was completely unexpected: in the past it was assumed that seagrasses, which flower under water, were all water-pollinated. This discovery perhaps makes more likely my speculative, and slightly tongue-in-cheek, suggestion that one day we might discover fish-pollinated flowers (Ollerton 1998).

Having dealt with the invertebrates, let's now consider the main vertebrate animals that act as pollinators.

Birds

Birds are the largest group of vertebrate pollinators. Of these, it is hummingbirds (family Trochilidae) that are the best known for their interactions with flowers (Figure 2.8), and are certainly the most diverse (Table 2.1). Other bird families have become more or less specialised to feeding on nectar, particularly the sunbirds in the family Nectariniidae (Figure 2.9) and honeyeaters (Meliphagidae). However, no

FIGURE 2.8 The royal sunangel hummingbird (*Heliangelus regalis*) is endemic to the elfin forests of the Andes of parts of Peru and Ecuador. This bird is visiting flowers of a *Brachyotum* species (Melastomataceae) in northern Peru. Photo by Jesper Sonne.

FIGURE 2.9 A Cameroon sunbird (*Cyanomitra oritis*) visiting flowers of a *Tabaernemontana* species (Apocynaceae) on Mount Cameroon. Photos by Francis Luma.

bird feeds exclusively on nectar – not even the most specialised of the hummingbirds. For example, the sword-billed hummingbird (*Ensifera ensifera*), which I was fortunate enough to observe closely when doing field work in Peru with one of my former postgraduate researchers, Stella Watts, has a beak longer than its body, an adaptation for feeding from nectar in very long-tubed flowers, but it will also take insects from other flowers (Moore 1947, Remsen *et al.* 1986). Sunbirds likewise will feed on arthropods and fruit as part on an omnivorous diet. This makes sense, because nectar as a food is rich in carbohydrates but deficient in most other nutrients (bees and other flower-specialist insects make up for this by consuming pollen).

There is an Old World/New World split here, as hummingbirds are confined to the Americas whereas those other groups are found in Africa, Asia and Australasia. I should really say that hummingbirds are *currently* confined to the New World, because fossil evidence shows that they occurred in Europe during the Oligocene about 30 million years ago (Mayr 2007, Mayr and Micklich 2010). This ancient bird was dubbed *Eurotrochilus inexpectatus* – 'the unexpected European hummingbird' – a fitting name indeed. This is not the earliest evidence of bird pollination, for a 47-million-year-old fossil bird of a previously undescribed group was found with pollen grains preserved in the gut area (Mayr and Wilde 2014). However, this could also represent flower eating rather than nectar feeding and legitimate pollination.

In addition to the bird families that specialise in visiting flowers, a number of generalist passerines have been shown, or are suspected, to be pollinators. These include species from families as varied as the leaf warblers (Phylloscopidae), the woodpeckers (Picidae) and the tits (Paridae) visiting flowers of Asian rhododendrons, African aloes, and Caribbean trees (Dalsgaard 2011, Huang *et al.* 2017, Diller *et al.* 2019, Ollerton *et al.* 2019b). This is clearly a global but under-appreciated phenomenon, and is perhaps best studied in the Canary Islands, where flowers from several families have evolved relationships with birds such as the Canary Islands chiffchaff (*Phylloscopus canariensis*), a common species found throughout the different habitats on these islands (Ollerton *et al.* 2009a, Olesen *et al.* 2012, Fernández de Castro *et al.* 2017). I'll say more about these flowers in the next chapter.

Bats and other mammals

In excess of 230 species of bats are known to act as pollinators (Table 2.1), servicing the flowers from more than 500 plant species in at least 67 families (Fleming *et al.* 2009). Bat pollination is confined to the subtropics and tropics and, as with the birds, there is a biogeographic split in the types of pollinating bats found in different parts of the world. In the New World it is the American leaf-nosed bats (Phyllostomidae) that visit flowers, while in most of the Old World it is the flying foxes and their relatives (Pteropodidae), which are found across large parts of Africa, Eurasia and Oceania. Finally, in New Zealand two species of short-tailed bats (Mystacinidae), one of which is almost certainly extinct, are not specifically adapted to flower feeding but are known to be pollinators of some native plants (Lord 1991). These three groups of bats are only very distantly related to each other, again like the birds, and have independently converged to exploit an abundant resource (nectar and, to some extent, pollen). Some species have evolved extraordinary adaptations to exploit flowers, including a tongue that is longer than its body in the case of one species from Ecuador (Muchhala 2006). However, as we saw with the birds, no large-bodied animal can subsist entirely on what flowers offer, and all species include fruit or invertebrates in their diet to some extent.

Among the mammals, bats are the most abundant and diverse group that pollinate flowers (Figure 2.10), and they have a clear advantage over other warm-blooded hairy animals: they can fly. However, a surprising variety of mammals belonging to only distantly related orders and families also visit flowers. These include various rodents, bush babies, genets, elephant shrews, monkeys, Madagascan lemurs, and marsupials such as Australian honey possums and South American opossums (Carthew and Goldingay 1997, Regan *et al.* 2015, Amorim *et al.* 2020). Many of these interactions with flowers occur at night, because most mammals are nocturnal. Thus, like moth pollination, mammal pollination has been less well studied than its diurnal counterparts and there are no doubt more mammalian pollinators to discover. For instance, it's only recently that we

FIGURE 2.10 Pollination of *Burmeistera glabrata* (Campanulaceae) by the bat *Anoura cultrata* in Ecuador. Note the streak of yellow pollen on the bat's head. Photo by Nathan Muchhala.

have documented proof that elephant shrews visit and pollinate flowers in Africa (Wester 2015). Although they look superficially like mice or true shrews, these small mammals in fact belong in their own taxonomic order (Macroscelidea) and are most closely related to elephants.

Of the monkeys, apes, bush babies and other primates, perhaps not surprisingly it is the smaller species, with a body weight less than 1 kg, that are more likely to visit flowers for nectar and thus effect pollination (Heymann 2011). However, there is one primate pollinator that is much bigger than most, a large ape with dextrous paws and a greater-than-average curiosity for exploring flowers: *Homo sapiens*. Humans actively pollinate a range of crop plants, and we know that they have been doing this for thousands of years: ancient Egyptian images from 2300 BCE show that farmers pollinated date palms, removing the inflorescences from male plants and placing them in the crowns of female plants to allow wind to carry the pollen to the fruit-bearing flowers. This is a practice that continues to this day. Vanilla provides another example of hand pollination. These orchids originated in Mexico but most cultivation occurs in Madagascar and the islands of the Indian Ocean where the natural pollinators (probably euglossine bees) do not occur. Self-pollination is therefore carried out by hand using small lengths of wood to manipulate the flowers and press the pollen mass against the stigma. The resulting vanilla pod is mature after about six months. We'll return to the question of human agency in Chapter 7 when we consider agricultural perspectives on pollinators and pollination. However, I can't leave this section without at least mentioning the robotic pollinators that have been the subject of a number of news articles in recent years. They are a wonderful example of technology looking for a problem to solve that doesn't exist, and may potentially do more harm than good (Potts *et al.* 2018 – see also p. 122).

Lizards

On first consideration, lizards seem to be the unlikeliest of pollinators. They are rather smooth-skinned and usually feed on arthropods and other small animals. However, flowers from a range of unrelated families, including bellflowers (Campanulaceae), screw pines (Pandanaceae), myrtles (Myrtaceae) and spurges (Euphorbiaceae), have evolved to attract and reward these reptiles (Olesen and Valido 2003, Godínez-Álvarez 2004 – Figure 2.11). The lizards themselves belong to at least 37 species in the lacertid (Lacertidae) and gecko (Gekkonidae) families, and lizard pollination is especially prevalent on islands (35 of the 37 records are from places such as the Balearic and Canary Islands, Mauritius, New Zealand, etc.) This association with islands may be due to the low diversity and abundance of insects, which restricts both the availability of other possible pollinators and of lizard prey, providing a dual reason for the establishment of mutualistic relationships between lizards and island plants (Olesen and Valido 2003). However, lizard pollination in continental habitats may be more common than we realise; it's only recently been confirmed in Africa, for instance (Cozien *et al.* 2019, Wester 2019).

FIGURE 2.11 The Mauritius blue-tailed day gecko (*Phelsuma cepediana*) pollinates a range of flowers, including *Trochetia blackburniana* (top) and *Roussea simplex* (bottom). Some of the lizard-pollinated flowers have brightly coloured nectar that is a visual signal to the geckos (Hansen *et al.* 2006). Photos by Dennis Hansen.

The origins of animal pollination

One often reads in textbooks, or sees presented in natural history programmes, a scenario of flowering plant (angiosperm) evolution that goes something like this. First of all there were the gymnosperms, relatives of conifers and cycads, and they were simple and primitive and were pollinated by the wind. Then the flowering plants evolved and they were much more complex and advanced and were pollinated by insects. That's why angiosperms are so successful today, while there aren't many gymnosperms left. It's an iconic story that's satisfying and tidy and seems to encapsulate the stories we humans tell ourselves about how evolution progresses (substitute 'monkeys' for 'gymnosperms' and '*Homo sapiens*' for 'flowering plants' and you'll see what I mean).

The problem is that this scenario is wrong in almost every respect. In fact the earliest flowering plants evolved, perhaps 170 million years ago, in ecological communities that were complex and contained many insect-pollinated gymnosperms (just as quite a number are insect-pollinated now, including plants such as cycads and *Welwitschia*). Recently fossil finds in Spain, Russia and, especially, China have shown us just how diverse insect pollinators were before the flowering plants evolved. These pollinators include mid-Mesozoic representatives from groups that still pollinate today, such as thrips, flies and beetles. However, it also includes groups that are rarely if ever pollinators at the present time, such as lacewings (Neuroptera) and scorpionflies (Mecoptera) (see my review of this in Ollerton 2017). The role of these insects as pollinators has been deduced by looking at the structure and probable function of their mouthparts, as well as from the presence of pollen grains (Ren 1998, Ren *et al.* 2009, Labandeira *et al.* 2007, 2016, Labandeira 2010, Peñalver *et al.* 2012, 2015, Peris *et al.* 2017).

One important implication of these discoveries is that formerly important groups of pollinators have been supplanted by other insect groups such as the bees, butterflies and moths discussed above, possibly because the plants on which they relied went extinct. That is a sobering story in light of current concerns about plant extinctions. There are lots of reasons why angiosperms are so successful, but evolving insect pollination as a key innovation is not one of them. However, the main period of flowering plant diversification and rise to dominance, beginning about 100 million years ago in the mid-Cretaceous, does coincide with the radiation of some of the main existing pollinator groups, such as the bees (Cardinal and Danforth 2013). Some of the evidence of this co-diversification is coming from amber, fossilised plant resin, which can preserve evidence of interactions, such as bees with orchid pollen masses (pollinia – see Chapter 3) attached, showing these interactions have existed for at least 15 million years, and that the orchids as a group probably began to evolve about 80 million years ago (Ramírez *et al.* 2007).

Conclusions

Bees are an important group of pollinators, but they represent only one major set of insects – and indeed even within Hymenoptera they are not as diverse as the wasps. In this chapter I hope that I have at least provided you with a sense of the diversity of the 'other' pollinators that are, collectively, just as important as the bees, but which receive much less attention and consideration. Although I've been based in Britain for my whole career, I've also been fortunate enough to carry out field work across the world and to engage with pollination ecologists from many countries. It's clear from this, and from reading the ever-growing literature on the topic, that the pollinators of any one region such as Europe, and the pollination biology of the native plants, are only partially representative of what is happening elsewhere in the world. Our ideas, or expectations, of what constitutes a pollinator have in the past been mainly considered from a point of view that's centred in the north-temperate zone, where bees, hoverflies and lepidopterans dominate. Pollination systems that involve unexpected groups such as flower chafers, spider-hunting wasps, lizards, rodents or elephant shrews are frequently considered amusing novelties of natural history. This despite the fact that such taxa can be locally dominant (Olesen and Valido 2003, Ollerton *et al.* 2003, Johnson 2004, Shuttleworth and Johnson 2009a) and may be much more widespread than is currently known. Not only that, but discoveries of new fossils, such as hummingbirds from Europe, make us rethink our understanding of the biogeography of species interactions.

Science is about explaining the patterns that we see around us: why certain things are as they are and not something else. Exceptions to general patterns provide us with important tests of our explanations, and that's certainly true when it comes to generalising about the biology of different groups of pollinators. Exceptions to virtually all of the generalisations I've mentioned above can be found across the world's pollinating fauna. There are parasitic bees, nocturnal bees, and bees that eat rotting flesh ('vulture bees' – see Camargo and Roubik 1991). Nocturnal butterflies and diurnal moths challenge our assumptions of lepidopteran biology, while the vegetarian masarines do the same for the wasps. Studying these exceptions allows us to consider how certain animal and plant lifestyles evolved and how their ecology functions. There's no doubt, however, that many discoveries about the biodiversity of species interactions remain to be made, not only in the tropics but also in apparently well-studied parts of the world such as Britain. Let's go back to Figure 2.1 for an example. The common carder bee (*Bombus pascuorum*), one of its nectar and pollen sources, greater knapweed (*Centaurea scabiosa*), and the specialist plant parasite knapweed broomrape (*Orobanche elatior*) were all described as species by the late eighteenth century. However, it wasn't until I studied their pollination biology with some of my students more than 200 years later that it was discovered that the common carder bee was the main pollinator of knapweed broomrape, a rare case of a plant host and its parasite sharing pollinators (Ollerton *et al.* 2007a). Familiar species often surprise us when we consider their interactions with other species. In the next chapter we will look at this from the plant's point of view – to discover what it means to be a flower.

Chapter 3

To be a flower

It's almost impossible to escape from flowers. They provide both a backdrop and a foreground to so many aspects of our lives, marking seasonal celebrations such as Christmas (poinsettias) and Easter (daffodils), and life events like weddings and funerals in many of the world's religions. Flowers are a cornerstone of human creativity, influencing countless poems and song lyrics. But it is in the visual arts and crafts where we see their particular power. Beautiful gardens are an obvious example, but floral emblems and designs have been used in jewellery, in clothing, and as decorative devices probably for as long as people have dressed and ornamented themselves, and sheltered within built structures. Flowers had a particular symbolism for the ancient Egyptians, and we know that they wore collars of both real and artificial blooms, especially cornflowers and water lilies. Three thousand years later the hippies were doing much the same; it was called 'flower power' for good reason.

Illustrations of flowers appear in murals and on ceramics throughout the ancient world. Later they feature as religious emblems in cultures as diverse as medieval Europe and Aztec Mesoamerica sometimes so stylised that we forget their origin: in heraldry, for instance, the *fleur-de-lis* is thought to derive from the flowers of a species of *Iris*. Hundreds of years later, during the art nouveau period, floral motifs were developed into semi-abstract shapes that were then re-imagined on the psychedelic shirts and posters of the 1960s. The current craze for houseplants in Britain, especially orchids, is just the latest demonstration of the timeless appeal of flowers.

An important reason why flowers are used as inspiration in writing, art and design is that, as well as being often beautiful, they are incredibly variable in colour, shape, size, odour and structure (Figure 3.1). This is what we term the *phenotype* of the flower, as opposed to the underlying genetic influences, the *genotype*.

Because flowers are so ubiquitous in our lives we rather take them for granted, in the sense that we don't really appreciate them on a deeper biological level. Ask most people what flowers are for, what their function is, and they will probably reply with something like 'to look beautiful'. If pressed they might add '... for pollinators' – but the central notion is that flowers are aesthetically pleasing for people. This may be true for those horticultural varieties that have been purposely bred to appeal to our senses, but what about their wild progenitors? Why do they

FIGURE 3.1 A small sample of the staggering diversity of flower and inflorescence size, shape and colour to be found in the flowering plants.

look the way they do? Clearly it's something to do with attracting pollinators, but therein lies a conundrum. Why should an over-brained bipedal ape like *Homo sapiens* be so fascinated by plant parts that have evolved to attract completely different groups of animals? Why are people not quite as enchanted by roots and stems and leaves?

If you believe that the natural world and all its diversity was created by a supreme being, then the answer is simple: God made flowers to look beautiful so that Adam and Eve (and their descendants) could enjoy them. That's certainly an argument I have heard Creationists use. Back in 2006 I (ill-advisedly) engaged in a public debate with an Australian fundamentalist who made the mistake of getting into an argument about flowers with me. 'Why is it', he asked, 'that sweet peas are so attractive? In my garden they are never visited by bees, because they self-fertilise. Clearly God made them for us.' I politely suggested he grow some different *Lathyrus* species, ones that had not been selectively bred by people.

If, on the other hand, you accept the scientific evidence that the biological richness of the planet has evolved over many hundreds of millions of years, and that flowers appeared long before people, then we must seek another explanation for why we find them so appealing. The simplest explanation is that the things that make flowers attractive to pollinators are also things that appeal to our senses and sensibilities in a number of ways: our colour vision means that we react to bright colours; our sense of smell responds positively to pleasant smells; the usually symmetrical arrangement of the parts of flowers appeals to something deep-seated within our psyches that finds such symmetry pleasing.

However, it's not actually true that flowers are beautiful to attract pollinators. Pollinators do not care if a flower is aesthetically pleasing. What a flower-visiting animal is concerned with is obtaining a reward, usually (though not always) nectar or pollen. We can demonstrate this using artificial flowers containing sugar-water: bees and butterflies will learn where they are and visit them regardless of their colour; they can be black or brown, it doesn't matter. The point is that the insects associate the reward with a certain colour or pattern and keep returning to seek that colour or pattern, and the reward it promises, until such time as the reward disappears.

The colour, scent and symmetry of flowers, and their huge variability, is a by-product of several things: the requirement of flowers to be distinguishable from one another; the interplay of long- and short-distance attraction of pollinators; the need for flowers to defend themselves from floral larcenists who would steal nectar or pollen; defence against flower and seed predators; and a host of other biological functions that have nothing to do with aesthetics.

These are the ideas that I want to explore in this chapter, to really get to grips with what it is to be a flower, and to understand how flowers function and how they manipulate the behaviour of pollinators. There's so much more to flowers than just their decorative or symbolic uses and the attractions of pollinators, but to really understand this we first have to get back to basics and think about the structure of a typical flower.

All flowers are the same, all flowers are different

There is a great German word for it: *Bauplan*. The literal translation is 'building plan' and it refers to the basic blueprint of an organism. For example, all tetrapods (amphibians, reptiles, birds and mammals) have a spinal column, a head containing a brain, and four limbs (except where these have been lost during evolution, as in snakes). In contrast, adult insects (among other features) always have three pairs of limbs and two pairs of wings, though the latter are often more or less reduced in certain groups. The *Bauplan* of a flower involves a rather predictable set of organs arranged in whorls (see Figure 1.1). The two outermost whorls (sepals and petals) are sterile in the sense of not playing a direct role in reproduction. The sepals form a protective sheath around the developing flower and are usually green; these both defend the flowers from herbivores and the weather, and they photosynthesise, fuelling the production of the rest of the flower. Next come the petals, which usually serve to attract pollinators both by being colourful and, in many cases, as the site of scent production. The next whorls are fertile. Pollen, containing the male sex cells (gametes) is produced by the stamens, specifically the anthers; ovules containing the female gametes are contained within the ovary, which forms part of the pistil. At the base of the pistil there is often a ring of tissue that secretes nectar, termed the nectary. Note that the terminology used to describe these different sets of organs varies a lot, with some authors favouring different terms, and there are further words that define collections of organs. For example, calyx is the collective term for all of the sepals, corolla for all of the petals. Calyx and corolla combined form the perianth. Androecium refers to the male organs, gynoecium to the female. A collection of flowers all on the same stem is termed an inflorescence, though sometimes this is an inflorescence of only one flower. Botanical terminology can be a bit confusing.

When we speak of pollination we are referring to the movement of pollen, containing the male sex cells, to the receptive female surface, the stigma. Subsequent events – involving the germination of the pollen, the growth of a pollen tube down the style, and delivery of the male gametes into the ovules – then result in the fertilisation of the ovule and the production of a seed. This is an important distinction, because while pollinators control much of the biology that is involved with the delivery of pollen to stigma, it is the plant that controls fertilisation, for example by rejecting pollen that has come from one of its own flowers (if it is self-incompatible) or from the flowers of a different species.

All flowers are built to this basic plan, and the order of the organs (sepals, petals, stamens, pistil) is almost invariant. Among the 352,000 or so extant flowering plants, only a few exceptions are currently known, involving species in the genus *Lacandonia* (Triuridaceae) and *Trithuria* (Hydatellaceae). In these species the flowers are 'inside out' with male parts in the middle surrounded by female organs (Rudall *et al.* 2016). The comments I made regarding generalising about pollinators at the close of Chapter 2 thus also apply to flowers. Most of the variation in flower structure is, however, based on that same *Bauplan*, and it includes: varying functions (the sepals of *Fuchsia* species are colourful to attract pollinators, for instance); loss of one or more sets of organs (e.g. in dioecious species with separate male and female individuals); the compression of flowers into inflorescences that look and function as if they were a single flower, for example in many members of the daisy (Asteraceae) and arum lily (Araceae) families. In addition, flowers occasionally produce organs that do not strictly conform to the sets outlined above, for example coronas such as the 'trumpet' of daffodils (*Narcissus* spp.), the radial filaments of passion flowers (*Passiflora* spp. – see Figure 9.3b) and the various coronal structures of species in the dogbane family (Apocynaceae).

The sizes of flowers can vary enormously too, from less than one millimetre in diameter in species of duckweeds in the genera *Wolffia* and *Lemna* (Araceae) to more than one metre in the parasitic corpse flowers of *Rafflesia arnoldii* (Rafflesiaceae). Although the latter is the largest individual flower, the inflorescence of the titan arum (*Amorphophallus titanum*) functions as a single flower and can reach over three metres in height. Like the duckweeds this is also a member of the family Araceae: it's incredible how diverse flowers can be even between closely related species, though their general form remains similar enough to be used for taxonomic purposes (see Chapter 1). This says a lot about the power of evolution to affect floral phenotype, and there is a bewildering variety of flower shapes, arrangements of petals, colours, and chemical makeup of scents.

The odours of flowers can be made up of dozens of individual chemicals, each of which must be manufactured by the plant. Recent advances in chemical analysis are allowing us to dissect these fragrances and assess their individual role in attracting particular types of pollinator. This in turn is giving us insights into how natural selection by different groups of animals has driven the diversification of some groups of plants as populations become adapted to different pollinators and ultimately evolve into new species (Johnson and Steiner 1997, van der Niet and Johnson 2012, van der Niet *et al.* 2014). These processes explain a lot of flowering plant evolution – but not all of it. For instance, wind-pollinated plant families can be very diverse but this is clearly the result of adaptation to aspects of the environment such as different soil types and climates rather than interactions with pollinators. Likewise the evolution of genera such as *Euphorbia* (see Chapter 1) and large groups of succulent Cactaceae and Crassulaceae seems to be driven more by abiotic than biotic factors.

All this variation is one of the reasons why I study flowers: there seems to be an endless diversity. I have a bit of an addiction to visiting botanical gardens and keep a life list of all of those I've been to; each time I visit a garden, without fail, I see a flower that I've not previously seen that surprises me in some way. What is especially fun is to try to work out what a particular floral phenotype 'is for'. How does it function? What pollinates it? How do those pollinators interact with the parts of the flower? The answers to those questions are often not as straightforward as we might imagine.

The functions of flowers

It seems natural to separate out flowers into their component parts and to consider the function (singular) of each part. But floral parts frequently play multiple roles across different flowers, or even in the same flower. Petals can be both attractive and restrictive, for example forming a long tube that prevents short-tongued insects from stealing nectar. Sepals can protect the developing flowers in bud and stop nectar robbing when the flower opens. Pollen clearly has a primary function in reproduction but can also play a role in attraction (through colour or scent), as well as offering a reward for pollinators. All these roles are integrated into flowers in such a way that they function as a single unit of reproduction: removing floral parts can severely hinder that function or change the mechanical relationship between flower and pollinator with the result that effective pollination is curtailed. In fact, this is one of the ways in which pollination ecologists test hypotheses about the roles of different floral strategies, by removing them or augmenting them to see how they influence pollinator effectiveness. One example of such an experiment, involving the manipulation of the flower heads of wild carrot (*Daucus carota*), is presented in Chapter 5.

A striking feature of the flowering plants, and one which has played an important role in their current ecological dominance, is the ability of species to grow under different conditions and to modify their growth to suit, something that we term 'phenotypic plasticity'. For example, many plants that grow in full sun will be low in stature and with relatively small leaves, but the same plant in shade will be taller, with larger leaves, in order to capture as much of the available light as possible. Likewise, roots will grow deeper into soil during drought conditions in order to hunt for water. But the same is not true of flowers. As a general rule, flowers do not increase or decrease in size or shape in response to environmental conditions to anywhere near the same extent as leaves, stems or roots. That's because they have to retain a high degree of integration of their component parts in order to function. However there is some variability in things like flower colour and petal size that is probably environmentally determined, and certainly the number of flowers that are produced and the timing of flowering can be hugely influenced by factors such as the weather, which we will consider in more detail in Chapter 6.

To examine floral functions in more detail it will be useful to consider a few case studies of how flowers actively participate in the process of pollination, rather than simply sitting and waiting for pollinators to do their thing. By behaving in certain ways, flowers manipulate the behaviour of their pollinators to the advantage of the plant, and to demonstrate this I will focus on three aspects of flower phenotype: scent, nectar and, to begin, colour.

Flower colour change in the Canary Island wallflower

The Canary Island of Tenerife is one of my favourite places in the world, and since 2003 I've made almost 20 trips there. In that year we picked it as a destination for an undergraduate field course, and it was exactly the right choice. Lying about 300 kilometres off the coast of North Africa, it's remote enough for a unique and diverse flora and fauna to have evolved, but close enough to the UK to be accessible with a short flight. It's also one of the tallest islands in the world, topping out at 3,718 metres. This steep altitudinal gradient, and a complex topography, allows a range of distinct plant communities to exist on a relatively small island that is only about 80 kilometres in extent across its long axis. All of this has drawn explorers and biologists to the island for centuries.

During an intense week of field work we explore many aspects of the island's ecology and biodiversity, and make a point of repeating some data collection from previous years to assess long-term variability in the systems we are studying. When we're on Tenerife I'm fond of telling the students the tale of how Charles Darwin was excited at the prospect of visiting the island in 1832 when he was voyaging on HMS *Beagle*. He had read the accounts of Tenerife published by Alexander von Humboldt and others, and was keen to see 'this long wished for object of my ambition ... perhaps one of the most interesting places in the world' (Darwin, *Beagle* diary, 6 January 1832). It was not to be. An outbreak of cholera in Britain meant that the *Beagle* and its crew had to be quarantined for 12 days, time the captain felt could not be wasted. The *Beagle* sailed south and Darwin never again had the opportunity to visit Tenerife. During the telling of the tale I refer to Tenerife

FIGURE 3.2 The Tenerife pine forest (Corona Forestal), the main habitat of the Canary Island wallflower (*Erysimum scoparium*), an individual of which can be seen bottom right.

as 'Darwin's Unrequited Isle' and point out how lucky the students are to be able to visit, easily and cheaply, such a wonderful natural experiment.

If Darwin had been allowed to explore Tenerife he would no doubt have seen the Canary Island wallflower (*Erysimum scoparium*). This endemic member of the cabbage family (Brassicaceae) is common in the pine forest and subalpine zone above about 1,500 metres (Figure 3.2). The species is unusual in that it has flowers that change colour: they are pure white when they first open, but from the second day onwards they darken to violet, then ultimately purple, as the flowers accumulate pigment (Figure 3.3). The purple flowers stay on the plant for up to 10 days and, from a distance, these are the flowers that are most noticeable: in the spring following a wet winter the steep hillsides can shimmer in a purple haze.

As the flowers age and change colour they also stop producing nectar. The pollinators, which are endemic solitary bees, learn to associate white flowers with more reward and focus their attention on the newly opened blossoms. This is clearly an evolved strategy, as it benefits the plant to have its most recent flowers preferentially visited, rather than the older flowers that have already received pollen. You can see this in Figure 3.4 – as the flowers age the bees visit them less often, until eventually they ignore them altogether.

This plant strategy of retaining flowers and changing their colour as they age is an unusual one. Most plants don't bother; they just shed the petals so that

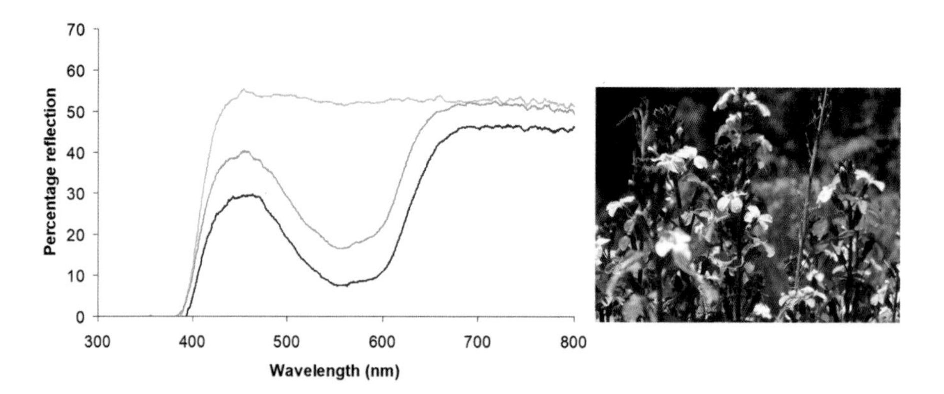

FIGURE 3.3 The flowers of *Erysimum scoparium* change colour with age. The graph shows the reflectance of the flowers at different wavelengths, measured in nanometres (nm), which are billionths of a metre. At short wavelengths, less than 400–500 nm, we move from ultraviolet into the blue end of the spectrum. Around 700 nm appears red to our eyes. As the flowers age they change from white (top line) through violet (middle line) finally to purple (bottom line). Note that the youngest flowers, at the top of each inflorescence, are white, and that the older flowers are darker.

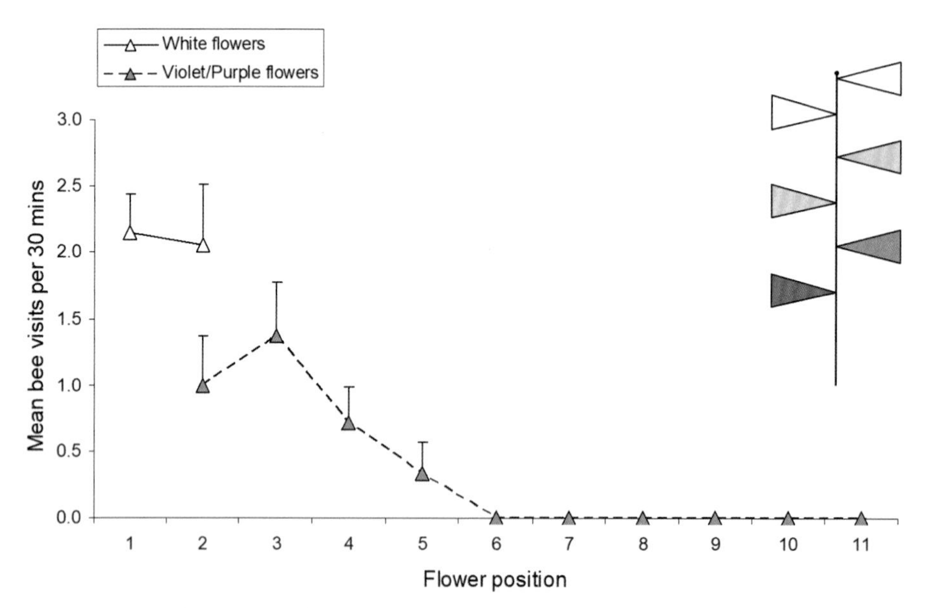

FIGURE 3.4 As the flowers of *Erysimum scoparium* age (left to right on the flower position axis) they receive fewer visits from bees, with the white flowers receiving far more than the flowers which have started to accumulate pigment.

pollinators ignore the old flowers. Only a small proportion of flowering plant species show floral colour changes, but it has repeatedly evolved and occurs in more than one-fifth of the angiosperm families (Weiss and Lamont 1997). So why does the Canary Island wallflower retain those old flowers? What's the advantage to the plant? The answer is that they act as a medium- to long-range advertisement to pollinators (remember, the plants look purple from a distance). In a paper I published with some of my students from the field course we demonstrated this by removing the purple flowers from experimental plants and seeing how it affected the bees' rate of visitation: plants without purple flowers were paid much less attention by the bees (Ollerton *et al.* 2007b).

It's a very intriguing system and a really nice example of how flowers can manipulate the behaviour of their pollinators, in this case using a phenomenon known as associative learning. The bees learn to associate white flowers with a reward (nectar) and purple flowers with little or no reward. This kind of pollination system, in which a plant influences the behaviour of its pollinators to its own advantage, has been demonstrated in other species too, such as lupines (Gori 1989). However, there are (broadly) three types of flower colour change, varying according to timing and position of the change in colour: post-anthesis, as just described, which happens after the flower opens; pre-anthesis, which occurs before flowers open (this is the type that gives red-hot pokers (*Kniphofia* spp.) their name); and change in colour of just parts of the corolla, for example the annulus around the centre of some forget-me-nots (*Myosotis* spp.). Each of these functions in a slightly different way, and there is still a lot that we don't understand about the circumstances under which they evolve.

The bird-pollinated plants of the Canary Islands

Another thing that we have studied on Tenerife is an amazing guild of bird-pollinated flowers involving about sixteen plant species in eight genera (Alayón 2013). This was not as part of the field course but in a project that involved one of my former PhD researchers, Louise Cranmer (whom you'll meet again in Chapter 12), and a long-standing colleague, Lars Chittka. As long ago as the 1950s it was noted that these flowers bear some of the hallmarks of classically bird-pollinated blooms such as bright colouration (reddish or orange), fairly sturdy construction, and the production of quite large amounts of nectar (Figure 3.5). However, there's a problem: the Canary Islands have no specialised flower-visiting birds, a paradox that was first noted by the late Stefan Vogel, one of the founders of the modern science of pollination biology.

Vogel suggested that perhaps the plants had originally been pollinated by sunbirds (Nectariniidae) which had subsequently become extinct in the Pliocene (between 5.3 and 2.6 million years ago), and that these plants were relicts of the former lush vegetation that once covered the Saharan region (Vogel 1954, Vogel *et al.* 1984). Although Vogel knew that the flowers were visited (and probably

FIGURE 3.5 One of the guild of bird-pollinated flowers on the Canary Islands, *Digitalis* (*Isoplexis*) *canariensis*, which grows in the laurel forest on Tenerife.

pollinated) by generalist passerines (see Chapter 2), he assumed that these birds were 'secondary' pollinators that had discovered the flowers after colonising the islands in the Pleistocene (2.6 million to 11,700 years ago).

It's a neat idea, but as soon as I read about it I felt that the story was wrong. For one thing, no sunbird fossils have ever been found on the island, though fossils of other birds such as finches and seabirds have been discovered. Absence of evidence is not evidence of absence, of course, and such fossils might be found in the future. More fundamentally, however, it seemed a large stretch of the imagination to think that these flowers could survive in their original, sunbird-pollinated form for thousands of years, waiting to be rediscovered by a different group of birds. It seems more likely that the plants would either go extinct or (fairly quickly) evolve to be bee-pollinated. Subsequent field work by researchers from Spain and Denmark confirmed Vogel's original observations that the flowers are now mainly visited by generalist passerines such as chiffchaffs and blue tits (e.g. Olesen 1985, Valido *et al.* 2002, 2004, Dupont and Olesen 2004), but up until that point there was little evidence of the effectiveness of these birds as pollinators. That was Louise's job – to actually quantify how much pollen the birds removed and deposited during their visits.

By observing the flower visitors, collecting flowers in different stages to look at pollen removal from anthers and receipt onto stigmas, and bagging flowers to exclude pollinators, Louise was able to show that these avian visitors are extremely effective pollinators of three species of Canary Island bird flowers. Furthermore, these plants show at least one quite specific adaptation to pollination by a group of birds that only rarely visit flowers: they have evolved rather long individual flower lifespans, up to 21 days in some instances (Ollerton *et al.* 2009a). In order to prevent insects from stealing the accessible nectar that's sitting around in these flowers, the plants have evolved two strategies. First of all, insects find them hard to see as they reflect little light in the ultraviolet part of the spectrum and do not show a strong contrast with background foliage when viewed through a bee's eyes. Visits by insect are therefore rare. However, if any do find the flowers, the plants have packed them with unpleasant chemicals to discourage further visits. The nectar of the Canary Island foxglove (*Digitalis* (*Isoplexis*) *canariensis* – Veronicaceae – Figure 3.5) for instance is bitter and causes one's tongue to go slightly numb, as I learned from personal experience.

Since we published that research, these plants have been the subject of further studies that have increased our understanding of these most intriguing flowers. It's now known, for instance, that generalist birds are also effective pollinators of other species within the guild, such as *Navaea phoenicea* (Malvaceae – Fernández de Castro *et al.* 2017) and that bird pollination evolved in one group, the genus *Lotus* (Fabaceae), about 1.7 million years ago (Ojeda *et al.* 2012). However, biodiversity is rarely simple and straightforward. Some of those species that are included within this Canary Island bird guild are in fact pollinated not only by birds, but also by insects and/or lizards (Olesen and Valido 2003, Valido *et al.* 2004, Ortega-Olivencia *et al.* 2012, Navarro-Pérez *et al.* 2013).

This generalist passerine bird pollination system, with or without other species being involved, is probably more common than we think. But it was probably much more prevalent in previous times when North Africa and the Mediterranean regions were wetter and covered in extensive vegetation. Evidence for this is the fact that it is found in the leguminous Tertiary relict tree *Anagyris foetida* (Ortega-Olivencia *et al.* 2005), which was the first confirmed example of bird pollination in the European flora. Since then plants from other families have also been shown to be bird-pollinated (Navarro-Pérez *et al.* 2013, 2017), and no doubt more are waiting to be discovered. It is interesting to consider that right up until the twenty-first century it was assumed that there were no native bird-pollinated plants in Europe, reckoned to be the best-studied flora in the world. As I noted in Chapter 2, however, once we start to consider interactions between species, even regions with a well-researched biodiversity can surprise us.

There's much else that remains to be discovered about bird pollination, despite it being one of the most carefully examined of all the pollination systems because of the huge amount of research interest in birds in general. One of the features associated with bird pollination, which is frequently repeated in textbooks and research articles, is that the flowers usually have no scent perceptible to the human nose. This is then interpreted as evidence that these birds have no sense of smell, which strikes me as circular reasoning at best, and reflects a persistent (and false) general notion that birds lack a well-developed olfactory system (Martin 2020). In fact this assumption has hardly been tested in bird pollination systems, and I know of only two studies that have assessed whether flower-visiting birds can smell. Both studies looked at hummingbirds, and both found that the birds could associate scents with food in artificial flowers (Goldsmith and Goldsmith 1982, Heringer *et al.* 2005). Many hummingbird-pollinated flowers are indeed scentless, at least to human senses, but there is evidence that at least some have an odour, though its role in attracting birds is unclear (Knudsen *et al.* 2004).

It surprises me that there's so little published on this topic, given how much research has otherwise been done on hummingbirds. Clearly vision is more important for hummingbirds when locating food, but that's not the same as stating that hummingbirds have no sense of smell. It seems to be one of those myths that won't go away, of which there are quite a few in pollination biology. But if birds can actually smell, why are bird-pollinated flowers (including those in Tenerife) so often scentless? Maybe it's because producing an odour is physiologically expensive for a plant and largely a waste of energy and resources when trying to attract an animal that mainly relies on its sense of sight to locate food. But could it be also that the flowers have evolved a strategy of olfactory inconspicuousness to avoid detection by nectar and pollen-robbing bees? It's a nice example of how, through evolution, plants can optimise their flowers to suit different kinds of pollinators, in the process saving energy and resources that can be used for growth.

Fly pollination in the family Apocynaceae

Fly pollination systems are much less well understood than interactions involving other groups of pollinators, such as birds or bees, despite the fact that the Diptera

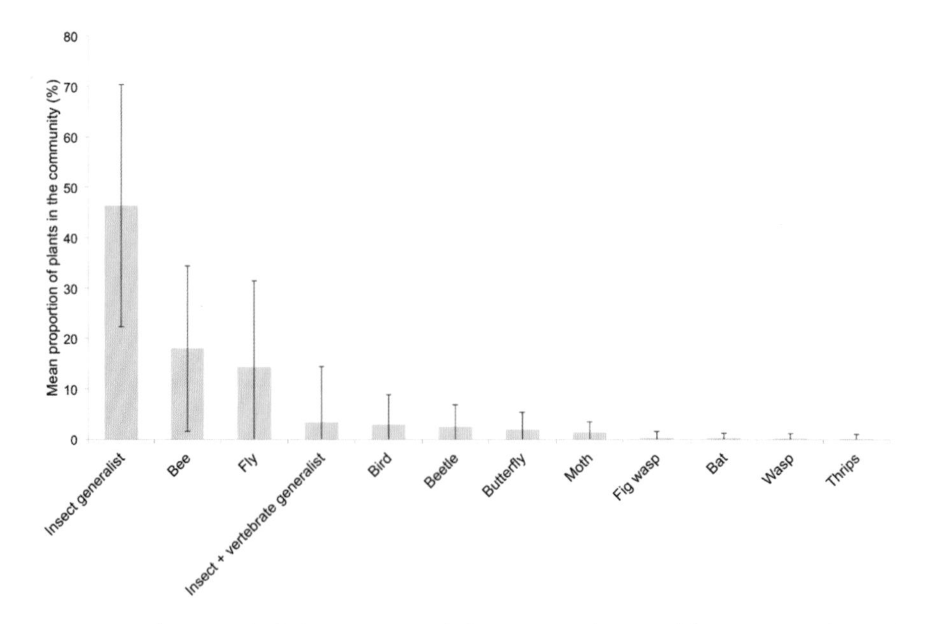

FIGURE 3.6 The mean (± SD) percentage of plant species that use different types of animal pollen vectors. Based on data from 32 published plant communities in which plants were classified into broad pollination systems by the original authors. From Ollerton (2017).

are far more diverse than those groups and that fly pollination is almost as common as pollination by bees (Figure 3.6). However, analyses such as these lose a lot of the biological detail of how plants and pollinators interact, and in fact we can distinguish between several different major forms of fly pollination. For example, some involve flies that behave more like bees and spend a lot of time visiting flowers for nectar, and sometimes pollen in the case of hoverflies (Syrphidae). Flies in this category can show extreme adaptations to flower feeding, with mouthparts which are longer than their bodies and accurately match the depths of the flowers that rely on them for pollination (Johnson and Steiner 1997, Goldblatt and Manning 2000). There are a limited number of fly families and genera that fit into this category, for example members of the Syrphidae, Nemestrinidae and Tabanidae.

Another category of fly pollination is much more diverse and involves the exploitation of flies that don't normally visit flowers (or if they do it's the occasional visit to an umbellifer or a daisy to recharge on nectar). These flies are frequently duped into visiting flowers because the plants tap into deep-seated, hard-wired behavioural traits like egg laying, mating or feeding. There seems to be no limit to the number of different types of flies that fit into this category. For an example I'll turn to a group of plants that has been the focus of much of my research, the dogbane and milkweed family Apocynaceae. With more than 5,300 known species (and many more being described every year), Apocynaceae is one of the top ten largest flowering plant families, with a global distribution, though most of its diversity is in the tropics and subtropics. Britain has no native species, though *Vinca major* and *V. minor* are naturalised, some *Asclepias* species are grown as garden ornamentals, and others are grown as house plants. This family has proven to be endlessly fascinating in its diversity of interactions with pollinators, and in the range of adaptations shown by its flowers. Most species produce their pollen not as free grains, as the majority of flowering plants do, but as coherent packages known as pollinia (singular: pollinium). Only one other family does this, the orchids (Orchidaceae), which are completely unrelated to the Apocynaceae. One outcome of producing pollinia which attach to an animal is that it's easier for pollination ecologists to determine which are the legitimate pollinators of the flowers, a point I'll return to in the next chapter.

As befits such a large family, the range of animals known to pollinate species of Apocynaceae is equally wide, and at the present time pollination by many types of bees, butterflies, moths, flies, wasps, beetles, hemipteran bugs and birds is known, while bat pollination is predicted for at least one species (Ollerton *et al.* 2019a). Since we completed that analysis we have added cockroach pollination to the list of confirmed systems in the family, which is one of just a handful of cases (Xiong *et al.* in press).

Fly pollination is especially common within the family and has evolved independently in diverse groups in different parts of the world. Some of these flowers are quite clearly mimicking animal corpses, for example the well-known 'carrion flowers' in the genus *Stapelia* and its relatives (collectively termed the stapeliads – Figure 3.7).

FIGURE 3.7 Representative stapeliad taxa, showing the range of flower colour, shape and size. Top: *Apteranthes europaea* var. *judaica*, *Tridentea dwequensis*, *Caralluma adscendens*; centre: *Hoodia gordonii*, *Echidnopsis dammanniana*; bottom: *Orbea schweinfurthii*, *Tromotriche longii*. Photos by Ulrich Meve, University of Bayreuth.

FIGURE 3.8 Representative taxa of *Ceropegia*, illustrating floral diversity in the genus. Top: *Ceropegia aristolochioides* subsp. *aristolochioides*, *C. radicans*; bottom left: *C. rendallii*; middle right: *C. rhyncantha*; bottom right: *C. yemenensis*. Photos by Ulrich Meve, University of Bayreuth.

So convincing are these flowers that pollinating flies lay eggs on them expecting the maggots to be able to feed once they emerge. In fact they starve to death. One might expect that this would act as a strong evolutionary incentive to change behaviour and thus avoid these flowers, but it seems that the flies cannot distinguish between the carrion mimics and the real thing, and the urge to lay eggs is just too great.

Closely related to these stapeliads are members of the genus *Ceropegia*, which employ a slightly different strategy. There's another great German word for these flowers – *Kesselfallen-Blüten*, 'kettle-trap-flowers', usually shortened to 'trap flowers'. Once again, the late Stefan Vogel was an early researcher on these plants (Vogel 1961). If you take a look at Figure 3.8 you can see how they got their German name, though personally I prefer their scientific name: *Ceropegia* translates as 'wax fountain' in Greek. It's a big genus of almost 200 species, all of them (as far as we know) fly-pollinated.

The flowers attract flies, which then fall into the corolla where they are trapped by the smooth internal walls and the downward-pointing hairs that line the tube. The flies are free to move around within the inflated base of the flowers, and at least some *Ceropegia* species (possibly all) provide nectar. The role of the nectar seems to be to position the flies' mouthparts correctly so that they pick up or deposit pollinia, rather than to reward them and ensure that they visit flowers of the same species. However this distinction and the role of nectar as a manipulator of animal behaviour is still debated by scientists, and we may not understand as much as we think we do (Pyke 2016). After a period of time, typically 24 hours, the hairs collapse and the flowers move from a vertical to a more horizontal position, allowing the flies to exit. Unlike the stapeliads, which mainly attract quite large flies related to bluebottles and flesh flies, *Ceropegia* species are usually pollinated by very small flies, typically between one and three millimetres in length, though some of the larger *Ceropegia* species attract bigger flies (Coombs *et al.* 2011). These are the sort of flies that most pollination ecologists would ignore, or slap at (some are blood-feeding midges) while looking at the more charismatic, but ecologically no more important, bees and birds.

The diversity of flies that pollinate these flowers is extraordinary and increasing all the time as data are gathered on previously unstudied species. For example, in the eight years between publishing our two large overviews of the pollination biology of *Ceropegia* (Ollerton *et al.* 2009c, 2017) we added five new families of confirmed pollinators. This diversity currently stands at sixteen families and is still rising. It's difficult to know how many genera and species are included, as many of these groups of small tropical flies have hardly been studied by taxonomists.

An important and largely unanswered question, then, is why these flies visit the flowers. Because most of the pollinating flies are female, it was assumed that the flowers were mimicking places to lay eggs such as rotting plant material, fermenting fruit or mushrooms (Ollerton *et al.* 2009c). This is probably true for some species, but certainly not for all. Recent research by Annemarie Heiduk in Germany has experimentally demonstrated that some species are mimicking the odour of the fluids that struggling insects exude when caught by spiders (Heiduk *et al.* 2010, 2015, 2016, 2017). These flowers are pollinated by 'jackal flies' from the family Milichiidae that, as the name suggest, swoop in and start feeding on prey caught by large predators (in this case spiders). It's very sophisticated and unprecedented – nothing like it had previously been reported in the flowering plants. However, a similar pollination system of 'kleptomyiophily' has recently been described in the unrelated trap-flower genus *Aristolochia* (Oelschlägel *et al.* 2015).

But for most other species of *Ceropegia* we have hardly begun to understand their interactions with pollinators. One species is frequently grown as a houseplant commonly known as 'hearts on a string' (*Ceropegia linearis* subsp. *woodii*). In its native habitat in South Africa it is pollinated by blood-sucking flies of the genus *Forcipomyia* (Ceratopogonidae – a group of flies that also pollinate cocoa – see Chapter 7). Bring the plant into cultivation and grow it thousands of kilometres away in Europe and the pollinators are, once again, species of *Forcypomyia* (Ollerton *et al.* 2009c). Clearly the flowers are producing a scent that is quite specific in attracting these flies, but as yet we don't know what chemicals are involved and why the flies react to it. No doubt there are many other surprises in store for anyone who studies *Ceropegia* and its relatives.

Things we know we don't know

The study of Apocynaceae pollination systems that I mentioned above (Ollerton *et al.* 2019a) was able to gather data for about 10% of the known species in the family. Similarly, a review of Cactaceae pollination by my Argentinian colleague Pablo Gorostiague found published data for approximately 10% of cactus species (Gorostiague *et al.* unpublished). Some smaller groups of plants have been much better studied. For instance, the 280 or so species of penstemons (*Penstemon*, *Keckiella* and relatives) have data available for 49 species, or 17.5% (Wilson *et al.* 2006); pollinators are known to at least fly family level for about one-third of the *c.*200 *Ceropegia* species (Ollerton *et al.* 2009c, 2017); and the genus *Dalechampia* has been intensively studied by Scott Armbruster, who has pollinator data for about two-thirds of the 120 species (Armbruster *et al.* 2009).

Of the super-large families, orchids (Orchidaceae) have been well studied but no one has brought together the scattered data in recent years to provide a summary, whereas the daisy family (Asteraceae) is really not well studied, probably because of the high number of generalists, which are somehow seen as less interesting than the specialists (see Chapter 5). But my gut feeling is that, for the 352,000 species or so of flowering plants as a whole, a figure of 10% is a reasonable estimate of the number of species for which we have anything like good data on their pollinators. In other words, for nine out of ten plant species we have only a limited idea as to what pollinates them. We can make inferences from the shape and colour and size and type of reward offered by many flowers. But even if we are correct in describing a flower as pollinated mostly by bees or flies or birds, there's

FIGURE 3.9 The unusual, waxy, emerald-green flowers of *Deherainia smaragdina* smell of cheese. When the flowers first open (left) they are in male phase, dispensing pollen. After a day or two they enter female phase (right) and the stamens move away from the stigma, which then becomes receptive to pollen.

still a lot of information missing. What type of bees? Are they collecting pollen or nectar or both? Does the flower change colour in the human or insect visual spectrum? How do its anthers move in relation to strategies for self-pollination? Is wind pollination involved? Do other types of animals contribute to pollination? Does any of this vary geographically or between different years?

Plants that are grown in cultivation are often the ones that are most familiar to us, but usually we lack any knowledge of their ecologies beyond the garden or glasshouse. This extends to their pollination requirements, where, in the absence of any information gathered within the native range of the plant, we must sometimes look for clues in their floral biology to predict the natural pollinators. Take the small, evergreen shrub with the wonderfully eliding name of *Deherainia smaragdina*, a Mexican member of the primula family (Primulaceae – Figure 3.9). It's often encountered in botanic gardens, and it was the smell of their flowers that I first noticed: strong and pungent like a ripe blue cheese, or unwashed feet, drawing me to it from the other side of the Palm House at the Royal Botanic Gardens, Kew. At first I couldn't see where the smell was coming from, then I spotted the blooms. They were larger than I was expecting (a couple of centimetres across) given that they were not immediately obvious, and very waxy and stiff to the touch. In fact (to the human eye) they were quite well camouflaged against the plant's own leaves, not at all what one expects from a flower ('smaragdine' means emerald-like, so a very fitting species name). However, camouflaged flowers that rely only on scent for attracting insects are not unknown in the plant kingdom, and probably under-recorded: see for example Adam Shuttleworth and Steve Johnson's work on wasp-pollinated flowers in South Africa, which are cryptically coloured to blend in with the background vegetation (Shuttleworth and Johnson 2009b).

One intriguing thing about *Deherainia smaragdina* is that the bisexual flowers are in a male phase when they first open, moving into female phase after a day or two (Figure 3.9). In the male phase the pollen-bearing stamens are centred in the flower, hiding the female stigma (which is probably not receptive at this stage); over time the stamens move outwards to expose the stigma and the flower goes into female phase. Visiting pollinators pick up pollen from flowers that have recently opened and deposit it on older flowers, allowing the plant to avoid self-pollination of the same flower (though it could happen to other flowers on the same plant). But why the flowers should smell of cheese is a mystery. It's probably attracting a particular type of pollinator, but what they are no one knows – it's never been studied, as far as I'm aware. We might predict from the scent that it's flies, but I think that wasps or beetles are also a possibility. Although widely grown in botanical collections, *D. smaragdina* is one of the 90% of plants for which we have no information about its pollinators.

Final thoughts: flowers as active players in the process of pollination

One of the most rewarding things about teaching new groups of students each year is opening their eyes to concepts and ideas that they have not previously considered. Take for example the idea that plants have behaviour. Students are usually at first sceptical, then amazed, when I suggest this and then give examples of the ways in which plants can react to stimuli and manipulate the actions of pollinating animals and other organisms, such as prey in the case of carnivorous plants, or the symbiotic fungi that inhabit the roots of most species.

As we have seen above and will discover in other chapters, flowers are not simply passive participants that sit and wait for their pollinators to turn up; they play an active role in the process of pollination. Amongst other things, they can time the opening of their flowers and the production of nectar and pollen to periods in the day or the year when their most effective pollinators are abundant (see Chapter 6). They can also change the colour of flowers and at the same time withdraw a nectar reward in order to train pollinators to visit the most recently opened flowers, as we saw with the Canary Island wallflower. And *Deherainia smaragdina* clearly exhibits what can only be described as behaviour when it adjusts the functional sex of its flowers, so changing the role of flower visitors (whatever they may be) from pollen receivers to pollen depositors.

As I hope I've shown in this chapter, to be a flower is to be an active participant on an ecological stage in which flowers and their pollinators are just two elements in a theatre packed with many, many players. How to make sense of all these diverse interactions is the topic of the next chapter.

Chapter 4

Fidelity and promiscuity in Darwin's entangled bank

> All the world's a stage,
> And all the flowers and pollinators merely players …
>
> With apologies to William Shakespeare,
> who wrote something similar in *As You Like It*

The idea of flowers and pollinators (and all the other organisms with which they interact) as actors on an ecological stage seems a little contrived, and like many metaphors it is rather fanciful, but if it's good enough for Shakespeare it's good enough for me. If we consider the daily and seasonal ecological changes that occur over time in all habitats (see Chapter 6) as the plot of a play, then within this time frame the storyline can change and be adapted. As for species, some of them have very fixed roles and stick to the script. These reliable troopers only ever engage with a limited set of other actors, while others are more flexible and improvise with many of the players they encounter, contingent on a variety of environmental and biological factors. There is a complex interplay between the fixed-role specialists and the improvising generalists that's been the subject of a lot of research in recent decades. And then of course there's the audience: those species within a community with which our players rarely, if ever, interact, but which provide structure, microclimate, complexity, and other indirect elements to the performance in which we are interested. But before we explore this further I want to take us back to 1859 and the publication of one of the most important books ever written.

Usually we think of Charles Darwin's *On the Origin of Species* (1859) as a book about evolution. And so it is. But there's also quite a lot of ecology in it. At the end of *Origin* Darwin muses on the complexity of relationships and interdependencies between species in a piece of writing that has become famous among ecologists:

> It is interesting to contemplate an entangled bank, clothed with many plants of many kinds, with birds singing on the bushes, with various insects flitting about, and with worms crawling through the damp earth … and dependent upon each other in so complex a manner …

Darwin used 'entangled' in the first edition and 'tangled' in subsequent editions, for reasons that are unknown. This section is therefore variously referred to as 'Darwin's Tangled Bank' or 'Darwin's Entangled Bank'. Both are correct.

The bank that Darwin was referring to is on his property at Down House in Kent, and it was one he observed many times during his walks through the garden. This notion of a tangle of species 'dependent upon each other in so complex a manner' gets right to the heart of understanding ecological interactions within communities, even if it is rather understated. Individuals are indeed dependent upon each other, but those dependencies are massively complex and range from the absolutely vital (for example, a bee that relies on a single species of plant for collecting its pollen) through to the relatively inconsequential, such as a flower that may be pollinated by dozens of different insects, each one of which plays an individually minor role. These plant–pollinator dependencies are in turn embedded within a set of all kinds of ecological interactions, involving predators of the pollinators (or their prey, in the case of carnivorous flower visitors), the plants and the herbivores that feed on them, parasites and hosts (see Figure 2.4, for instance), fungal symbionts, and many more.

Getting to grips with the complexities of these interactions is important if we are to fully understand the relationships between plants and their pollinators beyond just describing the reproductive interactions of single species of plants, as we did in the previous chapter. After all, this is how the world works: interacting groups of species determine ecosystem processes such as energy flow and nutrient cycling. The classical way of depicting such relationships is as a food web that shows the connections between species: the sun shines → the plants photosynthesise → herbivores eat the plants → predators eat the herbivores → top predators eat all the other animals → then everything that's been pooped or has died is decomposed by fungi and bacteria. It sounds straightforward enough. But for everything but the simplest of ecological communities such webs are always enormously complex and can involve hundreds to thousands of species, and thousands to millions of interactions. It's also never clear, outside of a laboratory setting, where such food webs begin and end. To give just one example, seabirds such as gulls connect the marine and terrestrial food webs of the planet. For that reason ecologists tend to focus on subsets of these webs that especially interest them, for example the predator–prey web, the plant–herbivore web or, of course, the flower–pollinator web.

Flower–pollinator interaction data

Ecologists who conduct ecological surveys in which they identify the plants, and catch and determine the pollinators that are visiting their flowers, are continuing a tradition that goes back many years. Nineteenth-century naturalists were fond of compiling lists of such species for their local area, and these types of compilations exist for quite a few sites, mainly in Europe and North America. The most comprehensive of all such lists, and still the most detailed set of plant–pollinator observations ever assembled, was compiled from surveys conducted at the end of the nineteenth and beginning of the twentieth century by Dr Charles Robertson, a physician based at Carlinville, Illinois, whom we encountered in Chapter 1. Robertson published a series of papers in natural history journals that culminated in his book *Flowers and Insects: Lists of Visitors of Four Hundred and Fifty-Three Flowers* (1928). This is an amazingly detailed set of observations that includes information about 1,428 species of flower visitor, their behaviour such as pollen and nectar feeding, whether they touched the reproductive parts of the flowers, and so forth. It's still considered an important scientific document by those of us who study these interactions, and the data that it contains have been used in quantitative surveys of plant–pollinator interactions in specific plant groups (Ollerton and Liede 1997, Ollerton *et al.* 2019a), to model how plants and pollinators respond when species go extinct (Memmott *et al.* 2004), and as a baseline for subsequent surveys of the plants and pollinators in that part of the USA (Marlin and LaBerge 2001, Burkle *et al.* 2013 – we will come back to these studies in Chapter 10).

Robertson's book is a rarity (I've never seen a copy for sale), but a scan of it can be viewed at the Biodiversity Heritage Library (www.biodiversitylibrary.org/item/43820). The title is slightly misleading, as in fact 456 plants are included, and interactions with hummingbirds are also documented. For ecologists, however, the real value is that this data set, as well as those from other studies, is available for free download from the Interaction Web Database (www.nceas.ucsb.edu/interactionweb) and other similar online repositories. In common with all such interaction networks, the Robertson data can be presented as a matrix with plants in columns and flower visitors in rows, with each cell in the matrix designated as either a 0 or a 1 (see Figure 4.1 for an example). A 0 means that the species do not interact, and a 1 indicates that they do. This can also be done as a quantitative matrix in which, rather than a 1, an interaction is given a value that represents how frequently the interaction takes place, for example the number of pollinator visits per minute. There are advantages and disadvantages to using both qualitative (0,1) and quantitative (0, $n > 0$) approaches. Qualitative networks only tell you that an interaction has taken place, and reveal nothing about how important that interaction was for the plant or the flower visitor. Quantitative networks can

Animal pollinators

Plants	Species 1	Species 2	Species 3	Species 4	Species 5	Species 6	Species 7	Species 8	Species 9
Species 1	1	0	0	1	1	1	0	0	0
Species 2	0	0	1	0	0	0	0	0	0
Species 3	1	1	0	0	1	0	0	1	1
Species 4	0	0	1	1	0	0	1	0	0
Species 5	0	0	0	0	0	0	0	1	0

FIGURE 4.1 An example of a plant–pollinator interaction matrix. In this case nine species of pollinator are interacting with five species of plants. A 1 in the matrix indicates that the plant–pollinator pair interact; a 0 indicates that they do not.

give you some of that information, but the relative importance of species to one another can vary hugely over time and space, and any one matrix may not be representative of what happens over large temporal and spatial scales. In their book *Mutualistic Networks* the Spanish ecologists Jordi Bascompte and Pedro Jordano (2014) explore these ideas in much more detail.

The Robertson data matrix is huge, containing more than 650,000 data cells (456 plants × 1,428 pollinators = 651,168). However, most of these contain a zero, and just 8,887 cells (or 1.4% of the matrix) designate an interaction between a flower and its pollinator. This is common: most plants and pollinators in a community have nothing to do with one another. But even for the small proportion of species in Robertson's community that do interact, how do we make sense of those relationships? We might scan through the more than 650,000 cells of the matrix looking for the pollinator visits to flowers that do occur and trying to see any patterns in those occurrences. But it's a long and tedious process that in any case doesn't really give us many insights into the ecology of these species. To assess these matrices more effectively we need to visualise them in some way, which brings us back to food webs. The advent of powerful computing technology since the late twentieth century has made it easier than ever to visualise networks of interacting species. What Robertson would have taken weeks to hand-draw we can achieve with a few key strokes.

Visualising flower–pollinator interactions

Networks that involve two trophic (feeding) levels such as flowers and their pollinators are termed 'bipartite graphs', and there are a few different packages out there for visualising and analysing that type of data, which vary in their complexity and user-friendliness. The most commonly used is the *bipartite* package in the open-source software R (Dormann *et al.* 2008). If you do a web search for R you can install it free of charge, and then install *bipartite*. There is also a more user-friendly system called Polinode (www.polinode.com) that was developed primarily to visualise business and social science data (the 'poli' part is nothing to do with pollination, that's purely coincidental). However, there's no reason why it can't be used for ecological data: from the point of view of the underlying theory and the patterns that are formed, interactions between different elements are fundamentally the same. This is true whether it's social relationships, disease transmission, online shopping, the way in which computers communicate with one another, or the encounters between pollinators and flowers. It's only the players (we're back to Shakespeare again) and the outcomes that are different.

Using *bipartite* I've plotted the Robertson data and, trust me, it's dense and complex. Visualising and making sense of it all in one go is impossible unless you print it in very large format, so I'm not going to present it here. Instead I'm going to focus on some of my own data, which are much simpler and more straightforward. The data come from surveys of flower visitors to nine species of asclepiads (Apocynaceae subfamily Asclepiadoideae – see Chapter 3) in the grasslands of KwaZulu-Natal in South Africa (Ollerton *et al.* 2003). The first plot (Figure 4.2) shows all of the flower visitors to the plants, regardless of whether they are proven pollinators or not. As you can see it's certainly an entangled bank of interactions. On the left are the plants, with lines linking them to the flower

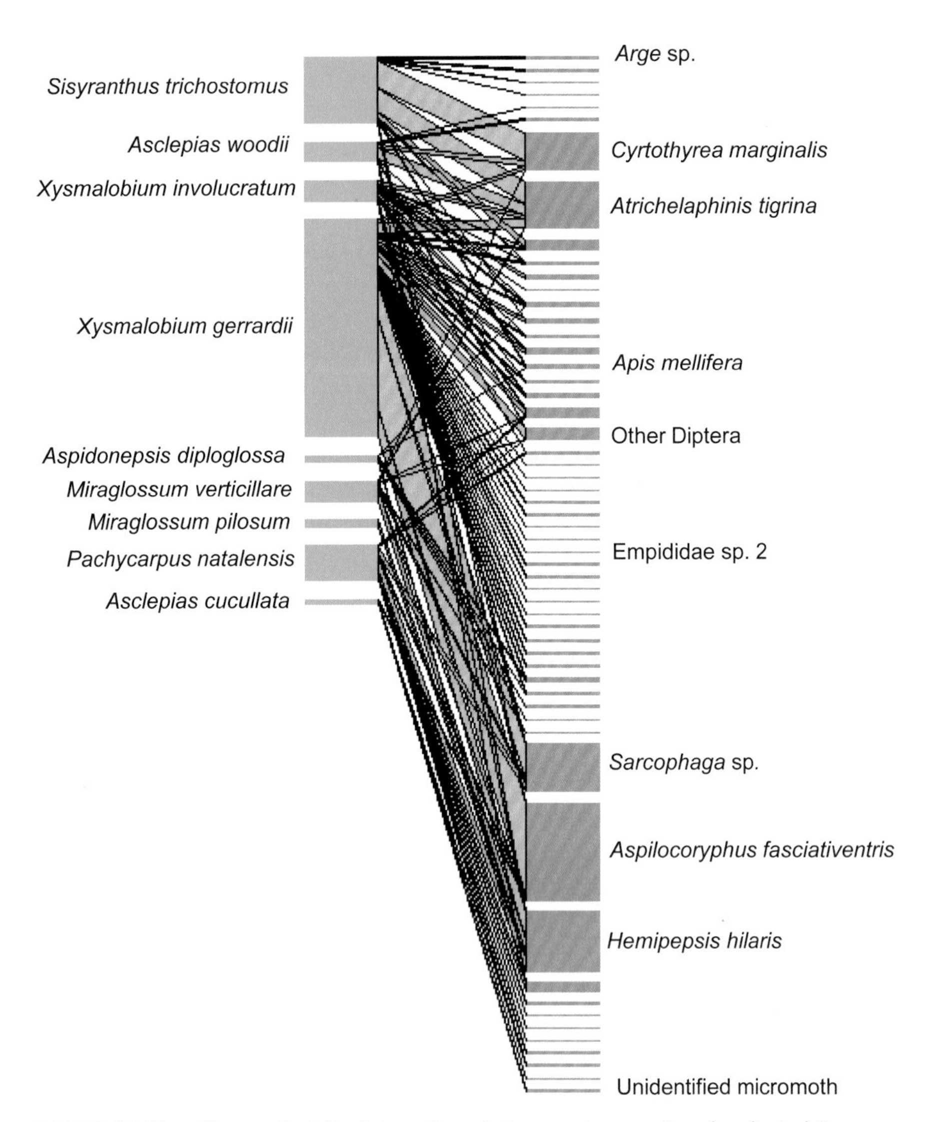

FIGURE 4.2 Bipartite graph of the interactions between nine species of asclepiad (in green on the left) and their insect flower visitors (in blue on the right). A connecting line indicates that the plant and the insect have an interaction, and the width of the line indicates how many insects were observed on the flowers during surveys. The size of the bar of a given species is proportional to the number of species with which it interacts and the frequency of the interactions. Note that most of the names of the insects have been omitted, to aid interpretation. Data from Ollerton *et al.* (2003).

visitors on the right. Each coloured bar denotes one species, and the width of the bar is proportional to the number of different types of insects that visit the flowers and their frequency (in the case of the plants) or the number of different types of plants visited by each insect. Thus, plant species such as *Xysmalobium gerrardii* are visited by many different insects, often very frequently. Other plants are visited by far fewer insects, just one species in the case of *Miraglossum pilosum*. Likewise, the insects vary in their fidelity to the plants: the spider-hunting wasp *Hemipepsis hilaris* (Pompilidae) visits six of the nine plant species, the two beetles *Atrichelaphinis tigrina* and *Cyrtothyrea marginalis* (both Scarabaeidae) visit five and four respectively, while many of the other species were seen on the flowers of only one plant. The hemipteran *Aspilocoryphus fasciativentris* (Lygaeidae) was found on the flowers of more species (seven) than any other insect, but its role as a pollinator is unclear. These insects are sap-sucking herbivores rather than nectar feeders, but they do carry pollinia of one of the asclepiads (*Xysmalobium gerrardii*) – though how often they move between the flowers of different plants is not known. Determining the role of animals as pollinators, and their importance relative to other visitors to the same flowers, is not at all straightforward in generalist species with a large number of visitors.

However, this plot shows only visitation, it does not give an insight into which of the insects are actually involved in pollination. As I mentioned in Chapter 3, the asclepiads package their pollen into pollinia which clip onto the insects and are readily identifiable. This means that the effective pollinators of these plants can be easily determined (Figure 4.3). If we take away the insects that do not carry

FIGURE 4.3 The beetle *Atrichelaphinis tigrina* is a confirmed pollinator of *Xysmalobium involucratum* in the grasslands of KwaZulu-Natal, South Africa. Note the pollen masses (pollinia) attached to the insect's body, legs and mouthparts. Photo by Adam Shuttleworth.

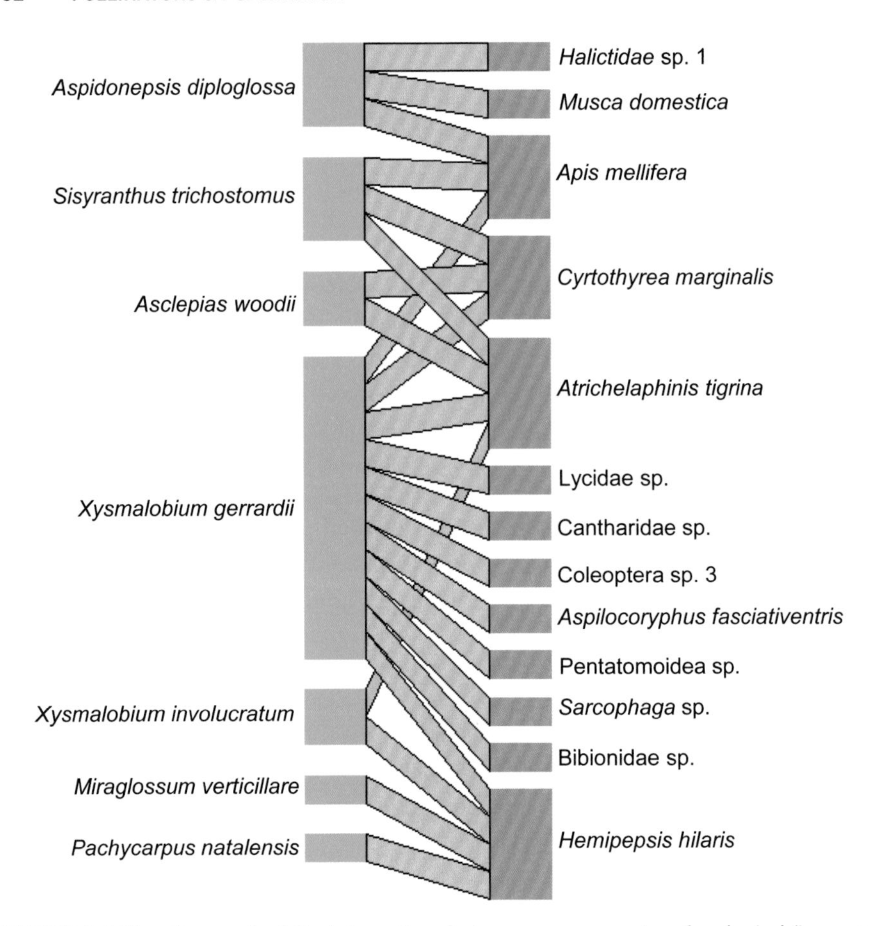

FIGURE 4.4 Bipartite graph of the interactions between seven species of asclepiad (in green on the left) and their insect pollinators (in blue on the right). A connecting line indicates that the plant and the insect have an interaction, and the width of the line indicates how many insects were observed on the flowers during surveys. The size of the bar of a given species is proportional to the number of species with which it interacts, and the frequency of the interactions. Data from Ollerton *et al.* (2003).

pollinia, and re-plot the bipartite graph, we see a similar pattern (Figure 4.4). Two things are immediately apparent. First of all, we've lost two plant species: the visitors to *Miraglossum pilosum* and *Asclepias cucullata* were never observed to carry pollinia. In the case of *M. pilosum*, even though we never observed it, we are 99% sure that the only visitors, those spider-hunting wasps, are the legitimate pollinators just as they are of the closely related *M. verticillare* with similar flowers. For *A. cucullata*, we're not so sure: it could be a generalist or a bee specialist, but the rate of insect visitation was too low to be sure. Secondly, we've also lost most of the flower visitors. In fact the pollination web is only 18% of the size of the flower-visitor web (matrix of 91 cells as opposed to 504 for the flower visitors). However, the overall structure that we saw in the flower-visitor web persists in the pollination web: *X. gerrardii* is still a generalist, and the more specialised species are even more specialised, in some cases having only a single pollinator. On the insect side, likewise, similar patterns are apparent.

I'd like to emphasise two important points here. Firstly, that this is a *minimum* pollination web, i.e. some of the insects that I've excluded will also be legitimate pollinators, it's just that we did not capture any with pollinia attached, as I noted for *M. pilosum*. Secondly, this interaction web is a really a sub-web of the much larger plant–pollinator community at the site, which is an eight-hectare area of moist montane grassland named Wahroonga. Our estimate of the plant diversity on this site is of the order of 300 species, with at least 25 of them asclepiads (Ollerton *et al.* 2003, 2009b, Adam Shuttleworth personal communication). The number of pollinating insects and birds must likewise be one or two orders of magnitude higher than we recorded. However, some of the flower visitors that are not pollinators of the asclepiads we were studying almost certainly pollinate other plants within the community. From that perspective these asclepiads are important because they provide resources that support insects that also have a pollinating role to play, just not in the pollination biology of these particular plants. Interaction network approaches to studying pollination ecology have been criticised in the past, because they often do not distinguish between flower visitors and pollinators, but most of those critics have failed to appreciate these important points.

The patterns of attachment between plants and their pollinators that we see when we visualise such ecological webs are not random in their structure. However, they are also not even in structure, as they would be if each plant had just one or two pollinators, for example. We can see this most clearly when we plot the data in a slightly different way, visualising the matrix as filled and empty cells to distinguish between interactions that do occur and those that don't (Figure 4.5).

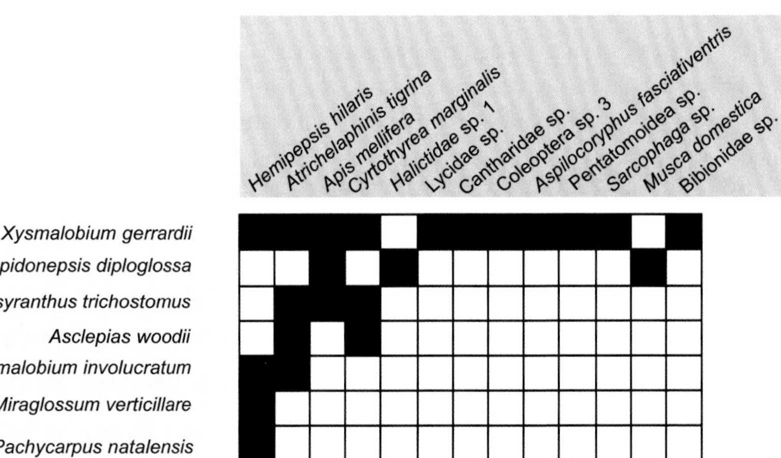

FIGURE 4.5 Interactions between the South African asclepiads and their pollinators visualised as a presence (black squares) – absence (white squares) matrix. The plants and pollinators have been ordered from most to least generalist along rows and columns.

What we observe is that generalist plants link to both generalist and specialist flower visitors, while specialist plants are pollinated by generalist flower visitors. There are no instances of specialist plants being pollinated by specialist pollinators. To some extent this pattern is driven by the abundance of species: *X. gerrardii* was a common species, as was *H. hilaris*, and therefore we are more likely to observe these species interacting. However, abundance of a species is an important ecological trait (some species are naturally rare, others naturally abundant) and so this pattern is not just an artefact. But abundance tells only part of the story: features of flower morphology and nectar characteristics suggest that each plant species either specifically attracts certain types of pollinators, or, conversely, does not filter out any flower visitors (in the case of the more generalist species). So overlying these patterns are restrictions on which flowers can be serviced by which pollinators, and which pollinators can gain resources from which flowers. The specific traits of these plants and animals, such as tongue length, flower depth, petal colour, nectar amount and concentration, and so forth, determine which interactions take place. This can result in the formation of 'modules' of species that interact more strongly with one another than they do with species outside of the module, and there are sophisticated ways of detecting these patterns statistically (Bascompte and Jordano 2014).

The interaction structure shown in Figure 4.5 is termed 'nested' because the pollinators that service the specialist plants are a subset of the pollinators that service the generalists; likewise, the flowers that are visited by the specialist pollinators are a subset of those visited by the generalists. In other words, specialised interactions are nested within generalised interactions. This nested structure was originally described in biogeography for the presence of species on islands: birds on small islands in an archipelago, for example, tend to be a subset of those on larger islands. Nestedness seems to be a universal biological pattern that occurs at a range of scales from molecules to ecosystems (Cantor *et al.* 2017). Almost all of the well-sampled flower visitation webs that I have seen published, from anywhere in the world, show more or less this pattern, and it's maintained even when we focus just on effective pollination (e.g. Ballantyne *et al.* 2015).

As we've seen, in these plant–pollinator communities some species play a major role, such as generalist pollinators that visit lots of different types of flowers, and that tend to be abundant. On the other side of the interaction, generalist flowers attract both these major pollinators and others that have only a minor role and visit only that type of flower. These specialist pollinators tend to be in low abundance, so visiting generalist plants that are pollinated by a lot of different types of pollinator and are common in the habitat makes good ecological and evolutionary sense: a rare pollinator that specialises on a rare plant increases its chances of becoming extinct if that plant itself disappears from the community. The same is also true for the plant, of course: specialising on a specialist pollinator could increase the possibility of its extinction.

Although this nested pattern is a fundamental insight into the structure of the entangled bank of plants and their pollinators, there are exceptions, though they tend to be special cases of two particular types. The first is plants that mimic other things, such as animal excreta, fermenting fruit, or carrion, and which attract pollinators that do not normally visit other flowers. Some of the *Ceropegia* species that I discussed in Chapter 3 are one such example. This is not purely a phenomenon of exotic tropical plants, however: closer to home in Britain cuckoo-pint (*Arum maculatum*) does a similar thing, attracting owl midges (*Psychoda phalaenoides*) with its smell of animal dung (Lack and Diaz 1991, Kite *et al.* 1998). The other exception is pollination systems that have evolved from seed predation interactions, as seen in the relationships between figs (*Ficus* spp. – Moraceae) and fig wasps (Agaonidae), and yuccas (*Yucca* spp. – Asparagaceae) and yucca moths (Prodoxidae). Seed predators (and other types of parasites) in general tend to be more specialised, and that's probably the reason why these exceptionally specific interactions have evolved in this way. I'll discuss plant strategies for being pollinated further in the next chapter. But before that I will consider what these patterns of nestedness and modularity mean for how such communities deal with disturbance.

Robustness and resilience in the entangled bank

In his comments about the entangled bank, Darwin was demonstrating a deep awareness of how nature works. However, compared to twenty-first-century ecologists, he had neither the data nor the analytical tools to properly study its workings. More than 150 years later we now understand that Darwin's insight that species in a community are 'dependent upon each other in so complex a manner' extends to *Homo sapiens*: we are, of course, dependent on the species with which we share the biosphere. At a time when we are making profound changes to the living world we need to understand how networks of interacting species such as plants and pollinators will react to disturbances. To borrow a phrase from Bascompte and Jordano (2014), the 'architecture of mutualistic networks may have profound implications' for how robust or resilient these communities are when subjected to human interference, or natural perturbations.

The terms 'robustness' and 'resilience' have a range of meanings in ecology, but broadly speaking we can think of robustness as being a feature that conveys stability on a community, whereas resilience in this sense refers to the ability of a community of plants and pollinators to rebound from disturbance, for example sudden wild fires or floods, hurricanes, unexpected freezing events, volcanic eruptions, or the impact of a disease. Not to mention the regular assaults that human societies throw at the natural world. In a recent review of how pollinators respond to extreme events Erenler *et al.* (2020) referred to them as SHOCKs – 'Single, High-magnitude Opportunities for a Catastrophic "Kick"' (Figure 4.6).

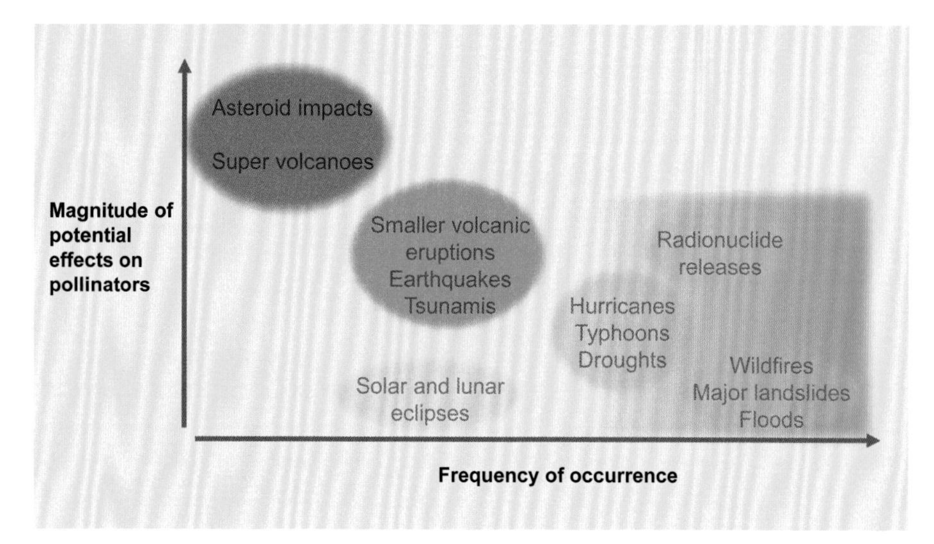

FIGURE 4.6 Large-scale events (SHOCKs) that can affect plant–pollinator communities, arranged along two axes: the likelihood of an event occurring and the magnitude of the event. In general, high-magnitude events are very low-frequency, and vice versa. The green shading indicates the SHOCKs that humans can influence (e.g. via climate change) then mediate directly; low-frequency SHOCKs are largely outside of human influence. The shading of the ovals reflects the scale of the impact on pollinators. Note that direct human transformation of landscapes is not included here as it's rarely a single, rapid process. I discuss human impacts in more detail in Chapters 10 and 12.

However, one of the conclusions of the Erenler *et al.* (2020) review was that we actually know very little about how the interactions between plants and their pollinators respond to such disturbances. The scientific literature is sparse, but growing. Bascompte and Jordano's statement implies that the structure of plant–pollinator communities, with its core of generalist species, and its nestedness and modularity, makes them less prone to disturbance. But it very much depends on the scale of the disturbance: events that happen towards the left side of the graph in Figure 4.6 are likely to wipe out whole ecological systems. Those towards the right will be much easier to recover from.

Until recently, the only work that had been done to test these ideas was to experimentally model what might happen using computer simulations (e.g. Memmott *et al.* 2004). Although this is a useful first step it doesn't capture all of the complexities of real communities, and the ability of pollinators to respond flexibly to events. With this in mind, in 2008 a couple of my postgraduate researchers (Sam Tarrant and Lutfor Rahman, whom you'll meet again in Chapter 12) and I devised an experiment that would test the ability of a plant–pollinator web to respond to a sudden disturbance. Initially we surveyed the plants and pollinators at a grassland site in Northampton to work out the patterns of interactions. From this we established that the most generalist plant was field scabious (*Knautia arvensis* – Caprifoliaceae), and that became the focus of our work. Then Sam,

Lutfor and I, together with a group of undergraduates whom we paid to help us (plus a young woman on her first day of work experience, for whom this was a baptism of fire – it was a very hot day!), systematically combed the site and removed the open flower heads from all of the field scabious that we could find: 14,500 of them in total. A few days later we returned to the site to re-survey the plants and pollinators, then did the same thing a few weeks later once the field scabious had re-bloomed. As a control we also surveyed a separate, nearby site at the same times.

This was the first time such an experiment had been attempted, to my knowledge. I jokingly refer to this approach as 'poking-it-with-a-stick-to-see-what-happens' ecology, but in truth a lot of scientific experiments are like that, from the Large Hadron Collider to artificial warming of forest soils. We may have vague hypotheses that we are testing, but in truth we can only be open-minded as to what to expect to find once the experiment has run.

The results from this experiment really surprised us. The removal of 14,500 flower heads, which we calculated would produce about 500 millilitres of nectar per day, *had no effect whatsoever on species richness or abundance of flower-visiting insects on this site.* All that happened was that the insects which had been feeding on field scabious moved on to the second most generalist flower, lesser knapweed (*Centaurea nigra* – Asteraceae), which thus had a boost in its diversity and abundance of pollinators. However, this effect did not propagate down through the rest of the interaction web, and other plant species were not affected. This was very surprising because half a litre is a lot of nectar and a big loss of resources for those pollinators. We might have expected to see a movement of insects to the adjacent site as the pollinators voted with their wings, but this didn't happen. Once the field scabious had regrown the insects moved back onto those flower heads and the species re-established its position in the network.

Although, in essence, very little changed following our removal of those flower heads, the experiment actually told us rather a lot about how plant–pollinator communities are structured. It told us that there is some level of resilience to disturbance in pollinator communities, that they are able to tolerate a measure of perturbation by locally redistributing themselves onto other plant species. Field scabious and lesser knapweed are similar to one another in that they have compact flower heads containing many individual florets that reflect light towards the blue end of the spectrum, so the movement of insects from one to the other, rather than to plants with different morphologies such as legumes, is not so surprising. However, it is surprising that other plants with similar morphologies, such as greater knapweed (*Centaurea scabiosa*) and creeping thistle (*Cirsium arvense*), did not also receive increased visits. I suspect that the main factor here is abundance of plants. Lesser knapweed was much more numerous than those other species, indeed its flower heads were almost as abundant as those of field scabious, and pollinators tend to select abundant nectar and pollen sources to visit. Thus, another thing that this experiment demonstrated is the flexibility of behaviour shown by flower-visiting insects, that they can quickly adjust their foraging when faced with a loss of favoured resources. This flexibility of behaviour no doubt has been an important factor in the evolutionary success of these groups of insects: individuals which are inflexible and cannot adjust their foraging patterns when faced with perturbations would no doubt be lost from a population very quickly.

The final thing that we learned from this experiment, in which insects were 'absorbed' onto lesser knapweed but not onto the other remaining species in flower, was that sufficient nectar was being produced in that community to support the insects present. Lack of nectar and pollen sources is often considered to be one of the main factors affecting pollinator diversity and abundance at a landscape and national level (see Chapter 10), but at a local level this may not be the case, and even a relatively low-diversity, grass-dominated area such as the site we used can produce more than enough nectar to support the pollinators of that community (though whether there are sufficient pollinators to service all of the flowers being produced, or whether some species or individuals are pollen-limited, is unknown).

Actually, we did learn something else from this study: novel, proof-of-concept experiments such as this are almost impossible to get published in the scientific literature. Although no one had ever tried such an experiment before, and the findings were (I think) very interesting, the fact that this was an unreplicated experiment (plus a control) meant that editors and reviewers did not like the study. I pointed out to one editor that the Large Hadron Collider is also, in effect, an $n = 1$ experiment, but to no avail. However, I have presented this work at a couple of conferences and it generated much interest and follow-on research. Some of this I collaborated with when Paolo Biella began his PhD work in the Czech Republic under the supervision of Jan Klecka. Paolo went further than we did in Northampton and sequentially removed the most generalist plants across replicated communities. Importantly, he also measured reproductive success in these plants by looking at pollen deposition and pollen tube growth. One of the most important findings, I think, is that the effect of such experiments on plant reproduction is highly idiosyncratic – species respond differently and unpredictably (Biella *et al.* 2019, 2020a).

Other pollination ecologists are taking slightly different approaches to manipulating networks. For instance Brosi *et al.* (2017) and others have set up field experiments in which pollinators, rather than plants, are removed, giving further insights into the resilience of such communities.

Plants that attract large numbers of flower visitors, such as field scabious, greater and lesser knapweed, and creeping thistle in British grassland plant communities, and some *Xysmalobium* species in South African grasslands, are hugely important. We consider them to be 'core generalists' as far as supporting pollinator communities is concerned. 'Generalist' because of the wide range of flower visitors, not all of whom are pollinators of that plant, but which may be pollinators of other plant species in the community. 'Core' because they are abundant and sit at the heart of a network of plants and pollinators, around which other, more specialised plant species cluster and benefit from the supporting role the core generalists provide for their pollinators. They sit top left in a nested matrix like those shown in Figures 4.5 and 4.7, as do pollinator species such as the more common bumblebees (*Bombus* spp.), which are likewise core generalists because they service a wide range of plants. Note that we would not refer to these as 'keystone' species (a much misused term) as they are abundant, whereas keystones, by definition, are in low abundance.

A huge range of different species can be core generalist plants; it largely depends on the type of habitat and where in the world it is located. However, they tend to have some things in common. They usually belong to plant families that present their flowers in dense flower heads, such as the daisies (Asteraceae), the scabiouses (Caprifoliaceae) and the umbellifers (Apiaceae and Araliaceae), or produce masses of small flowers over a short period, like the rose family (Rosaceae). The flowers individually produce rather little nectar at a time, but collectively they produce a lot, and (importantly) they have a high rate of nectar recharge. That nectar is also accessible to flower visitors with both long and short mouthparts. Finally, and probably most importantly, the plants tend to be abundant, and it is their great abundance as much as anything else that determines their role as core generalists.

Even within one region the importance of core generalists varies between communities and has a seasonal dimension, with certain species and families playing a role at different times of the year, and frequently in sequence. In Britain, for example, early in the season rosaceous species such as blackthorn (*Prunus spinosa*) and then hawthorn (*Crataegus monogyna*) are important; somewhat later it is umbellifers such as cow parsley (*Anthriscus sylvestris*), then hogweed (*Heracleum sphondylium*). Later still the daisies (Asteraceae) come into their own, though dandelions (*Taraxacum officinale* agg.) are hugely important throughout the season, especially in grassland. In the autumn common ivy (*Hedera helix*) is by far the most important species. However, in a high Arctic community in Canada, on Ellesmere Island at more than 81 degrees north, it is species of *Dryas* (Rosaceae), *Saxifraga* (Saxifragaceae) and *Salix* (Salicaceae) that are the core generalists (Figure 4.7a), while in a tropical palm swamp community in Venezuela the core generalists are species of *Ludwigia* (Onagraceae) and *Hyptis* (Lamiaceae) (Figure 4.7b). If you compare the structure of these two plant–pollinator communities, however, you will see similar patterns too: both are nested in structure, for one thing. These contrasting and similar patterns between arctic, temperate and tropical plant–pollinator communities brings us neatly to the final section of this chapter, in which we consider the extent to which tropical plant–pollinator communities differ in their ecology from those in the rest of the world.

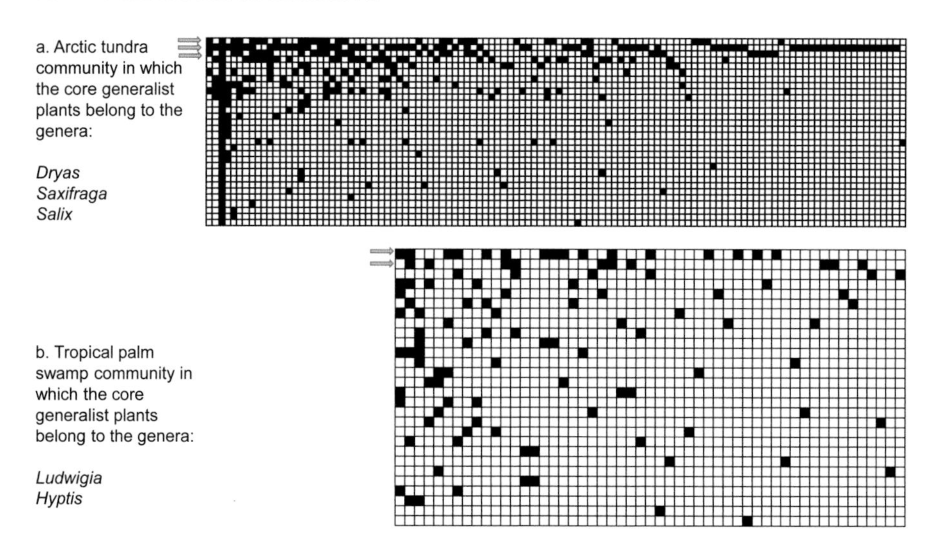

FIGURE 4.7 (a) The plant–pollinator interaction web of an Arctic tundra community in Canada (data from Kevan 1970); (b) The plant–pollinator interaction web of a tropical palm swamp community in Venezuela (data from Ramirez and Brito 1992).

Tropical specialisation: getting to grips with a myth

In hushed tones the narrator describes the intricate details of yet another highly specialised relationship between two species of indescribable beauty and complex behaviour. Slowly the camera view pulls back to reveal the green cathedral of a tropical rainforest. 'The tropics', continues the narrator, breathlessly, 'are special.'

How many times has this scene played out, in one form or another, in television documentaries about the tropics? And of course the narrator is correct: the tropics are indeed special. But how special? Or more to the point, how different are tropical communities from temperate or arctic, or even subtropical communities? The very fact that some scientists describe themselves as 'tropical ecologists' suggests that they see the tropics as different (in contrast, I know of no one who calls themselves a 'temperate ecologist') – even though the world does not suddenly change at 23.4 degrees north and south of the equator. The overwhelming diversity and abundance of species to be found between the latitudes of Cancer and Capricorn impressed even the earliest European explorers of these regions. Biogeographers soon recognised that there were many more species to be found in tropical grassland or woodland than in most equivalent habitats in temperate areas. This latitudinal trend of species richness declining from the tropics to the polar regions is well researched, though we now know that its strength varies greatly between different groups of organisms and between hemispheres, and the causes of these patterns are still hotly debated (Hillebrand 2004, Kinlock *et al.* 2017).

Lagging far behind the description of such first-order patterns in biodiversity, our understanding of latitudinal trends in species interactions is far from complete. Indeed, when I first became interested in how interactions change with latitude back in the early 2000s I could find only a handful of research papers that had even considered the question worthy of study. Perhaps this was because 'everyone knows' that the tropics host more specialised interactions than other parts of the world? Well, yes and no. It's true that there are more *examples* of specialised interactions in the tropics, but it's also the case that there are more *species*. So this is the real question: is a greater *proportion* of tropical species more specialised? There are sound ecological reasons to believe that this might be the case: if you have more species in a community, as you do often in the tropics, then there are fewer resources available per species, such that specialisation may become inevitable. However, the empirical data show that this is not always the case, and that it varies enormously depending on the type of interaction. There is evidence that some interactions are indeed more specialised in the tropics (reviewed by Ollerton and Cranmer 2002, Moles and Ollerton 2016), but for plant–pollinator relationships there is no consistent trend. We also have examples of species from some plant families that tend to have fewer pollinators in the tropics, and therefore to be more specialised – for example the cactus family (Cactaceae) – but the trend is weak (Gorostiague *et al.* unpublished). On the other hand the asclepiads are not more specialised on average in the tropics, and neither were the plant communities that we analysed (Ollerton and Cranmer 2002). In fact the data showing tropical species as more specialised were biased by the fact that tropical plants were under-sampled compared to those in the temperate zone. If you spend less time observing the pollinators of a plant you're going to observe fewer different types of pollinator – it's that simple. Later studies have confirmed this, and in fact one analysis of interaction networks by Schleuning *et al.* (2012) suggests that plant–pollinator interactions are *less* specialised in the tropics.

The question of latitudinal trends in specialisation of interactions has proven to be a surprisingly controversial topic, and by no means all ecologists (especially tropical ecologists) accept these findings. In 2014 Australian researcher Angela Moles and I wrote a guest post on the Dynamic Ecology blog called 'Is the notion that species interactions are stronger and more specialized in the tropics a zombie idea?' It generated a lot of discussion and comments, some supportive and some pushing back on our arguments, and we were invited by the editor of the journal *Biotropica* to expand on this as a commentary (Moles and Ollerton 2014, 2016).

The concept of a 'zombie idea' in ecology was first promulgated by American ecologist (and Dynamic Ecology founder) Jeremy Fox, to refer to a long-held idea that actually has no empirical basis but which refuses to die. A few people have suggested that framing such concepts in this way, and the manner in which Angela, myself and others have presented them, sometimes results in a combative or strident rhetorical tone. But if we were guilty of being overly aggressive in this case, it's because we care very deeply about this question, because it matters. Not just matters in the sense of 'let's understand the fundamental ecological patterns and processes of latitudinal trends in species interactions', but matters in terms of real conservation.

There is so much propaganda (and I use that word deliberately) being pushed out about how saving tropical ecosystems should be a priority because they are special, and contain intricate, irreplaceable interactions. Sure, they are, and they do. But so are other parts of the world, outside of the tropics, and especially outside of tropical rainforest. But they hardly get a mention and we are in danger of losing, for example, huge areas of grassland in South Africa that contain both astonishing plant diversity *and* incredible, co-evolved sets of interactions between those plants and their pollinators. In fact, some of the most specialised relationships that occur between flowers and their pollinators occur in South Africa (Ollerton *et al.* 2006, Johnson 2010), most of which is not in the tropics: it ranges from 34.8 degrees south and extends into the tropics only at its far northeastern corner. Why plants and pollinators in southern Africa should be so specialised is an intriguing question, and it may in part be due to the great age of some of the ecosystems there – which has allowed these relationships to evolve and persist over time.

Clearly, species of plants and pollinators differ hugely in how loyal they are to their partners within the entangled bank of interactions in which they are caught up. This matters for the ecology of the species concerned, and may have profound implications for the evolution of these relationships. It's this topic that we turn to in the next chapter.

Chapter 5

The evolution of pollination strategies

A strategy can be thought of as a purposeful plan of action that is designed to achieve an aim. So it may seem strange to consider plants as having strategies to ensure that their flowers are pollinated; after all, they cannot plan their actions in advance, nor has their biology been designed (unless we believe in an omnipotent creator of all species – see Chapter 3). However, in a biological sense we can consider a strategy as a set of traits that has *evolved* to fulfil a particular function. In that case, 'pollination strategies' make sense with respect to how flowers work and how pollinators interact with those flowers (see Chapters 3 and 4). In this chapter I will explore how those functions and interactions have come about, how they have evolved to be what they are, and what that might mean for their current and future evolution. I will focus on two aspects of pollination strategies: first, what determines the number of pollinators that service a flower; and second, how the flowers have adapted to be pollinated by those pollinators.

The specialised–generalised continuum

As we saw in Chapter 4, interactions between flowers and their pollinators can be thought of as a continuum. At one end of that range are the very specialised relationships in which one species of pollinator services one species of plant, such as pollination of *Yucca* species by yucca moths or, as recounted in Mike Shanahan's excellent book *Ladders to Heaven* (2016), some of the fig–fig wasp interactions. At the other end are much more generalised interactions in which many pollinators (sometimes hundreds of different species) interact with a plant species, and many of those pollinators, in turn, service other plants in a community. Along this continuum we find many moderately generalised species such as the Canary Island *Sonchus* species in Figure 5.1.

Until about twenty years ago most of the attention of pollination ecologists was focused on the specialised end of this spectrum of interactions, and for good reasons. As Darwin himself had noted (see below), these interactions provide good examples of evolution by natural selection. Not only that, but specialised

FIGURE 5.1 One of the woody sow thistles (*Sonchus* spp.) that are endemic to the Canary Islands. This species, like many Asteraceae, has a generalist pollination strategy and attracts a diversity of nectar-foraging insects. On Tenerife the endemic Canary Island bumblebee (*Bombus terrestris canariensis*), and the Canary Island large white butterfly (*Pieris cheiranthi*) are regular visitors.

interactions are amenable to dissection and study by scientists wishing to understand the fine details of these processes and how they have evolved. The relationships between flowers pollinated by hummingbirds and hawkmoths with long mouthparts, or between figs and fig wasps, became the textbook illustrations of how flowers evolve relationships with the animals that carry their pollen from plant to plant.

However, in the mid-1990s things began to change and some pollination scientists started to question this state of affairs. Not the legitimacy of the textbook examples of coevolution – clearly these existed, and were wonderful stories of natural history – rather, they began to question how frequent such intricate, specialised interactions are in nature, how loyal pollinators are to just a single species of plant (and vice versa), and whether the textbook examples really represented the majority of relationships between flowering plants and their pollinators. In Figure 5.2a I've ranked 388 species of plants from the family Apocynaceae (see Chapter 3) according to the number of confirmed species of pollinator each plant is known to have. Most species have more than one pollinator; indeed, more than half have at least three, and it's not uncommon for plants to have double-digit pollinators recorded. As I noted in the previous chapter, ecological data such as this need to be treated with caution. Most of these plants have only ever been studied in single populations for a limited amount of time, and the data therefore show the minimum number of pollinators for each species.

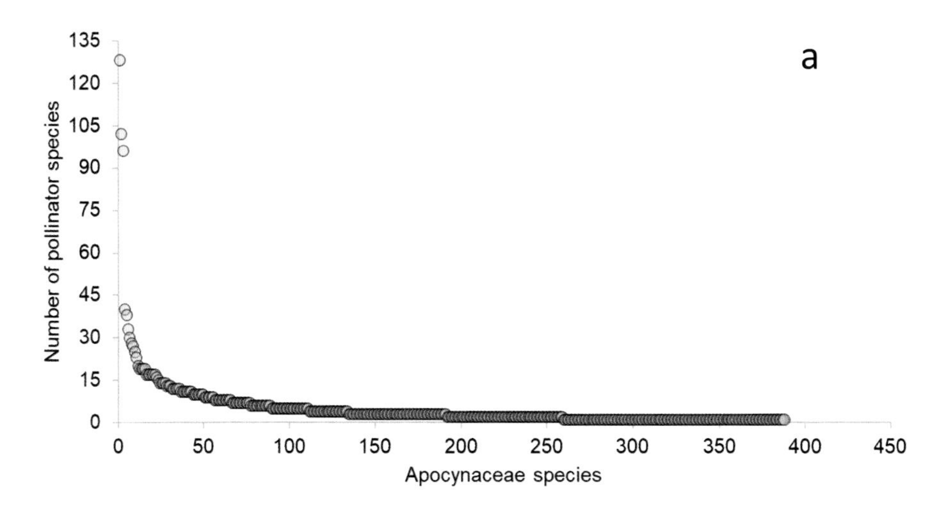

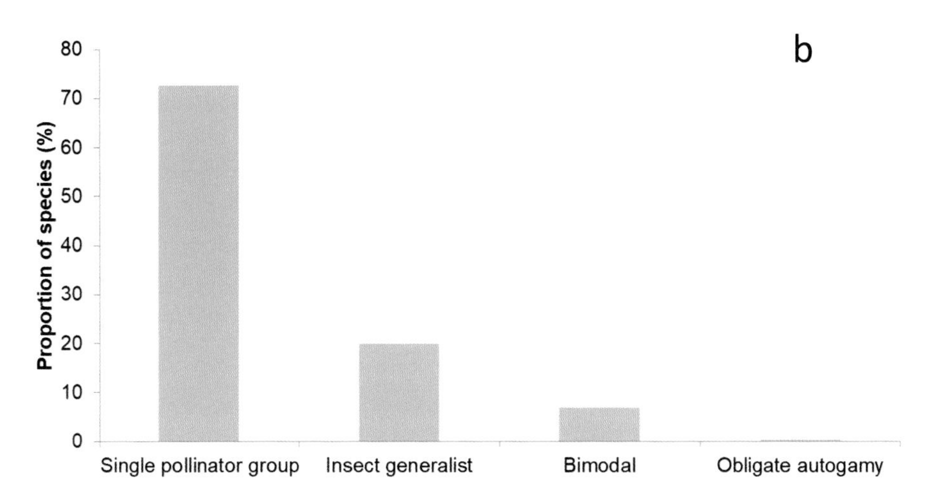

FIGURE 5.2 (a) The number of pollinators of 388 species of Apocynaceae ranked by their level of generalisation. (b) The same species categorised into whether they use a single group of pollinators or are generalists or use two distinct types of pollinator (bimodal). Obligate autogamy, in which the species only ever self-pollinates, is exceedingly rare in the family and is included for the sake of completeness only. Data from the Pollinators of Apocynaceae Database (see Ollerton *et al.* 2019a).

Figure 5.2b shows another way to consider the same data, by categorising the plants by whether they use a single type of pollinator (birds, bees, moths etc.) or are pollinated bimodally by two distinct groups of animals (moth + butterfly or bee + bird, for instance) or are true generalists. Although pollination by a single type of pollinator dominates, many of these species can be pollinated by a diversity of that type of pollinator. In the past I've described these patterns in terms of 'ecological' specialisation for the number of pollinating species a plant has; 'functional' specialisation for the number of different groups of pollinators; and 'phenotypic' specialisation referring to specific adaptations for pollination by a functional group (Ollerton *et al.* 2007c). That terminology has been quite widely adopted by other pollination ecologists, though my friend and colleague Scott Armbruster has suggested some alternative descriptors, including 'evolutionary specialisation' (evolving towards more specialised relationships), and prefers 'functional group specialisation' (see Armbruster 2017). Such discussions serve to bring clarity and understanding to a complex and evolving field of science.

Note that ecological, functional and phenotypic specialisation do not have to be correlated. Thus, bird's-foot-trefoil (*Lotus corniculatus* – Fabaceae), which we will return to in Chapter 6, can be described as bee-pollinated (functionally specialised), by a wide diversity of bumblebees and larger solitary bees (ecologically generalised); however, the flowers show specific adaptations for bee pollination (phenotypically specialised). To give another example, Rocío Pérez-Barrales at the University of Portsmouth has studied *Narcissus papyraceus* (Amaryllidaceae) across the Strait of Gibraltar. Her work has shown that this species is phenotypically specialised for moth pollination: it has long-tubed flowers and a nocturnal fragrance. In some populations long-tongued nectar-feeding moths are the main pollinators, whereas in others it's moths plus hoverflies and solitary bees. The species therefore has adaptations for phenotypic specialisation, but is ecologically generalised to varying degrees across its range (Pérez-Barrales *et al.* 2007, 2009).

To understand how the field of pollination ecology broadened its view and shifted away from considering (mainly) interactions that are highly specialised ecologically, functionally and phenotypically, we need to consider some of the history of these developments.

From specialisation to (including) generalisation

As so often in science, diverse individuals were independently asking questions about the specialisation–generalisation continuum in different parts of the world: for example, Pedro Jordano and Carlos Herrera in Spain (Jordano 1987, Herrera 1996). In the early 1990s I'd also been thinking about this during my PhD work at Wytham Woods in Oxford, thanks to an introduction to the world of flowers and their pollinators from my main supervisor, Andrew Lack. The plants in the limestone grasslands of Wytham often attracted many different types of pollinators, particularly the thistles and other members of the daisy family (Asteraceae), and umbellifers (Apiaceae) such as wild parsnip (*Pastinaca sativa*). The focus of my PhD was flowering patterns, pollination, reproductive output and seed predation in *Lotus corniculatus*, a legume (Fabaceae), the pea-like flowers of which excluded all but a few species of bumblebees from fully accessing the nectar and pollen, though butterflies could rob nectar without pollinating. Bird's-foot-trefoil was more specialised than the thistles and umbellifers, but any large bee could be a pollinator, so it lay somewhere along that spectrum from extremely specialised to highly generalised.

Following completion of my PhD in 1993 I did some field work in Australia, and on the way back called in to see two Californian ecologists, Nick Waser and Mary Price. There we discovered that we had been thinking about similar questions regarding the importance of generalised pollination systems, as had two other postdoctoral researchers, Lars Chittka and Neal Williams, and we agreed to write a paper on the topic. This was subsequently published as 'Generalization in pollination systems, and why it matters' (Waser *et al.* 1996). The opening of the abstract set out our stall:

> One view of pollination systems is that they tend towards specialization. This view is implicit in many discussions of angiosperm evolution and plant–pollinator coevolution and in the long-standing concept of pollination syndromes. But actual pollination systems often are more generalized and dynamic than these traditions might suggest.

As soon as it appeared the paper began to generate discussion and debate, both in the ecological literature and more informally in seminars and conferences, and in exchanges between colleagues. Waser *et al.* (1996) seemed to polarise opinions, and still does. Some in the field thought we were stating the obvious and that the paper didn't need writing in the first place, while others believed the whole basis of our thesis was wrong and that specialised interactions were far more frequent and important than we had stated. Gratifyingly, there were also many researchers who saw our publication as marking an important shift in how we think about plants and pollinators. We could never have predicted the impact that our paper would have had on the field and the way in which it stimulated researchers to reconsider what they thought they knew about plant–pollinator interactions. In particular it encouraged a more critical assessment of pollination syndromes and what they are telling us about how nature is organised, a topic that we turn to next.

How much do we really know about pollination syndromes?

As you'll recall from Chapter 1, I make a distinction between pollination *systems* (flowers plus pollinators and how they interact) and pollination *syndromes*, which is the expression of the phenotype of the flowers, in terms of scent, colour, shape, reward and so forth. Pollination syndromes provide us with some excellent examples of a phenomenon that has long fascinated ecologists and evolutionary biologists: convergent evolution. A lot of science is about documenting repeated patterns in nature. Scientists seek to understand the processes that determine these patterns and to make predictions about how and when they are going to be observed in the future or in other parts of the world. There are many examples of such patterns, including cyclical population dynamics of species such as lemmings; the occurrence of specific types of plant communities (e.g. rainforest, grasslands) in areas with particular climates; and convergent evolution of unrelated species with similar ecological niches, such as large, predatory placental and marsupial mammals (e.g. the dog and wolf family compared to the thylacine or Tasmanian 'wolf').

Pollination syndromes are an example of convergent evolution that has fascinated botanists since the nineteenth century. The underlying process behind these patterns is that some flower characteristics have repeatedly evolved in different plant families, owing to the convergent selection pressures applied by some groups of pollinators. Thus, red, scentless flowers producing lots of nectar are typical of many hummingbird-pollinated plants in the New World, while white, night-scented flowers often signify moth pollination. Plant species possessing these archetypical flower traits have been used as textbook examples for decades to illustrate the predictable and specialised nature of some plant–pollinator interactions, as I noted above.

The ideas behind the pollination syndromes originated with nineteenth-century biologists such as Federico Delpino and were later developed by a number of researchers including Stefan Vogel (Waser *et al.* 2011, Johnson and Wester 2017) and Knut Faegri and Leendert van der Pijl in their seminal book *The Principles of Pollination Ecology* (1966). The problem is that until recently the pollination syndromes had rarely been subjected to critical tests of their frequency and predictive value (Waser *et al.* 1996, Hingston and McQuillan 2000).

There is also a degree of misleading inaccuracy to the names used for the classical syndromes – which is most apparent when we consider the syndrome 'sapromyiophily', which means flowers that 'love' flies that feed on decaying material. The huge range of flies that are attracted to such flowers (Chapter 3), and the diversity of scent compounds, means that we have to split sapromyiophily up into finer and finer categories (Ollerton and Raguso 2006). In addition, at least some of these flowers are pollinated mainly or solely by beetles, for example the infamous titan arum (*Amorphophallus titanum*).

It's been tacitly assumed that (after more than 150 years of study) we clearly know all there is to know about the syndromes, even though there have been criticisms levelled at them since their inception. However, in the last couple of decades biologists have begun to seek answers to questions such as: How often do plant species conform to the expectations of the classical pollination syndromes? How good is our ability to predict the pollinators of a plant based just on its flower characteristics? What is the role played by flower visitors that do not conform to the predictions of the pollination syndromes? Similarly, what is the role of animals that steal nectar or pollen, or act as herbivores, in shaping flower traits? What new examples of convergent evolution of flower traits remain to be discovered, generated by groups of pollinators that were never considered in the classic literature, such as cockroaches (Xiong *et al.* in press)?

Research conducted in many different parts of the world has addressed questions that most biologists had assumed were already answered or which were not worth asking in the first place. And the answers to them are proving to be both surprising and controversial. For example, the largest field test of the frequency and predictive power of pollination syndromes that has been conducted to date (Ollerton *et al.* 2009b) concluded that only a small proportion of the 352,000 species of flowering plants could be categorised into the pollination syndromes as classically described. Likewise, we estimated that the average predictive power of the pollination syndromes was about 30%, and that this varied enormously depending on where in the world a plant community was located. Other studies have shown that 'secondary' flower visitors can pollinate flowers at least as effectively as the 'primary' pollinators predicted by the syndromes (e.g. Waser and Price 1981, 1990, 1991, Duffy *et al.* 2020); that floral antagonists can play an important role in shaping flower traits (e.g. Junker and Parachnowitsch 2015 and references therein); that bimodal pollination systems involving two very distinct groups of pollinators are widespread (Shuttleworth and Johnson 2008, Dellinger *et al.* 2019, Ollerton *et al.* 2019a); and that there are still examples of convergent evolution to 'unexpected' pollinators waiting to be discovered in less well-researched parts of the world (see Chapter 2 and Ollerton *et al.* 2003).

In this context 'less well-researched parts of the world' actually equates to *most* of the world. Plant–pollinator interactions have hardly been studied across huge swathes of the globe, especially in Africa, Asia and Australasia, as shown by assessments of the global distribution of plant–pollinator network studies (Vizentin-Bugoni *et al.* 2018), the data used in published meta-analyses (Archer *et al.* 2014), detailed studies of the pollination ecology of plant species in the family Apocynaceae (Ollerton *et al.* 2019a), and text analysis of the published pollination literature (Millard *et al.* 2020). To get a sense of the extent of the geographical gaps in our knowledge, I suggest that you take a look at the map of the Web of Life Ecological Networks Database, selecting just the 'pollination' networks (www.web-of-life.es/map.php). It's in the nature of any field of research that these sorts of biases will reduce over time, and certainly we are starting to see more work come out of Asia and Africa in particular. However, this general pattern will persist for a long time: much of the world is unexplored from the perspective of interactions between plants and their pollinators.

As so often is the case in science, opinions differ, and not everyone accepted these critiques of pollination syndromes. A comprehensive push-back that challenged the challengers was by a joint Mexican–Argentinian team (Rosas-Guerrero *et al.* 2014). Using a statistical technique called meta-analysis, underpinned by a review of the available literature, they suggested that pollination syndromes are much more reliable than Ollerton *et al.* (2009b) concluded, with a predictive power perhaps as high as 75%. Although all contributions to this debate are welcomed, some of my collaborators and I saw problems with this study that biased and undermined the robustness of their conclusions. Thus we wrote a response (Ollerton *et al.* 2015).

I won't go into the technical details of what we perceived as problems in Rosas-Guerrero *et al.*'s approach to testing pollination syndromes, but in summary they relate to how the literature review was conducted (failing to include all of the studies that could have provided data for their meta-analysis); the significant bias in the current literature because plant–pollinator interactions are not studied randomly (biologists are often drawn to large-flowered plants possessing those archetypical, classical flower traits associated with particular syndromes – see also Johnson and Wester 2017); the variation in how different researchers determine the effectiveness of the pollinators in their system, meaning that these studies are not always comparable; and issues around annual variation in pollinator identity and presentation of data.

The authors of course responded to our critique (Aguilar *et al.* 2015), and we in turn responded to theirs (Waser *et al.* 2015). And so it goes. It's important to stress here that all of the large-scale tests of pollination syndromes so far published can be criticised, and that they all possess a range of weaknesses, including our own. This is work in progress – and I hope to see more studies published in the future that learn from these earlier attempts.

In an era when we are more and more concerned about loss of pollinator diversity, including extinction at both a species and a country level (see Chapter 10), and the plants that they rely on and which in turn need the pollinators, do these debates really matter, or are they of purely academic concern, of interest only to a few botanists and ecologists? As you might expect, I'd argue that they do matter: there are still some fundamental aspects of pollination ecology that we don't completely understand, or that we have only recently begun to seriously address. These include the number of flowering plants that require animal pollination (Chapter 1); the diversity of pollinators at a global and regional level and the relative importance of different types of pollinators (Chapter 2); whether or not plants and pollinators are more specialised in tropical than in temperate communities (Chapter 4); and the fact that there are still many geographical and taxonomic gaps in our understanding of the full diversity of plant–pollinator interactions. Without some of this fundamental knowledge we are unable to make effective arguments, policies and strategies for conserving not just plants and pollinators, but also their interactions. Despite providing a focus and framework for understanding pollination biology for over 150 years, the pollination syndromes continue to surprise us, and to provide a vital antidote to scientific hubris. We really do not have nearly as much knowledge about plant–pollinator relationships as we assume.

In fact these debates about pollination syndromes echo arguments that have been going on since the nineteenth century (Waser *et al.* 2011). More than 130 years ago the Darwinian biologist Hermann Muller criticised Federico Delpino for ignoring the 'wrong' flower visitors. Interestingly, Delpino was fundamentally a teleologist who saw purpose in nature, expressed through (as he perceived them) the highly ordered relationships between flowers and pollinators. The historical roots of pollination syndromes are indeed strange when judged by modern standards. As we discuss in the Waser *et al.* (2011) paper, Stefan Vogel was another prominent pollination biologist, and advocate of the importance of pollination syndromes, who was also fundamentally teleological in his thinking. I was fortunate enough to meet Stefan at a symposium in honour of his eightieth birthday at the International Botanical Congress in Vienna in 2005. He graciously signed my copy of his book *The Role of Scent Glands in Pollination* (1990) and said, with a twinkle in his eye, 'you and I have probably got a lot to discuss.' Unfortunately we never got the opportunity, and Stefan died in 2015, still active in his ninetieth year. Although I suspect we would have disagreed about some fundamental aspects of pollination syndromes, I dedicated our 2009 paper on *Ceropegia* (Ollerton *et al.* 2009c – see Chapter 3) to him 'in honour of his pioneering work on pollination' in the genus. Stefan's legacy of research, particularly in the tropical regions of South America and in South Africa (Johnson and Wester 2017), is a fitting tribute to his memory.

If pollination syndromes refer to the phenotypic traits of a flower and how they may, or may not, predict the pollinators of a species, the term 'pollination systems' can be used to refer to those traits of the flower plus the pollinators themselves, as I noted in Chapter 1. How pollination systems evolve and diversify is the topic of the next section, and we're going to focus once again on my favourite plant family: Apocynaceae.

Diversity and evolution of pollination systems in large clades

Interactions between flowering plants and the animals that pollinate them are known to be responsible for part of the tremendous diversity of the angiosperms, currently thought to number some 352,000 species. This relationship between plant diversity and pollination biology has attracted the interest of researchers for almost 150 years (Darwin 1862, Crepet 1984, Johnson 2006, Kay and Sargent 2009, Vamosi and Vamosi 2010, van der Niet and Johnson 2012, van der Niet *et al.* 2014). But the diversity of different types of pollination system (bird, bee, moth, fly etc.) is unknown for most large, related groups of plants that have evolved from a common ancestor (groups that systematists term 'clades') such as families and subfamilies. In addition, we know little about how these interactions with pollinators have evolved over time and in different parts of the world. Only a handful of groups of flowering plants have been studied with respect to questions such as:

- What is the diversity of pollination systems in large clades?
- How is that diversity partitioned between the smaller clades (e.g. subfamilies, tribes, genera) of a family, and what are the evolutionary transitions between the major groups of pollinators?
- Do these pollination systems vary biogeographically across the clade's range?

These sorts of questions have been a source of fascination to me for over twenty years, since Sigrid Liede and I published a study of what was then known about pollination systems in the family Asclepiadaceae (the asclepiads – see Chapter 3). In that paper we said that the research 'is intended to be ongoing ... [and we] ... hope to re-review asclepiad pollination within the next decade' (Ollerton and Liede 1997). At the time I didn't think it would actually take more than twice that long. However, over those two decades (and more) a lot has changed, not least the fact that Sigrid acquired a husband and a hyphenated surname (Sigrid Liede-Schumann). For another thing, the family Asclepiadaceae no longer exists, broken up and subsumed within a much larger Apocynaceae. Also, over that period I did a lot of work in the field and in the herbarium on some of the smaller groups within the family, such as *Ceropegia*. Others had also been working on different genera in various parts of the world, increasing the amount of data that we had access to. Finally, the level of sophistication of the analyses we are now able to perform has increased beyond recognition compared to what we could achieve in the mid-1990s. At the International Botanical Congress in Shenzhen, China, in summer 2017, Sigrid and I decided that the time was right to push ahead with a study of pollination systems in Apocynaceae, and I devoted most of the rest of 2017 and into early 2018 to doing just that. It was a monumental effort, and along the way we picked up another 73 (!) co-authors who provided a huge amount of previously unpublished data and/or contributed state-of-the-art data analyses (Ollerton *et al.* 2019a).

The first thing to say is that Apocynaceae is massive: it is one of the top 10 or 11 largest angiosperm families with more than 5,300 described species. It's also globally distributed, with particular centres of diversity not only in the tropics but also in some subtropical and temperate regions such as South Africa. Given the size and distribution of the family, our Pollinators of Apocynaceae Database was only ever going to be a sample of the plants and their pollinators, and in fact we ended up with information about the pollination systems of just over 10% of the species in the family. That sounds small, I realise, but in fact it's the largest such compilation that's ever been brought together for such a large clade. So lots more data on plant–pollinator interactions needs to be collected before we can say we fully understand how pollination systems have evolved in this most remarkable family, and any conclusions must be provisional. But that's fine: as I noted above, science is always a work in progress. Indeed, as I was writing this book in 2019 I was invited to collaborate on a study of the first example of cockroach pollination in the family (Xiong *et al.* in press).

What did our data tell us? We found that Apocynaceae is characterised by an enormous diversity of pollination systems involving almost all of the major pollen vectors and some that are nearly unique to the Apocynaceae (Ollerton *et al.* 2019a). Some examples of Apocynaceae species and their pollinators are shown in Figures 4.3 and 5.3, and some of the flowers that are adapted for fly pollination can be seen in Figures 3.7 and 3.8. The one notable pollination system

FIGURE 5.3 Examples of the diversity of pollination systems in the family Apocynaceae. Top, left to right: the hawkmoth-pollinated *Schubertia grandiflora* with *Manduca florestan* visiting the flowers (photo by Felipe Amorim, Brazil); the large bee-pollinated *Calotropis gigantia* with a pollinating *Xylocopa* sp. Bottom, left to right: *Dregea sinensis* is a generalist pollinated by bees such as *Apis cerana* as well as other insects (photo by Zong-Xin Ren, China); the Madagascan endemic *Cynanchum obovatum* may be a scoliid wasp specialist but requires further study (photo by Ulrich Meve).

missing from our survey is pollination by bats, which has never been recorded in this family. However, based on what in this case may be a predictable pollination syndrome, another apocynologist colleague, Mary Endress, suspects that it occurs in at least one South American species. If we consider how the family has evolved, then older, earlier-diverging clades have a narrower range of pollination systems than those that evolved later. Thus, pollination systems such as wasp and fly pollination seem to be more recently evolved. There is also significant convergent evolution of pollination systems, especially fly and moth pollination, by geographically and phylogenetically distinct clades: plant species can have flowers that look remarkably similar even though they are only distantly related and grow naturally on different continents.

Transitions from one type of pollination system to another have happened frequently during the evolutionary history of Apocynaceae, which dates back at least 50 million years (Magallón *et al.* 2015). Some of these transitions are evolutionarily constrained, and rarely or never occur, whereas others have taken place much more often, such as transitions between wasp and beetle pollination.

Broad perspectives on the diversity of pollination systems in related groups of plants such as this tell us a lot about the deep-time evolution of those groups and the present-day biogeographic patterns of relationships with pollinators. However, a problem with categorising pollination systems in this way is that it misses much of the subtlety and complexity that can only be obtained by a very detailed study of the ecology of plant–pollinator interactions. One way of thinking about this is that interactions with pollinators, the timing of flower opening and closing, aspects of floral attraction and reward, and so forth, comprise the 'pollination niche' of the plant (Parrish and Bazzaz 1978, Ollerton *et al.* 2003, 2007a). Ryan Phillips and colleagues have recently written a very comprehensive review of this topic (Phillips *et al.* 2020). In the next section I give an example of how complex and difficult it can be to fully understand a plant's pollination niche – an example that goes all the way back to speculation by Charles Darwin in the nineteenth century.

Evolutionary hangovers? Darwin and the wild carrot

For many flowers with adaptations to particular groups of pollinators it's relatively easy to assess how certain characteristic of those flowers, such as scent or colour or shape, might relate to the type of pollinator. This is the basis of the pollination syndromes that I discussed above. However, there are also lots of less straightforward examples where constructing a story of how the plants have evolved is more difficult. The umbellifers (Apiaceae) provide a good example of this and, as is so often the case, we can thank Charles Darwin for providing an initial insight. Darwin had been observing wild carrot (*Daucus carota* – Figure 5.4) and noted that in most plants there is a larger, dark-coloured flower at the centre of the inflorescence. Speculating on this structure, Darwin wrote in 1877 (p. 8, my emphasis):

> In the Carrot ... the central flower has its petals somewhat enlarged, and these are of a dark purplish-red tint; but it cannot be supposed that this one small flower makes the large white umbel at all more conspicuous to insects ... That the modified central flower *is of no functional importance* to the plant is almost certain. It may perhaps be a remnant of a former and ancient condition of the species.

The idea that some features of flowers had a past adaptive function but are now evolutionary 'hangovers' from earlier times might seem to contradict Darwin's views about the evolution of adaptive traits as set out in *Origin of Species*. But logically it makes sense – it seems unreasonable to expect every aspect of an organism's biology to have a current purpose, when we know that evolution is a complex process. Inspired by Darwin's claim, and by the later work of some other pollination ecologists who had studied the plant, notably Dan Eisikowitch in Israel, I set out to test this idea experimentally. To do this I recruited a final-year undergraduate, Ellen Lamborn, and between us we designed a complicated experiment that involved various treatments, including removing the dark central flower and

FIGURE 5.4 The inflorescence of wild carrot (*Daucus carota*) is visited by a wide diversity of small bees, flies, beetles and wasps. Note the dark floret in the centre of the inflorescence.

adding varying numbers of artificial florets (made from black insulating tape) in order to test Eisikowitch's 'fly catcher effect' (Eisikowitch 1980). This is the notion that the central floret serves as a false fly that encourages other flies to land on the inflorescence. For each of these treatments we observed the number and diversity of pollinators, and the resulting seed set of the umbels.

Field work was carried out at a local nature reserve with a good population of wild carrot, during the summers of 1996 and 1997. The results were intriguing. First of all, the relative numbers of different pollinators varied enormously between years, and this became especially obvious when we included data from the same site in later years (Figure 5.5).

In some years soldier beetles (*Rhagonycha fulva*) were very common, whereas in other years they were not present at all. Likewise, sawflies (*Tenthredo* sp.) and various flies and parasitic wasps varied in their relative abundance. However, none of these groups responded in any systematic way to our experimental treatments, and there was no clear pattern to the seed-set results. It didn't seem to matter whether or not the dark central floret was present. Scientists often don't like these 'negative' results, and such findings can be difficult to publish, but they supported Darwin's hypothesis and we thought they were worthy of publication, so we submitted a manuscript to the journal *Functional Ecology*, where it was duly published (Lamborn and Ollerton 2000).

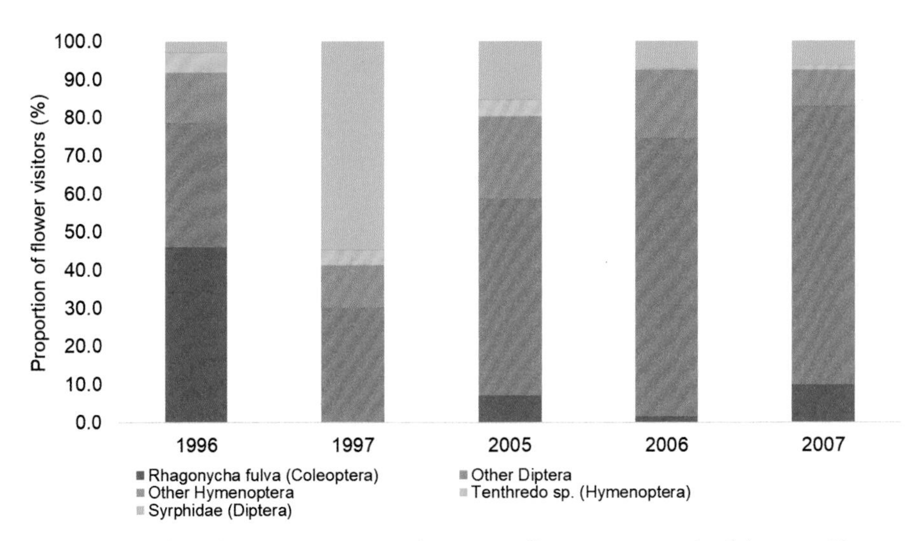

FIGURE 5.5 The relative proportions of major pollinator groups of wild carrot (*Daucus carota*) at a site in Northampton in different years. Data from Lamborn and Ollerton (2000), Ollerton *et al.* (2007c) and Ollerton (unpublished).

That wasn't, however, the end of the story. A few years later my friend and colleague Dave Goulson published a research paper in which he and some students had carried out experiments on wild carrot in Portugal at a site where the main pollinator was the varied carpet beetle (*Anthrenus verbasci*), a species that occurs in Britain but which we had not seen visiting the flowers at our site (Goulson *et al.* 2009). Dave and his students had noted that the beetle is of a similar shape and size as the dark central florets of wild carrot, and that the beetles responded positively when they added more florets to an inflorescence. They went on to manipulate the umbels as we had done, but this time adding various numbers of dead beetles. In contrast to our findings, Dave and his co-workers discovered that the presence of the dead beetles attracted other beetles (presumably to mate with them). Although they could not be sure of the role of these beetles as pollinators, this (together with the findings of some earlier researchers) suggested to them that the florets do indeed function as insect mimics, and that Darwin was wrong in his conclusion.

The plot thickened when Sabrina Polte and Klaus Reinhold (2013) addressed the same question in Germany. Using video recordings, they found that the presence or absence of the dark floret made no difference to the behaviour of pollinators landing on the umbels. However, they did find that umbels with a dark central floret suffered less parasitism from a gall midge called *Kiefferia pericarpiicola*. This is an uncommon species in Britain, and certainly we had not observed it. Polte and Reinhold proposed that the dark central floret mimics the presence of a gall, making the fly less likely to oviposit on that umbel. Such counter-warfare examples are not unknown in the plant world – for example, some tropical passion flowers produce leaves with spots that mimic the eggs of butterflies. But its occurrence in the European flora is rare, and this could be an intriguing example. More recently still Gonzalez *et al.* (2018) also showed that the presence of the central floret did not affect pollinator visits, and that the heights of the inflorescences may be important in determining how many insects land on them.

It is possible that all of these studies are correct, and that different things are going on in different parts of the range of wild carrot, explaining the contrasting results found by different researchers. It's certainly a subject that deserves further research, and I hope that future pollination ecologists pay it some attention. But it also illustrates a wider point: truly understanding the functions of flowers, and how those might vary between populations and over time, requires hours of patient observation and (preferably) experimental manipulations. It's no wonder then that only a small proportion of the flowering plants have been studied in anything like the depth of wild carrot. So much more remains to be discovered, even in the European flora, as I discussed in Chapter 4. There are also many myths require dispelling, even for apparently well-known groups such as the orchids, as we'll see in the next section.

Are orchids a special case?

The plant family Orchidaceae (the orchids) has been a particular focus for biologists interested in understanding how flowers evolve since even before Darwin wrote a book on the subject (Darwin 1862). It was no accident that this was the book which Darwin chose to write after *Origin of Species* was published: the evolution of orchid flowers under the agency of their pollinators gave, he believed, profound insights into the process of natural selection that he had laid out three years earlier. It is an important and ground-breaking book. However, aspects of it have been misunderstood in ways that resonate to the present day, particularly with respect to how specialised the orchids are in their relationships with pollinators. There's no doubt that orchid flowers are specialised in the sense that most produce pollinia of various types (analogous to the unrelated asclepiads – see Chapter 3) and they can sometimes engage in bizarre relationships with insects, such as mimicking female bees and wasps, thus tricking the males to visit them. However, one sees statements time and again that this evolution is the result of orchids having exclusive relationships with their pollinators, such that each orchid has only one type of pollinator. I've seen this repeated in textbooks, on horticultural websites, in grant proposals, and on Wikipedia. But it's not true: it's a myth.

Orchids certainly have some fascinating and often quite intricate floral mechanisms to ensure pollination, but these have not necessarily evolved to attract and exploit just one species of pollinator. Even in the case of sexually deceptive orchids that fool their male pollinating insects into believing that they are mating with a female of the same species, it is sometimes the case that more than one insect species is involved. For example, in the well-studied bee orchids of the genus *Ophrys*, Vereecken *et al.* (2011) note that 'flowers are pollinated by a narrow taxonomic range of pollinators, from a single species to up to five closely related species', and that this is not the same as the mythological 'extreme case of one orchid/one pollinator'.

Likewise different species of euglossine orchid bees (Euglossini) may pollinate the same orchid flowers as they visit to collect scent compounds; for example in the Brazilian species *Dichaea pendula*, species from at least two different bee genera act as pollinators (Nunes *et al.* 2016).

The fact that 'one orchid/one pollinator' is a myth is not new knowledge; it's been widely discussed in the pollination ecology literature for decades. For example, Waser *et al.* (1996) showed data from the late nineteenth/early twentieth century that clearly indicated a range of specialisation in European orchids. Even earlier than this Tremblay (1992) showed that only about 62% of species for which he could find data had a single pollinator, and that this varied considerably between different subfamilies of Orchidaceae, with some subfamilies being more specialised than others.

More recently, I looked at this question and compared orchid specialisation using data from southern Africa, North America and Europe (Ollerton *et al.* 2006). Orchids are more specialised in southern Africa than in Europe and North America (as are a number of other plant groups including the asclepiads, which we compared them with). But even in southern Africa, only about 65% of the orchids studied have a single pollinator species. It's worth pointing out, though, that many of the species included in this analysis, and in Raymond Tremblay's paper, have been studied only at single sites and often in single years, meaning that often we have no idea if there is any spatiotemporal variation in the pollinators an orchid species exploits.

Why does this myth persist? I think it's for the same reason that myths of all kinds are retold from generation to generation: they are great stories that fascinate the teller and the audience. Indeed, orchids are very special plants with some amazing floral and vegetative adaptations, intricate relationships with fungi, and incredible diversity. But we don't have to mythologise their relationships with their pollinators to try to make orchids more special than they already are.

Why does the specialisation–generalisation continuum exist?

To rephrase the title of this concluding section, under what circumstances do some plants evolve pollination strategies with one pollinator, and other plants with many pollinators? Why do some species evolve into what could be considered an evolutionary dead end of hyper-specialisation? And why are these patterns different from others, such as host–parasite relationships, where specialisation is the norm? Or predator–prey interactions, where generalisation is so much more common? This is a question that was addressed by Waser *et al.* (1996), by some chapter authors in the book that Nick and I co-edited a decade later (Waser and

Ollerton 2006), and by subsequent researchers. They are important questions that relate directly to the ecological structure of plant–pollinator assemblages and how they persist over time and space (Chapter 4).

Opportunities for plants to specialise require that pollinators are predictable, that they will occur year after year at more or less the same time and in the same place. That kind of predictability can only occur when climate and geography are relatively stable or when pollinators are so abundant that fluctuations in their numbers have little impact on the plants they pollinate, as is probably the case with the microdiptera that pollinate *Ceropegia* species (Chapter 3). The stability argument is certainly one possible explanation for why plant–pollinator interactions in southern Africa tend to be more specialised than elsewhere in the world, as these habitats were relatively unaffected by the kinds of ice ages that affected the northern hemisphere (Johnson and Steiner 2000, Ollerton *et al.* 2006).

There is a long-standing notion that when species evolve along a very specialised trajectory, they somehow become evolutionary 'dead ends' that cannot further evolve. The panda, with its specialised bamboo diet, is perhaps the prime example of this. However there's little evidence of evolutionary dead ends among the flowering plants. In fact there is a growing number of well-documented cases of plants that have evolved generalist pollination strategies despite having rather specialised ancestors. The first published example was by Scott Armbruster and Bruce Baldwin, involving a handful of species of *Dalechampia* (Euphorbiaceae) in Madagascar that are more generalised than their African ancestors. These have switched from being bee-pollinated and resin-rewarding (the bees use the resin to build their nests) to using a wide range of different insects that they reward with pollen (Armbruster and Baldwin 1998). Similarly, in the genus *Miconia* (Melastomataceae), generalist nectar and pollen rewarding strategies have evolved within a clade of plants that predominantly uses a more specialised, buzz-pollinated strategy involving just bees (de Brito *et al.* 2017).

Generalist plant species such as these tend to exploit pollinators that are unpredictable in their abundance in time and space, as the example of wild carrot (Figure 5.5) shows. Some of this variability is caused by weather conditions in particular years affecting the flowering time of plants and the likelihood of peak flowering intersecting with the main activity period of pollinators. The role that weather and climate play in influencing pollinators and plants is the topic of the next chapter.

Chapter 6

A matter of time: from daily cycles to climate change

In these spring days,
when tranquil light encompasses
the four directions,
why do the blossoms scatter
with such uneasy hearts?

Ki no Tomonori (*c.*850–*c.*904)

The daily and yearly cycles of light and dark, warmth and chill, rainfall and drought that ultimately determine climate (the longer-term patterns across years and geography) have a huge influence on the natural world and on human societies. Understandably, these cycles and patterns also have an effect on when plants flower and when their pollinators are active. In this chapter I will explore in some detail the influences of these hourly, daily, seasonal and long-term changes in factors such as light, temperature and water availability. We will begin at a small scale (hourly changes in one place, for example) and go on to consider seasonal and climatic variability, including global climate change. Sometimes I joke that the reason I became a pollination ecologist was that I don't like doing field work in cold weather. There's some truth in that statement, and in fact it's hard to do any kind of work with pollinators when the weather is inclement (though not impossible – see the examples below). Air temperature and rainfall, in particular, change the way that both flowers and their pollinators behave. So let's start by considering day-to-day patterns.

Daily patterns of flowering and pollinator activity

The flowers of most species open once in their lifetime, often first thing in the morning, and remain open until they have fulfilled their role of attracting pollinators and setting seed. However, some flowers, for example members of the family Aizoaceae (Figure 6.1) open only when the sun shines, closing when it's cloudy or at nightfall, an adaptation that probably protects the pollen from water

FIGURE 6.1 The flowers of this species of Aizoaceae (*Carpobrotus edulis*), like many in the family, open when the sun is shining but close when it is overcast (inset photograph).

damage. Other flowers open in the evening, producing scents that attract their crepuscular or nocturnal pollinators, usually moths or bats, though bear in mind that there are some nocturnal butterflies and bees (Chapter 2) that have hardly been studied as night-time pollinators. Likewise, cockroaches can also visit flowers that open and produce their scent at night (Xiong *et al.* in press).

The timing of opening of such flowers, and the presentation of a nectar reward, can be very precise and predictable. During a month of teaching and field work with the Tropical Biology Association in Tanzania in 2011, we studied this in a putatively moth-pollinated rainforest species of Apocynaceae called *Dictyophleba lucida* (Figure 6.2). Over several evenings we scored the time that flower buds opened and recorded illumination levels using a light meter. As you can see, almost all of the flowers opened during a narrow time window of less than 90 minutes as light levels dropped rapidly, a constant feature of the tropical dusk. At the same time the air was filled with a strong, sweet scent that drifted through the forest to lure in the moths that are the most likely pollinators of these flowers. On one evening we captured moths that visited the flowers but could not confirm them as pollinators, though it seems highly likely that they are, given the nocturnal flower opening and the nectar reward that we detected. Note that it is the drop in light levels that seems to be the trigger for opening these flowers, not the complete

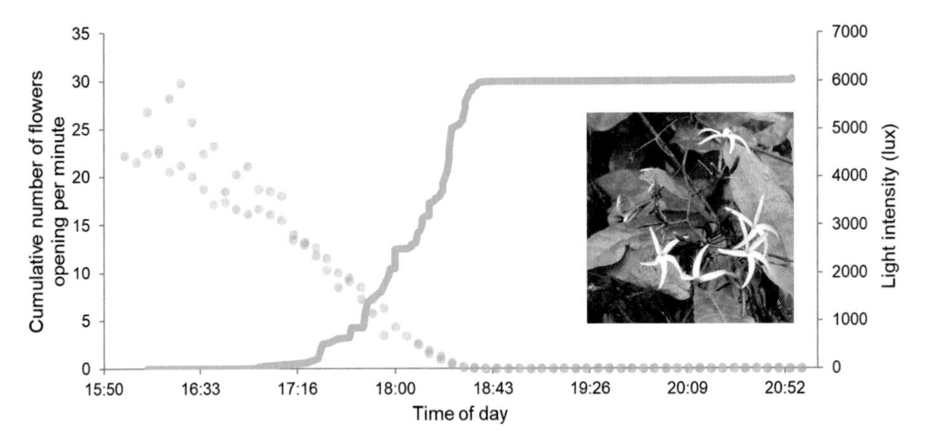

FIGURE 6.2 Daily flower opening in the moth-pollinated species *Dictyophleba lucida* (Apocynaceae). The green line shows the cumulative number of flowers opening per minute. The brown data points show light intensity, measured using a light meter. Data collected by Fidel Chiriboga, Evalyne Muiruri, Chediel Mrisha, Jeff Ollerton and Clive Nuttman, in Amani, Tanzania, 2011.

absence of light; by the time it's completely dark (lux of about zero) almost all of the flowers are already open.

Rainforest climbers in Tanzania are all very well and exotic, but similar patterns can be seen in wild habitats and gardens all over the world, in species such as evening-primrose (*Oenothera biennis* – Onagraceae) and honeysuckles (*Lonicera* spp. – Caprifoliaceae). Evening-primroses typically close their flowers by noon the next day, but the flowers of most other species remain open all day and night, and in all weathers, taking advantage of pollinator visits whenever they may occur. These two patterns (strictly nocturnal versus all-day opening) represent different floral strategies in the sense that I explained in Chapter 5. They may result from natural selection imposed by pollinators that vary in their effectiveness or by the damage that can be done to flowers by the 'wrong' visitors. For example, there may be an advantage in a nocturnal moth-pollinated flower closing up during the day to prevent its pollen being stolen by female bees. On the other hand, if visits by those moths are rare, the additional pollination services provided by diurnal flower visitors could more than compensate for pollen loss. Thus, all of this floral behaviour is timed to coincide with the sometimes patchy, sometimes predictable, activity of the animals that pollinate the flowers. If we compare the activity of different types of pollinator on a typical summer day in Britain, we see that some (for example bumblebees) will become active much earlier in the day than others, such as butterflies. These patterns reflect physiological differences between the various groups; unlike most insects, bumblebees can generate their own internal heat by quickly vibrating the muscles of the thorax. The insect physiologist Bernd Heinrich has studied this in detail (Heinrich 2004), and I've observed queen buff-tailed bumblebees (*Bombus terrestris*) flying in the winter in my garden when air temperatures have been as low as 3°C.

The plant kingdom is full of species with predictable flower opening times, colour changes, and other floral behaviour. In Chapter 3 I've already shown examples such as the Canary Island wallflower (*Erysimum scoparium*) and the trapping behaviour of *Ceropegia* species. The daily cycle of the latter is highly predictable. Flies are temporarily trapped by the flowers, during which time they deposit and/or pick up the pollen packets (pollinia). Eventually the flower has to release the flies, and this typically happens after 24 hours: the hairs lining the inside of the floral tube collapse and the flower moves from an upright to a more horizontal position, allowing the insects to walk out. However, this has been studied in only a limited number of *Ceropegia* species; it's a large genus, and it's likely that not all species behave in quite this way. Generalising from one member of a group of organisms to all of the others is not advisable, as traits and behaviours can vary a lot even in closely related species.

There are also flowers with highly predictable timings of pollen or nectar release, which pollinators can learn and subsequently time their activities to periods of peak reward. An example is seen in the coexisting species of African *Acacia* (Fabaceae) studied by Graham Stone and colleagues that release their pollen at different times of the day. This staggered release is learned by the pollinators (mainly megachilid and honey bees, and pollen-feeding flies), which move from one species to the next (Stone *et al.* 1996, 1998). The suggestion is that this pattern of pollen release has evolved because it reduces competition for pollinators between *Acacia* species.

For other groups of pollinators, the timing of when they arrive on flowers may be a matter of opportunism and competition, even in apparently predictable pollination systems. Night-flowering cacti in North and South America that are pollinated mainly by bats and moths, for instance, also receive some pollination from early-morning bees, which ensure reproduction when the less reliable nocturnal visitors are absent (Fleming *et al.* 2001, Gorostiague *et al.* unpublished). Another example can be seen in Figure 6.3, which shows data from a short study of pollinators of a species of *Asphodelus* (Asphodelaceae) on Mallorca. The only visitors first thing in the morning were flies and beetles, when the volume of nectar per flower was at its maximum (probably because some was still available from the day before) but before honey bees were active. Soon after the honey bees appeared, however, the numbers of flies and beetles dropped off considerably as the bees aggressively dominated the flowers and monopolised the nectar. When nectar volume fell at about 1 pm, following intensive exploitation by the honey bees, those bees departed and flies and chafers returned. This departure was temporary, however, and the honey bees came back in even greater numbers (presumably recruited by nest mates) in response to rising nectar volumes, displacing the flies and chafers. By late afternoon all of the visitors had departed even though nectar was still available. Although we didn't record it, I would not be surprised if night-flying moths came after dusk to these flowers.

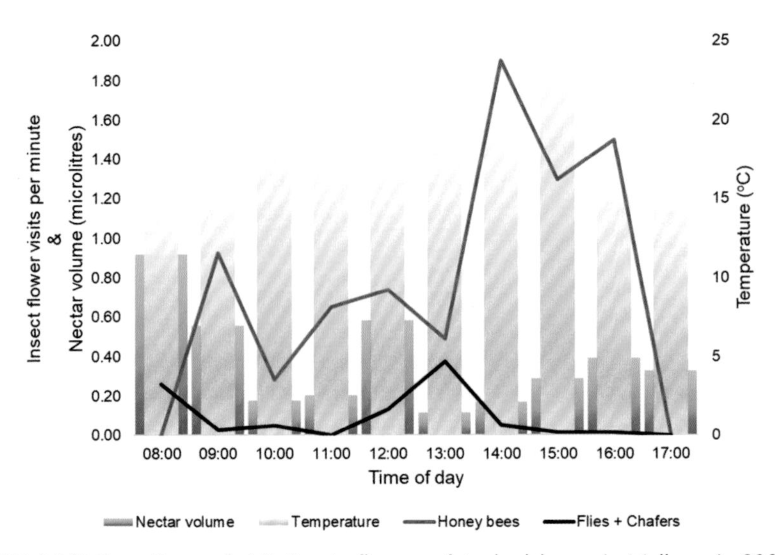

FIGURE 6.3 Daily patterns of visitation to flowers of *Asphodelus* sp. in Mallorca in 2002, by honey bees, and flies and chafer beetles, in relation to temperature and nectar production. The only other visitors were a very few native bees later in the day, not shown in the figure for reasons of clarity. Data collected by J. Ollerton with students during a field course.

These kinds of complex patterns are commonly observed by pollination ecologists: not everything is as predictable as we might expect and much depends on local circumstances. For example, if there had been no bee hives in the vicinity, the numbers of other species might well have been very different. Annual patterns of flowering and pollinator activity are also a mix of the predictable and the contingent, and it's to these that we turn our attention next.

Annual patterns of flowering and pollination

The word 'phenology' is used to describe the seasonal changing of events in the natural world. We talk about 'flowering phenologies' of plant species, or 'emergence phenologies' of hibernating insects. As I tell my students, there's a couple of other words that are sometimes confused with this: 'phonology', the study of speech sounds, and 'phrenology', the Victorian obsession with how bumps on heads allegedly reflect character traits. I've had students use both in their answers to exam questions ... perhaps I should stop mentioning it?

Annual patterns of flowering are some of the most striking phenomena in the natural world, but also something that we rather take for granted. In northern Europe we assume that snowdrops (*Galanthus nivalis* – Amaryllidaceae) will appear early in the year, daffodils (*Narcissus* spp. – ditto) around March, bluebells (*Hyacinthoides non-scripta* – Asparagaceae) a bit later, mayflower (*Crataegus* spp. – Rosaceae) in, well, May, and common ivy (*Hedera helix* – Araliaceae) in the autumn. These patterns (and those in other parts of the globe) are embedded in our collective subconscious, and we hardly question that there's a natural order to such events. For peoples all over the world, the timing of flowering has significance. For our early ancestors, and for modern hunter–gatherer communities, it is important to have a strong sense of what foods appear at different times of the year, a knowledge that is literally life and death for these societies. Even in agricultural communities in more recent years the additional nutrition provided by wild plants and animals could make the difference between surviving periods of famine, dying, or having to migrate. In some societies, understanding and recording of flowering and fruiting times have become ritualised. In Kyoto (Japan) there is a written record of cherry blossom flowering over twelve centuries, since the year 812 (Aono and Kazui 2008, Aono and Saito 2010). That such records have been kept for well over a thousand years reflects a Japanese obsession with the timing of flowering of those trees, and the associated spring custom of *hanami* or 'flower viewing', a tradition that goes back to at least the ninth century CE, as the verse by Japanese poet Ki no Tomonori at the start of this chapter illustrates.

In temperate climes such as occur in northern Eurasia, southern South America, and northern North America, the winter period where most plants cease flowering and most pollinators hibernate as eggs, pupae or adults, is a huge disruption to local plant–pollinator networks. Yet, each spring, these relationships re-establish themselves. Some overwintering adults may have remembered the flowers they visited the previous year, but for newly emerging individuals the process of finding food from flowers must be done on the basis of trial and error. It's perhaps easy to dismiss this as 'just what nature does', but re-establishing the links in Darwin's entangled bank (Chapter 4) is a complex ecological process underpinned by many generations of natural selection.

These local patterns, although they can vary in their specific details between years (e.g. Alarcón *et al.* 2008), scale up to form predictable patterns of flowering and pollinator activity that occur year after year across whole regions. Natural history records of plant flowering times and pollinator foraging, much of them collected by well-informed amateurs, have huge scientific importance in this regard. One of the values of such records to ecology is that it allows us to document where these species occur in space and when they are active in time. This can be done at a range of spatial and temporal scales, but large-scale patterns (for example at a country level) are, I think, especially useful because they provide scientific evidence that can inform national conservation strategies.

During 2017 I collaborated with a young early-career researcher, Nick Balfour at the University of Sussex, on an analysis of the phenologies of British pollinators and insect-pollinated plants (Balfour *et al.* 2018). This involved assessing more than one million records that document the activity times of aculeate wasps, bees, butterflies and hoverflies held in the databases of three of the UK's main insect recording organisations: the Bees, Wasps and Ants Recording Society (BWARS), the UK Butterfly Monitoring Scheme (UKBMS) and the Hoverfly Recording Scheme (HRS). Information on flowering times was taken from a standard flora (Clapham *et al.* 1990). As well as looking at annual flight periods and flowering trends for these organisms, we also focused on pollinator and plant species that were endangered or extinct, building on work I'd previously published (Ollerton *et al.* 2014 – see Chapter 10). The results were fascinating.

About two-thirds (62%) of pollinator species peak in their flight times in the late summer (July and August). There was some variation between the different groups, however, as you can see from Figure 6.4 where I've presented the data for bees and hoverflies. Particularly noticeable was the double peak of the bees, with the first peak denoting the activity of many early-emerging solitary bees, such as species of the genus *Andrena*, while the second peak is other solitary bees, plus of course the bumblebees, which by that time have built up their colonies. This pattern is well known to naturalists, but it was good to see it confirmed and illustrated with data. The plants were more surprising, though, as an unexpected phenological pattern with respect to different life forms was also apparent: insect-pollinated trees tend to flower first, followed by shrubs, then herbaceous species. This might be because larger plants such as trees and shrubs can store more resources from the previous year, which will give them a head start in flowering the following year, but that idea needs to be formally tested. However, the outcome is that trees tend to be pollinated by those earlier-emerging bees and hoverflies, whereas the herbs are pollinated by species that are active later.

When looking at the extinct and endangered pollinators, the large majority of them (83%) were species with peak flight times in the late summer, a much bigger proportion than would be expected given that 62% of all species are active at that time. However, this was mainly influenced by extinct bee species, and the same pattern was not observed in other groups. The obvious explanation is that historical changes in land use have led to a dramatic reduction in late-summer-flowering herbaceous species, and the consequent loss of floral resources has been highly

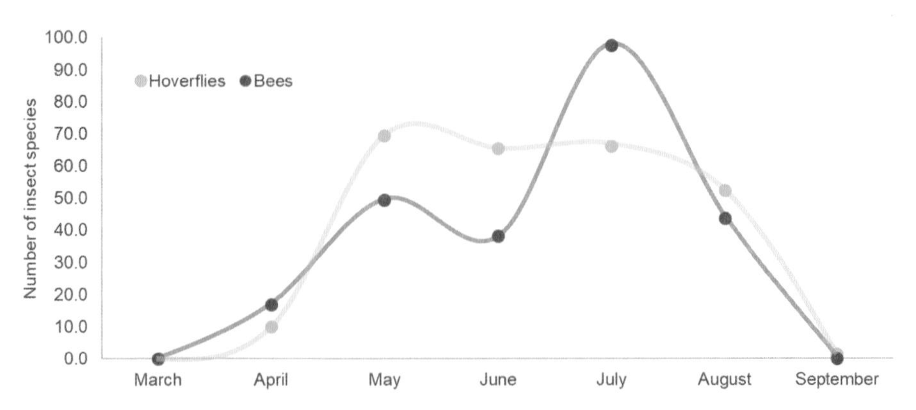

FIGURE 6.4 The phenologies of two important groups of British pollinators: bees and hoverflies. The plotted data are the number of species which peak in their observation frequency in each month, based on Balfour *et al.* (2018).

detrimental to those bees. The reasons why pollinators become rare or extinct are complex, a point that I will return to in Chapter 10.

The lack of late-summer floral resources for pollinators is a contentious issue. Some plant conservation groups in the UK have in the past recommended that meadows and road verges are cut in late summer to maximise plant species richness. Mowing road verges once or twice a year certainly benefits plant diversity, as a recent review by Jakobsson *et al.* (2018) demonstrated. But such mowing can also reduce the flowers available on which pollinators can forage, and studies have found that reduced frequency of mowing can increase pollinator diversity and abundance (e.g. Balfour *et al.* 2015, Garbuzov *et al.* 2015). This was amply demonstrated in the PhD work of one of our former postgraduate researchers, Sam Tarrant, who studied the pollinators and plants on restored landfill sites compared to nearby nature reserves. Sam showed that on restored landfill sites the abundance of pollinators in autumn surveys (conducted September–October) was just as high as for summer surveys. On nature reserves, which are routinely cut from mid-July onward, this was not the case – there was a much lower abundance of pollinators in the autumn (Tarrant *et al.* 2013). Clearly what is good seasonal management for plants may not be good for pollinators, and this is an area that needs to be studied more critically, as I'll discuss in Chapter 12.

Patterns beyond north-temperate regions

The ways in which plants flower and pollinators behave over the course of a year varies enormously with latitude. In the far north, studies in Greenland and Arctic Canada have shown that there is a narrow window of opportunity for plants to flower and for pollinators to forage and reproduce. The period can be as short as just a few weeks (though of course there is the benefit of 24 hours of daylight for some or all of that time). Similar patterns are found at high elevation, where alpine species must grow and reproduce before the weather gets too cold and they become dormant.

Moving towards the equator, the seasonal availability of resources and suitability of the weather for flowering and foraging becomes longer, and is more likely to be constrained by extremes of rainfall and heat than by cold weather, except on high mountains. During the time I've been taking trips to Tenerife with my students, always between April and May, we have encountered very variable weather conditions up in the mountains, ranging from baking heat to freezing snow. But one of the dominant bee species (*Anthophora alluaudi*) remains a constant – we see it all the time in all kinds of habitats – and is active twelve months of the year (Figure 6.5). I have even seen this species foraging on flowers in cold, foggy conditions at around 2,000 metres elevation. It's a tough bee! Clearly data like these are sensitive to when scientists are making observations (which is why I haven't added my own observations to Figure 6.5), but it seems likely that the numbers of these bees do indeed peak around April when most plants on the

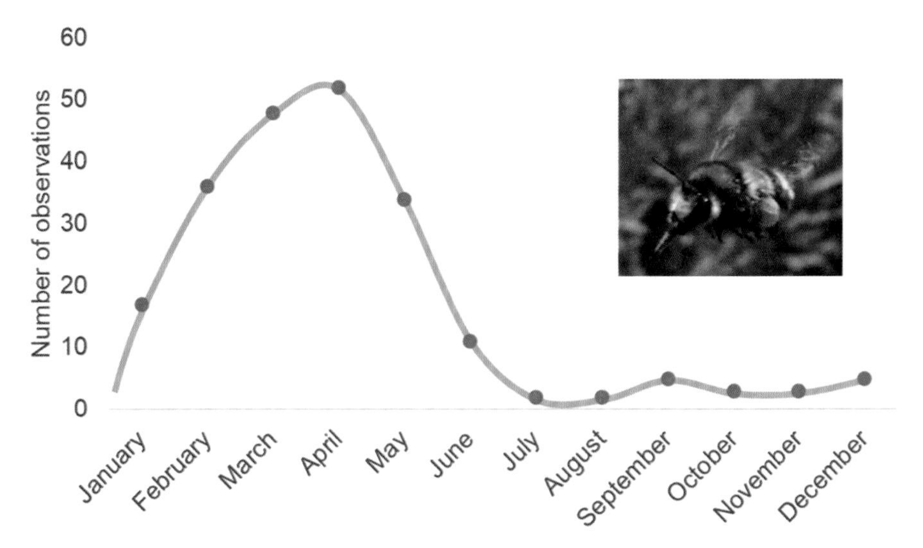

FIGURE 6.5 The abundance of the solitary bee *Anthophora alluaudi* on Tenerife over the course of the year. Data are based on records from bee specimens presented in Hohmann (1993). Inset photo by Victoria Price.

island are in flower, declining sharply in the summer when conditions are at their hottest and driest. However, it's also clear that the species does not hibernate and can be found flying throughout the year, as can other bee species including the endemic bumblebee (*Bombus terrestris* subsp. *canariensis*). This is due to a relatively clement, subtropical climate and the availability of food all year round. Closer to the equator things are similar but different.

One of the myths about the tropics is that they 'have no seasons'. Nothing could be less true. The tropics certainly are seasonal, it's just that the seasons are rather different to what we would expect in a temperate context. For example, there are frequently distinct wet, very wet, and drier periods in the 'wet' tropics, and plants in the rainforest in these regions may respond to these weather patterns by producing flowers. The patterns of flowering of individual species can be highly variable, however, and range from species that are more or less constantly in flower, to those that flower on an annual cycle, and to species that reproduce on a supra-annual basis (Bentos *et al.* 2008). Likewise, tropical pollinators may have distinctive peaks of activity, though there is a tendency for them to be active for most of the year and to move between different floral resources as they become available. In the Atlantic Rainforest of Brazil, for example, hummingbirds will migrate between different areas depending upon what is in flower (see Chapter 9).

In arid parts of the subtropics and tropics, water availability has a huge influence on the abundance of flowers: we see far more plants in flower following a wet winter on Tenerife than if the winter has been dry, for instance. This phenomenon can be seen most clearly in the occasional mass-flowering 'super blooms' that occur in deserts such as in southwest North America or the winter rainfall regions of South

Africa. In rare years there is enough water available to trigger large-scale flowering events that carpet the landscape with blooms. These may only last a few weeks, and then the flowers vanish. Pollinators of these plants must be able to respond to similar weather cues and quickly emerge to take advantage of the floral resources.

This correlation between flowering time and pollinator activity is clearly important if both groups of organisms are to reproduce successfully. Nowhere is this more apparent than in the case of the figs (*Ficus* spp. – Moraceae, Figure 6.6). This group of predominantly tropical plants is pollinated exclusively by tiny wasps in the family Agaonidae that are completely reliant on the figs to complete their life cycle. Because of this close relationship, populations of fig trees must produce their

FIGURE 6.6 Two species of fig (Moraceae): *Ficus macrophylla* from Australia (top) and *F. dammaropsis* from New Guinea (bottom).

inflorescences year-round in order to support the fig wasp population. A significant break in the phenology would kill the wasps and doom the trees to a slow decline as they fail to reproduce. Such a risky strategy has clearly been successful – there are about 850 described species of *Ficus*, and they have persisted for perhaps as long as 90 million years (Shanahan 2016). Despite being no more than a couple of millimetres in length, fig wasps can disperse easily to newly founded fig populations (Shanahan 2016). This means that figs and their pollinators can quickly recolonise areas that have been devastated by volcanic activity, as was seen in the re-establishment of the flora of Krakatoa during the twentieth century (Compton *et al.* 1988, New and Thornton 1992). It used to be thought that each fig species was pollinated by only a single species of fig wasp, and each wasp was supported by only one species of fig, but we now know that's not the case for many species (Machado *et al.* 2005, Yang *et al.* 2015). It is possible that this flexibility in relationships also helps to maintain the fig–fig wasp relationship over long time periods.

The flowering times of plants are sensitive to local weather conditions, especially temperature and water availability, and pollinator activity is likewise timed to the availability of floral resources. It follows, therefore, that alterations to weather patterns due to global climate change have the potential to affect these relationships, which is the topic of the next section.

Climate change, pollinators and pollination

There is no doubt that the Earth's climate is changing, that our atmosphere is getting warmer overall, and that this is affecting local weather patterns and regional climates. There is also no doubt that most of this change is driven by human activities such as burning of fossil fuels.

Ecologists have been studying how these changes to the climate are affecting pollinators and their flowers for some time. Early work was done with butterflies (e.g. Parmesan *et al.* 1999, Warren 1999) because of the large volume of data available, the result of long-standing interest in recording these striking and conspicuous insects. Bumblebees (*Bombus* spp.) have been studied more recently and, as they are one of the most important genera of pollinators across large parts of Eurasia and North America, they are worth considering in detail. In their *Climatic Risk and Distribution Atlas of European Bumblebees*, Rasmont *et al.* (2015) showed that, as Europe gets warmer, just over one-third of the 56 *Bombus* species studied are at risk of being lost from more than 80% of their current distribution, and 41% could be lost from 50–80% of their range. Just three species (about 5% of the total) were likely to increase their distributional range across the continent.

Bumblebees not only generate heat but also conserve it with their thick body hair and relatively large size, both of which incidentally help make them such effective pollinators. Clearly, they have evolved in regions with cooler climates, and so it's not surprising that they are especially affected by future European climates that are predicted to be hotter and drier. Spain and Portugal may have only one bumblebee species left by 2050, with potential knock-on effects for fruit and seed production in both wild and crop plants. Globally, bee diversity peaks at mid-latitudes rather than in the tropics, and is especially high in areas with warm, dry climates (Michener 2007, Ollerton 2017). So the bumblebees of southern Europe, for instance, may be replaced by other genera that are more tolerant of heat and drought. However, we currently know too little about the distribution and life histories of many of these groups to make firm predictions as to how climate change will affect them.

Research published in February 2020 suggests that climate change has already had a massive effect on the abundance of bumblebee species in North America and Europe. The title of the paper by Soroye *et al.* (2020) rather gives away its findings: 'Climate change contributes to widespread declines among bumble bees across continents'. This study shows that, for 66 species of *Bombus*, there had been a decline in species diversity in 100 × 100 km quadrats of, on average, 46% in North America and 17% in Europe. This loss of diversity has occurred in the period 2000–2014, relative to a baseline of 1901–1974. Using some sophisticated analyses, the authors show that climate change has been the main driver of these losses, and has been more important than factors such as changes in land use and use of pesticides. Which is not to discount those other contributors to pollinator loss: they can interact with climate change and are all part of the assault that we are imposing on the environment (see Chapter 10). The most significant finding of the Soroye *et al.* (2020) study is that it's extreme heat which seems to be the driving factor in determining *Bombus* declines. Bumblebees are large, hairy insects because they are adapted to cooler conditions: they are not, by and large, tropical insects, except in mountainous areas. Not surprisingly, then, it is the number of days of temperatures higher than those historically encountered by particular bee species that is the main driver of their loss from a region. This is the result of human-enhanced SHOCKs (see Chapter 4), and for heat-sensitive species like bumblebees, they are occurring more often than we had imagined when we wrote our review (Erenler *et al.* 2020). I fear that the coming years will see more examples of this as the effects of anthropogenic climate change continue to play out and our world experiences more extremes of weather events that are hotter, wetter, colder, drier, windier and more combustible than we have previously known.

Phenological observations that have been collected by naturalists and scientists over the last 200 years or so indicate that spring is generally earlier now in much of Europe than it was 20 years ago. With that shift, autumn has likewise been brought forward and is lasting longer, as shown also by changes in fungi fruiting patterns. Phenology scientists usually express these changes quantitatively, as number of days difference between events, such as bird migration dates or plant flower times, across a period of years. But any person with an interest in the natural world can see these changes for themselves, even in gardens, where for example snowdrops and daffodils bloom earlier than they did a generation ago. However, it is high-latitude regions that are expected to be hardest hit by anthropogenic climate change, and long-term ecological observations show that this is the case. For example, between 1996 and 2009, the flowering seasons within the Arctic became shorter and pollinators less abundant (Høye *et al.* 2013). One of the groups of pollinators most affected were flies of the family Muscidae, a dominant set of pollinators in a region where bees are few and far between. They are the principal pollinators of *Dryas* spp. (Rosaceae), an important source of nectar and pollen in the Arctic (Tiusanen *et al.* 2016). Similarly, Prevéy *et al.* (2019) using a database of 42,689 observations of flowering time in tundra plants, showed that late-flowering species were flowering earlier, thus shortening the overall community flowering time. The long-term effects of fewer pollinators and shorter flowering seasons are currently unknown but are unlikely to be beneficial for the ecology of high northern latitudes.

The responses of pollinators to climate change will depend on whether species can adapt to the new climate or move to areas that better suit them, or survive in smaller range areas. For the pollinators of eleven species of cacti in the Andes, future climate change is predicted to increase their spatial overlap with the plants despite a decrease in the range sizes of the plants (Gorostiague *et al.* 2018). Whole shifts in ranges might be more difficult, however, as being able to move depends on there being appropriate habitat into which to move. Conservation programmes aimed at pollinators are now considering how to 'future-proof' landscapes and regions to allow species to move to more suitable areas (Samways 2015). We'll consider this in more detail in Chapter 12, though it's worth stating here that this strategy may not be possible in the places most vulnerable to climate change such as high mountains and the Arctic, because there are limitations on where pollinators can actually go. Likewise, highly developed countries in Europe and elsewhere currently contain too little natural habitat to facilitate these movements: nature reserves and other protected areas need to be extended and linked together.

In their review of the impacts of climate change on pollination, Settele *et al.* (2016) noted that as well as mismatches in the phenologies of species, some of the other negative effects of even gradual climate change include alteration of habitats, and spread of non-native species and pathogens. There are also predicted impacts of extreme weather events such as droughts, periods of heavy rainfall, and unseasonable freezes. These include death of the pollinators themselves, reduced floral resources, and restricted opportunities to forage. All of this sounds depressing, and there's no doubt that climate change is going to have huge effects on human society and natural ecosystems in the future. However, I am optimistic that many plants and pollinators, at least in the temperate zones, will be able to cope with some degree of climate change. That optimism is based on our knowledge of the flexibility of plant–pollinator interactions (see Chapter 4), and also on how robust species can be to daily and seasonally unpredictable weather, as we will see in the next section. This is not to say that we should be relaxed about global climate change: we urgently need to reduce its predicted impacts and plan for the continuation of changes that are already happening. We also need to consider the other impacts that climate change is having on ecosystems, such as increased fire frequency (see the final section of this chapter).

Responses of flowers and pollinators to extremes of weather

Sudden changes in the weather in Britain, especially early in the season, are not at all uncommon, and climate change seems to be making such events more frequent and more extreme. Take 2018 as an example: my wife Karin Blak spotted the first queen buff-tailed bumblebee (*Bombus terrestris*) in the garden on 6 January, investigating a camellia flower just outside the kitchen window. Over the course of the next few weeks I saw a few more at different sites, plus occasional butterflies that had come out of hibernation such as the red admiral (*Vanessa atalanta*). The various social media were reporting similar things, and it looked as though we were going to have an early spring. Then at the end of February the 'Beast from the East' hit us, a weather system from Siberia that brought some of the coldest weather and heaviest snow the country had experienced for several years. That persisted for over a week before temperatures became much milder. On 16 March I was in the garden and spotted the first male hairy-footed flower bee (*Anthophora plumipes*) of the year, plus an *Andrena* species, and a brimstone butterfly (*Gonepteryx rhamni*), a queen bumblebee and another red admiral. Great, I thought, spring really is here! The next day it snowed: a 'Mini Beast from the East' had arrived. The photographs of our garden making up the composite image in Figure 6.7 were taken two days apart.

What happened to all of those insects that we had seen? Were they killed by the cold weather? Or did they survive? We have no firm data to answer that question – as far as I'm aware no one has ever tagged early-emerging pollinators and followed

FIGURE 6.7 Changeable weather in our garden and those of our neighbours in Northampton, March 2018. These two photographs were taken only two days apart.

their progress to see if they can survive these abrupt temperature changes. It would make an interesting, though labour-intensive, project but could be done using non-toxic paint of various colours to individually mark insects. I suspect that some were killed, but most were not and simply went back into hibernation for a short period, hunkering down in safe, sheltered spots. That makes much more evolutionary sense: any insects in Britain that cannot survive sudden changes in the weather would have gone extinct long ago. Another clue to support this idea is the fact that plants in flower early in the season, and in some cases the flowers themselves, usually survive the cold weather and come back as if nothing had happened. If the flowers can do it, and they have to stay where they are, surely the mobile pollinators can also do it.

As I noted above, there is often a tendency to think that the timings of things like flowering, insect emergence and bird arrivals are rather fixed, and that these phenologies are highly predictive of the seasons in northern temperate areas. That is true up to a point, but there is also a lot of variation in those timings between years. One example can be seen in Figure 6.8, which shows the flowering time of the same population of cuckoo-pint (*Arum maculatum* – Araceae) in two consecutive years. In 1996 a cold spring meant that the flowering was delayed by about one month relative to a typical year. However, it made no difference to the pollination of these plants and fruit set did not differ between years because the main pollinator, a small psychodid fly (*Psychoda phalaenoides*), is present all summer long (Ollerton and Diaz 1999).

This is certainly not the only species to be so flexible in its flowering; I first encountered such variability during my PhD field work (see Chapter 5) when the population of bird's-foot-trefoil (*Lotus corniculatus* – Fabaceae) at Wytham Woods, Oxford, likewise varied by a month in its flowering time in different years, though

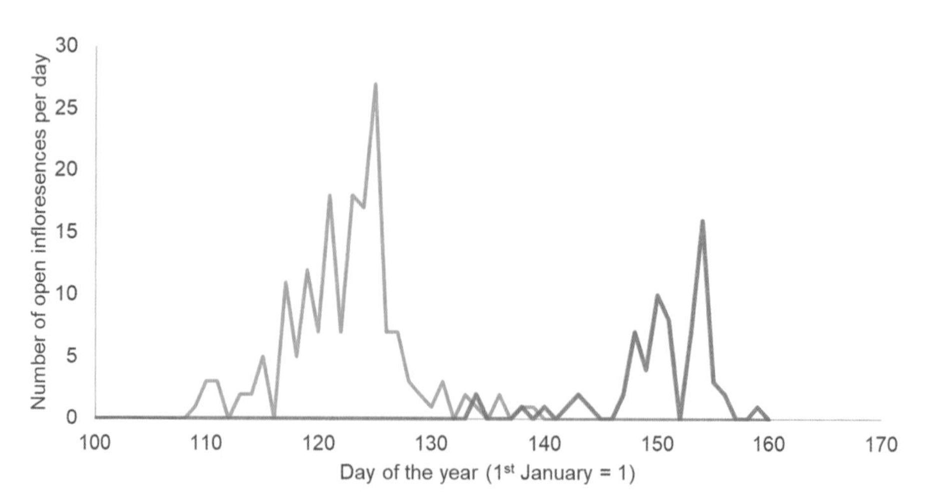

FIGURE 6.8 The flowering phenologies of a population of *Arum maculatum* at Park Campus, University of Northampton, in 1996 and 1997.

drought was the most probable cause. Similar to *Arum maculatum*, this change in flowering time did not affect average pollination success (the proportion of flowers that produce seeds) because the bumblebee pollinators were available through the flowering period. Whether climate change will cause phenological

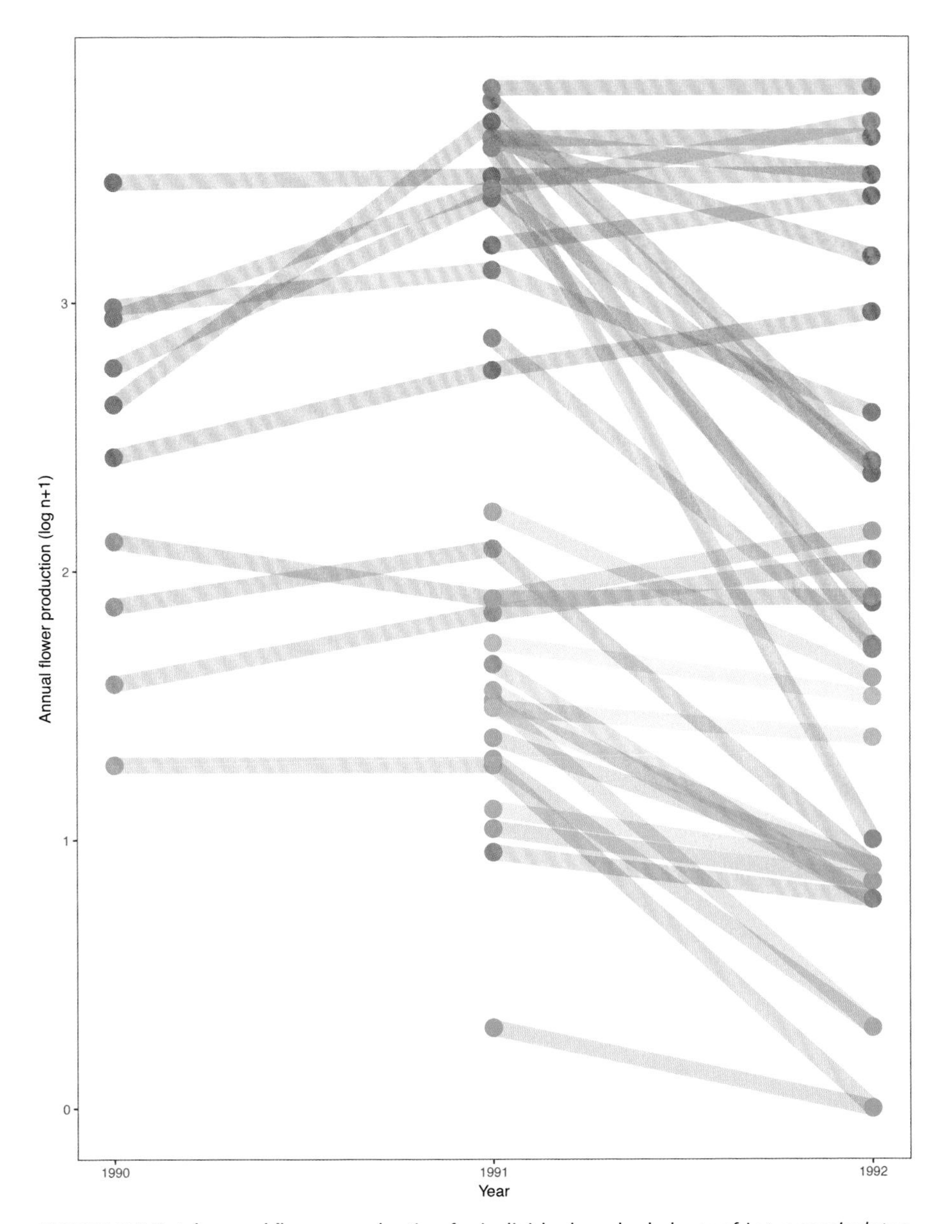

FIGURE 6.9 Total annual flower production for individual marked plants of *Lotus corniculatus*, Wytham Woods, Oxfordshire, 1990 to 1992. Note that flower production on the *y*-axis has been logged as $n + 1$ to account for individuals that produced no flowers in 1992. Data from Ollerton (1993).

mismatches between plants and their pollinators is unknown for most species, but the limited evidence to date suggests that such mismatches are not (currently) common (Forrest 2015).

Drought did have a profound impact on the number of flowers that many of those *L. corniculatus* plants produced (Figure 6.9). Each line in this plot is a different individual plant whose reproduction had been followed for two or three years (*x*-axis) with data on total flower production for each year on the *y*-axis (logged). Some plants have data for all three years, others for only two. The first two years were fairly average in terms of rainfall; 1992 was the drought year. As you can see, most plants produced fewer flowers in 1992, sometimes a *lot* fewer: individuals that had produced thousands of flowers in 1991 produced only tens or single digits the following year. But there were a few plants that produced just as many or even more in the dry year. I suspect that these were individuals whose microhabitat (or perhaps depth of tap root) meant that they had access to sufficient water. It's also possible that some individuals were genetically predisposed to be more drought-tolerant. This is quite a long-lived grassland species with a deep tap root; individuals probably experience multiple episodes of extreme weather over their lifetime and so are robust to these events. However, at a community level the reduction in number of flowers produced, and the amount of nectar available per flower, could have effects on pollinator populations (Phillips *et al.* 2018a). This has more impact in desert areas, where lack of water can result in many plants ceasing to flower and young bees suspending their development and remaining in their nests until such time as conditions change (Minckley *et al.* 2013).

The long-term effects of climate on plant–pollinator interactions

The Earth's climate has always changed, albeit at a slower rate than we are presently experiencing, and has always varied geographically. That change and variation has had profound effects on the distribution of different pollination systems and the level of specialisation of interactions across the globe. The relative proportions of wind and animal pollination in plant communities at different latitudes that I presented in Chapter 1 (Figure 1.5) are constrained mainly by current rather than past regional climates (Rech *et al.* 2016). Areas with higher average temperatures tend to have a larger proportion of animal-pollinated plant species, as do wetter regions, though only in forest habitats. These patterns could be due to there being fewer limitations on nectar production in parts of the world that have more warmth and water availability, and because wind dispersal of pollen is less effective in humid air. Thus the latitudinal trend of greater animal pollination in the tropics shown in Figure 1.5 can be linked directly to current climate rather than to the legacies of past climates, as can local ecological patterns such as pollinator diversity (Martín González *et al.* 2009).

In contrast, other studies have shown that the climate over the 10,000 years or so since the last ice age can have legacy effects on the structure of plant–pollinator networks, including their modularity and levels of specialisation (Dalsgaard *et al.* 2011, 2013). So if we wish to predict how the plant–pollinator interactions of a region will change over the coming centuries we need to consider both what the climate was like in the past and what it's going to be in the future. That's as well as understanding how habitat degradation, destruction and restoration is affecting these relationships (Chapters 10 and 12). Whatever the future holds for plants and their pollinators, however, it's not just natural ecosystems that will be affected; as we'll see in the next chapter, global agriculture owes a lot to the animals that pollinate many of our crops.

Extreme events and climate change: the view from Australia

During December 2019 and January 2020 I was based in Australia as a Visiting Research Fellow at the University of New South Wales. As well as working on a project there with colleagues, I was making final corrections to the manuscript of this book. The bushfires that are a normal feature of the Australian environment were, during that austral summer, unprecedented in New South Wales and Victoria. Exacerbated by a long-term drought and higher-than-usual temperatures, the fires had started earlier, become more intense, and burned a much larger area of habitat than is usual. This included areas that do not normally burn, such as rainforest. Climate change clearly had a role to play in this. The full impact of these fires on Australian animals and plants is yet to be assessed, but there is no doubt that it will be huge. Whole populations, perhaps even whole species, of some plants and pollinators may have been eliminated from the Australian environment. For example, there is particular concern about the green carpenter bee (*Xylocopa aerata* – see Santos *et al.* 2020 and www.aussiebee.com.au/carpenterbees.html). But this is a distinctive and relatively large species; most bees (and other pollinators) are much smaller and less conspicuous, and we know little about what has happened to their populations. The Australian bushfire season of 2019/20 may be one of the most significant tests yet of how plants and their pollinators can cope with large-scale, climate-change-induced SHOCKs (see Chapter 4).

Chapter 7

Agricultural perspectives

It is almost impossible to read anything about pollinators without coming across impressive statistics about their importance to agriculture. However, it's much more complicated than the oft-quoted statement that 'one in every three mouthfuls of food comes from bees', which holds little water for anyone tucking into a plate of fish and chips, where most of the vegetable component is either wind-pollinated (wheat) or propagated via tubers (potatoes). Mushy peas are a different matter of course, but regardless, these sorts of clichés – widely bandied around by NGOs and well-meaning individuals – are a huge oversimplification of a complex and important topic: food security and world hunger (Bailes *et al.* 2015). In this chapter I will explore the role of pollinators in agriculture, providing some case studies on how important they actually are, and asking which of the many different groups of pollinators are the most significant.

The economic value of pollinators

At a global scale we know that 75% of the 115 most productive crop plants are dependent to some extent on animal pollinators, and those crops account for 35% of worldwide crop yield (Klein *et al.* 2007). That's probably where the 'one in three mouthfuls' comes from, but it's clearly an odd interpretation of the facts. 'To some extent' means that crops range in their reliance from those that need pollinators to produce any kind of a harvest to those where perhaps only a small percentage of the seeds or fruit are the outcome of animal pollination, with the rest from wind- or self-pollination. This differs both between crop species and between varieties of the same species; for example, canola or rapeseed (*Brassica napus* subsp. *napus*) varieties vary in their requirement for insect pollination. The estimated value of this pollination service changes depending upon the author and the year it was calculated, but the most recent and accurate assessment suggests that animal pollinators account for between 5% and 8% of global crop production, with a market value of US$235 to US$577 billion per year (IPBES 2016). However, these gross figures hide considerable geographic variation in the importance of pollinators to the agricultural economies of different countries, an importance that has grown enormously for many regions since the 1960s (Potts *et al.* 2016).

In the UK, insects directly contribute an estimated £603 million a year to the economy through the pollination of crops (Vanbergen *et al.* 2014), though this is certainly an underestimate of the true value of pollinators. The figure does not include produce from home gardens and allotments and in fact, as far as I can discover, there are no published data for how much home-grown food is worth to household economies. But it must be considerable, given that many British gardeners grow at least one food plant in their patch. In his book *The Garden Jungle: or Gardening to Save the Planet* (2019), Dave Goulson points out just how much food can be grown in a domestic garden: yields can be far greater than from farmland of equivalent area (see also the recent study from Dave's group that looked at food production in the English city of Brighton and Hove, framed around the United Nations Sustainable Development Goals – Nicholls *et al.* 2020). Our small urban garden in Northampton measures just 10 × 20 metres, but nonetheless we grow a wide variety of fruit and vegetables, many of which require some level of pollination by insects. We are hardly self-sufficient, but it helps reduce the food bill throughout the year as what we don't eat immediately we preserve or freeze. I'll discuss this further in Chapters 8 and 9 when we consider the importance of urban pollinators for urban agriculture.

The UK figure cited above also doesn't take into account the estimated £1.8 billion it would cost farmers to pay labourers to hand-pollinate those crops, a situation that does occur in some parts of China, and in the vanilla orchid industry in Madagascar. The statistics on the value of insect pollination also fail to include some minority crops that, while in total value a small proportion of UK agriculture, nevertheless make a contribution to local economies (see *Thank the insects for Christmas*, below).

At a local scale in the UK the figures are no less impressive. For example, as part of the Nene Valley Nature Improvement Area (NIA) project, Jim Rouquette in our research group calculated the value of insect pollination of crops within the NIA (which encompassed most of Northamptonshire) to be £7.8 million per year (Rouquette 2016). This is the amount that would be lost from the local farming economy if pollinators were not present. However, the agricultural requirement for pollinators varies hugely across the county, depending on the crops being grown. Areas that grow crops which require pollinators, such as oilseed rape, apples, field beans and linseed (flax), have a high demand, areas growing predominantly cereals or with mainly pasture have no demand at all.

As I mentioned in Chapter 1, insect-pollinated crops are in a minority in Britain, where arable agriculture is dominated by cereals and root crops. However, these crops have become more important over time (Figure 7.1). Although the trend is clear, there are significant fluctuations in the area of different crops being grown over this period, owing to external factors such as variations in wholesale prices, the introduction of new crops, or an increased demand for others. Linseed was hardly grown in England prior to the late 1980s, for example, and farmers now cultivate four times the area of field beans that they did forty years ago. The recent

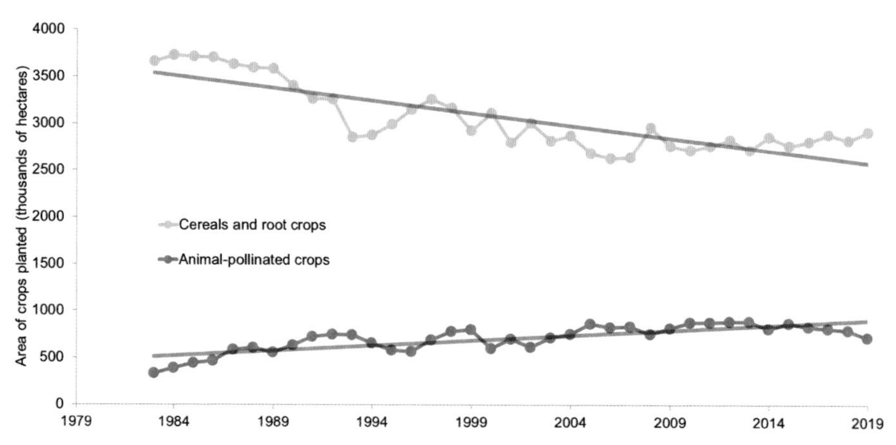

FIGURE 7.1 The areas of the major animal-pollinated crops (oilseed rape, linseed, field beans, peas, orchard and soft fruit) and non-animal-pollinated crops (cereals (excluding maize), potatoes, and sugar beet for human consumption) grown in England since the 1980s. Data from the June 2019 survey of agriculture in England (Defra 2020). Linear trend lines are shown in grey.

downturn in animal-pollinated crops, which is mirrored by an increase in the hectarage of the other crops (Figure 7.1), is largely due to less oilseed rape being grown. There has been a suggestion that this was caused by the banning by the European Union in 2018 of three of the main neonicotinoid pesticides used to control insect pests (see Chapter 13). However, Defra data show that the decline in planting of oilseed rape clearly started earlier than the ban, from 2012 in fact, and was likely due to a combination of economic reasons (NatWest 2019).

Since the 1980s the area of animal-pollinated crops has more than doubled in the UK, mostly due to the greater emphasis on rapeseed (canola) cultivation. In contrast, the non-animal-pollinated crops have declined in area by one-fifth, with reductions across all three of the main crops (cereals, potatoes and sugar beet). This pattern reflects the worldwide trend described by Potts *et al.* (2016) towards growing more and more animal-pollinated crops, ironically at a time when agriculture is seen as a major factor responsible for reducing pollinator abundance and diversity (Chapter 10).

Measuring the area or weight or financial value of a crop, however, only gives a crude insight into its importance, because we also need to consider the nutritional contribution that the crop makes to our diets (Ellis *et al.* 2015). While cereals and root crops provide most of the energy (principally carbohydrates) in the human diet, animal-pollinated fruit and seeds provide high proportions of many of our essential minerals and vitamins, including vitamins A, C and E, and various carotenoids. The relative importance of animal pollinators to providing these nutrients varies enormously, depending on the country and the overall balance of diets, but it is particularly pronounced in developing countries, where a higher proportion of the crops are animal-pollinated. Indeed, in a study published in *The Lancet* by

Smith *et al.* (2015), it was proposed that following a complete loss of all pollinators across the globe, as many as 262 million people in developing countries could find themselves deficient in vitamin A. In addition, up to 2.5 billion people who currently have diets with less than the recommended amounts of vitamin A would experience further declines. Similar findings were reported for the B vitamin folate. This is because the loss of all pollinators would result in a global reduction of fruit supplies by almost 25%, vegetables by more than 15%, and nuts and seeds by more than 20%. Such a loss of pollinators would, it was suggested, increase the number of global deaths from malnutrition-related diseases by around 1.4 million per year.

This is a very worrying prospect. Clearly the planet is not going to lose all its pollinators in the near future, but even losses at a local or regional scale may impact on crop quality in terms of the micronutrients being provided by that crop, and affect levels of malnutrition of the human population in those areas (Eilers *et al.* 2011, Brittain *et al.* 2014, Chaplin-Kramer *et al.* 2014).

This should concern governments, businesses, and civil society at large. The impacts of climate change that I discussed in the previous chapter, plus loss and degradation of habitat (which we will explore in Chapter 10), means that regional losses of a high proportion of pollinators are not just speculative fictions dreamt up by conservation NGOs and doomsayers. This has important implications for food security and human health, particularly at a time when we are being urged to buy local produce to reduce the carbon footprint of the food that we eat. Thus work referred to above gives a vital insight into how important pollinators are to agriculture, both nationally and across the planet. To explore this further, the following sections present three case studies of pollinators and their importance for agriculture, two as global in scope as it's possible to be, the other much more local to the UK, though with implications for northern Europe. These case studies highlight three things I personally enjoy very much: the three Cs of coffee, Christmas and chocolate.

How many bees does it take to wake up in the morning?

In the diary that Charles Darwin kept during his time aboard HMS *Beagle*, an entry written on a farm near Rio de Janeiro, where he stayed in April 1832, notes:

> On this [farm] are cultivated the various products of the country: Coffee is the most profitable: the brother of our host has 100,000 trees, producing on an average 2 lb per tree, many however singly will bear 8 lb or even more.

Like many academics I drink quite a lot of coffee, and it's my go-to drink first thing in the morning. It's also a hugely valuable commodity that is traded internationally. Charles Darwin seems to have been interested in the crop too, though doesn't appear to have appreciated the importance of pollinators to its production.

Coffee is represented in the global commodities market by different varieties of two species (*Coffea arabica* and *C. robusta*) in the family Rubiaceae. The flowers are pollinated by a variety of insects, predominantly wild bees and managed honey bees (Ngo *et al.* 2011). Because of its huge economic value (estimated to be in excess of US$80 billion per year) it supports millions of subsistence farmers throughout the tropics. It also supports a vast supply chain, and not a small number of baristas in the developed world. One sometimes reads statements about coffee being 'the second most valuable global commodity after oil', and in fact I repeated this notion in a recent review paper (Ollerton 2017). However, even careful scientists get things wrong sometimes, and I later discovered that it's a myth (Greenberg 2017). Nonetheless, regardless of its ranking, it's hugely valuable. In 2018/19 coffee production worldwide amounted to 170.937 million standard hessian coffee bags, each weighing 60 kilograms (International Coffee Organization 2020). One coffee seed (what we refer to as a 'bean') results from a single fertilisation event following the deposition of pollen on a coffee flower's stigma, and there are two ovules per flower to be fertilised. The weight of a single coffee bean is about 0.1 grams on average (I know because I spent an hour in the lab weighing a sample a few years ago) so there are approximately 582,524 beans in one of those standard 60-kilogram export bags. The total coffee bean production in 2018/19 was therefore 170.937 million bags multiplied by 582,524 beans per bag, which equals 99,574,752,191,955. Put into words, that is over 99 trillion coffee beans. That's a lot of coffee.

However, coffee is on average 50% self-pollinating, though outcrossed coffee is better quality (Klein *et al.* 2003, Classen *et al.* 2014), and one visit to a flower by a bee might result in the fertilisation of both ovules. Therefore we can divide our bean total by four to give us the minimum number of pollinator visits to flowers required to sustain global coffee production: almost 25 trillion visits. Think about that figure for a minute: at this very moment, somewhere in the world, billions of bees are going about their business, collecting nectar from coffee flowers and providing the basis for an industry that spans the world, generates jobs and prosperity for producers and the employees of coffee companies, and fuels a good deal of the intellectual output of universities.

The importance of the global coffee industry, and therefore of the pollinators it relies on, has risen by orders of magnitude since Darwin's day. Thanks to the availability of historical data it is possible to show how the role of bees in sustaining coffee production has increased exponentially over the past 170 years or so (Figure 7.2). We can't directly answer the question posed in the title of this section, as we simply don't know enough about tropical bee ecology or the number of visits each bee makes to coffee flowers in its lifetime. At the least we'd need to know the number of individual bees found in coffee plantations of different sizes. But what is clear is that billions of bees support the livelihoods of the farmers and other workers in the coffee supply chain. These bees in turn require semi-natural

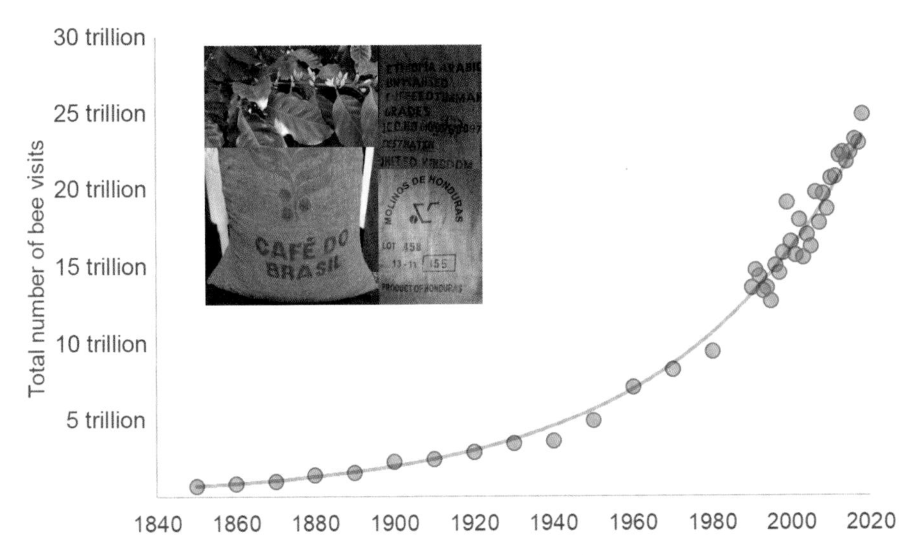

FIGURE 7.2 The historical increase in the number of bee flower visits required to pollinate the world's coffee crop, based on data of total production of coffee (one trillion = 10^{12}). Pre-1990 data are averages per decade; post-1990 data are per year. The grey line is an exponential fitted trend. Data pre-1990 are from Bacha (1992); those from 1990 onwards are from the International Coffee Organization (2020). Inset images show a coffee bush with a ripe fruit and opening flowers (top left) and a selection of standard 60 kg hessian coffee bags.

habitat to survive, as they have to find forage other than what is available within the coffee flowers, as well as places to nest and other resources (see Chapter 12).

It's unfortunate, then, that some of the front-line coffee retailers don't seem to appreciate how important pollinators are to their industry. Take Starbucks, for instance, a company founded on coffee drinking, with more than 28,000 outlets worldwide, employing over 230,000 people, and worth in excess of US$14 billion (though no doubt the COVID-19 crisis has had an impact on its profits in 2020). Some of those outlets have text on the walls describing where and how coffee grows, its cultivation, harvesting and distribution (Figure 7.3). Although this describes how coffee plants flower once a year, and that the flowers are jasmine-scented, there is no mention of pollinators. It's as if some magic happens … and nine months later you get coffee seeds for processing! Starbucks is not alone in this respect. I've yet to see a large coffee retailer highlighting in their outlets the role of these insects and how vital wild and managed bees are to their profits. Which is extraordinary, given the way that awareness of pollinators has risen up the political agenda (Chapter 13). Apparently Starbucks *et al.* don't want to acknowledge the role of these bees in supporting their (very lucrative) industry, at least not in the cafes themselves. If you Google 'Starbucks pollinators' then you find some information online about how the company values bees, but this ought to be on the front line: let the public know in the places where the public goes!

FIGURE 7.3 There ain't no b(ee) in Starbucks. In large coffee outlets such as this, it's rare to see any mention of the pollinators that are responsible for supporting the global coffee crop. Photograph taken September 2018.

I'll come back to coffee later in this chapter, but let's move on to Christmas, because whenever in the year you are reading this book, it's always 365 days or less to the next one.

Thank the insects for Christmas

The Northamptonshire 'peasant poet' John Clare, writing in *The Shepherd's Calendar* of 1827, beautifully captured the cultural value of plants at Christmas:

> The shepherd, now no more afraid,
> Since custom doth the chance bestow,
> Starts up to kiss the giggling maid
> Beneath the branch of mistletoe
> That 'neath each cottage beam is seen,
> With pearl-like berries shining gay;
> The shadow still of what hath been,
> Which fashion yearly fades away.

We'll return to Clare in Chapter 10, as many of his insights are environmental as well as cultural. Although I'm not at all religious, I do appreciate the Christmas holiday as an opportunity to relax and take a break from work, to share time with family and friends, and to unwind during the cold and gloom of a British winter. It's become commonplace to write about the importance of pollinators to a traditional Christmas dinner, but when I first started doing it on my blog in the early 2010s it was quite novel. Pollinators are important for seed production of many of the typical vegetables such as carrots, parsnips and sprouts, as well as herbs and spices, and forage plants that feed livestock. Not to mention soya beans and some of the nuts that go into vegetarian alternatives. However, the importance of pollinators also extends to determining the market value of two of the most emblematic of

Christmas plants: holly (*Ilex aquifolium*) and mistletoe (*Viscum album*). Holly and mistletoe are two seasonal crops that play a culturally important role as symbols of Christmas across the world, though both also have pre-Christian pagan roots. A few years ago, together with my University of Northampton colleague Jim Rouquette and Tom Breeze from the University of Reading, I investigated the role of insect pollinators in determining the commercial value of these plants (Ollerton *et al.* 2016). To do this we used the sales records from Britain's largest annual auction of holly and mistletoe, held every year in Tenbury Wells, Worcestershire, by auctioneers Nick Champion. The holly and mistletoe are collected locally and represent useful cash crops for the people who harvest them.

Since publishing our findings I've continued collating these statistics each year. Analysis of the sales records shows that insect pollination raises the sale price of these crops by on average two to three times (Figure 7.4). This is because holly and mistletoe with berries is more sought after than material without berries, with wholesale buyers paying higher prices at auction. These berries in turn are the result of pollination by insects such as flies and bees: both holly and mistletoe are 100% dependent on insect pollination, since they have separate male and female plants. There is some annual variation to the prices, and in years when berries are scarce (possibly because of low insect numbers or, in 2018, the very hot dry summer) the difference in price between material with and without berries can be even greater.

As far as I know this is the first study showing that insect pollinators play a large part in determining the value of culturally symbolic, non-food crops. We could live without decorating our houses with holly and mistletoe and other ornamentals, though it would be a duller world and they clearly have an economic value. But other non-food crops, including plants producing fibres, construction materials and pharmaceuticals, have huge significance – and many are insect-pollinated. Almost all of the economic valuations of insect pollination to agriculture, however, have focused on human and animal food crops; much less is known about how the value of non-food crops is enhanced by insects (Gallai *et al.* 2009, IPBES 2016). There is also the question of the long-term sustainability of 'crops' such as timber. A recent report by the International Union for the Conservation of Nature (IUCN) concluded that over 40% of the endemic trees of Europe, i.e. those native only to that part of the world, were threatened with extinction (Rivers *et al.* 2019). I estimate that about 67% of native British tree species are insect-pollinated, and the proportion in the rest of Europe will be at least that high. However, loss of pollinators, and subsequent effects on the number and quality of seeds being produced, was not one of the factors that the report identified as a threat. It may be that loss of pollinators is less important than, say, diseases and other pests, or clearance of habitat for urban or agricultural development, but I suspect that the main reason it was not highlighted in the report is simply that we know less about the reproductive biology of these trees than we do about those other factors.

Going back to the theme Christmas, however, where would our winter festival be without our final case study, chocolate?

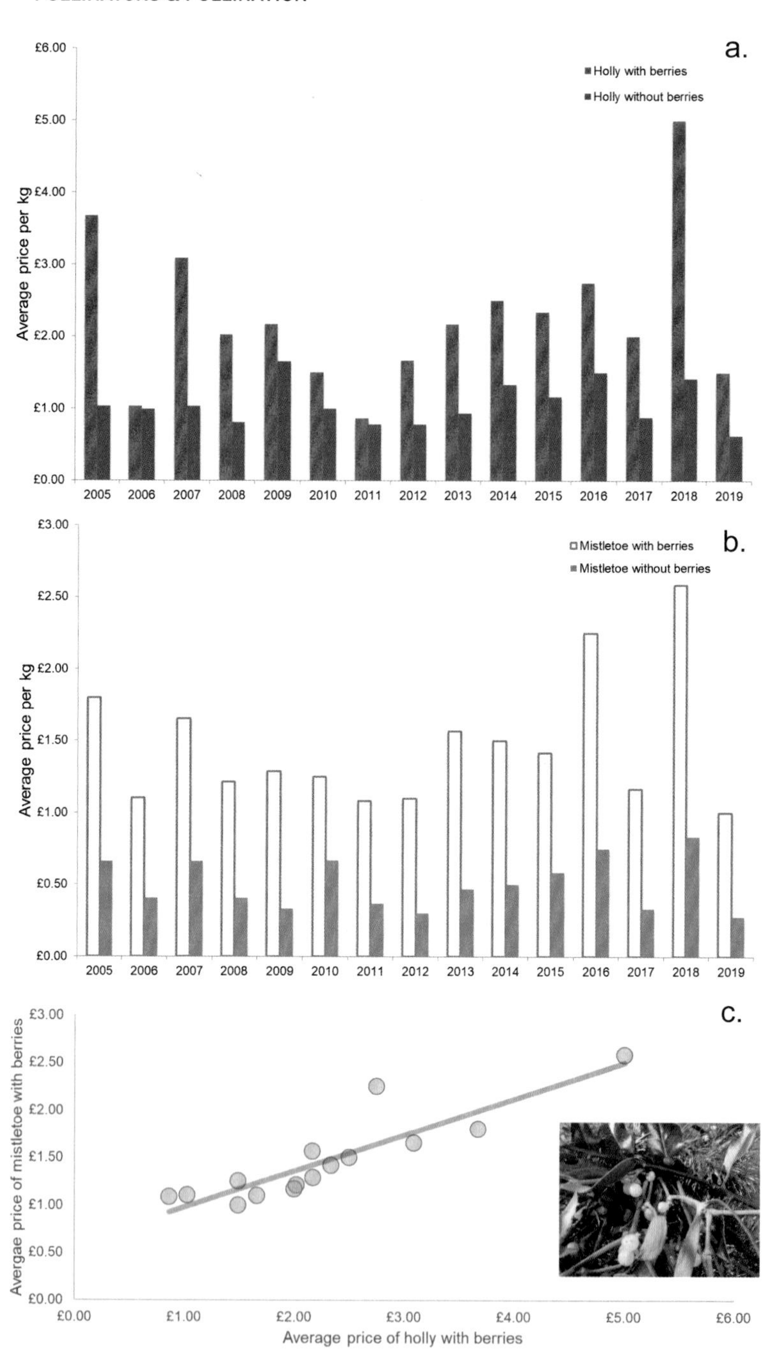

FIGURE 7.4 Annual wholesale auction average prices of (a) holly and (b) mistletoe with and without berries, 2005–2019. Note that for mistletoe, 'without berries' is a shorthand: the material is actually graded as first and second quality and all auction lots have berries, but first-quality material has more berries and is greener, less leggy, and so forth. (c) The average prices of material with berries are strongly correlated in each year. Data courtesy of Nick Champion Auctions, Tenbury Wells.

Flies and cocoa: a complex story

In 2018 the Cambridge Conservation Initiative published a report entitled *The Pollination Deficit: Towards Supply Chain Resilience in the Face of Pollinator Decline*, which was 'aimed at companies with agricultural supply chains who wish to gain a better understanding of the potential risks to their business posed by the decline of wild pollinators and how this translates to a business case for action'.

One of the businesses on which the report focused was the confectionery company Mars, perhaps not surprisingly given that a primary ingredient in their products is insect-pollinated. Chocolate is manufactured from cocoa (*Theobroma cacao* – Malvaceae), a small tree native to Central America but now widely grown across the tropics. There's a long history of cultivation of the crop going back perhaps 4,000 years, and the cocoa beans were originally used as both a form of currency and to produce a ritual drink, the precursor of modern chocolate drinks. Like coffee, it's an extremely valuable commodity, globally traded, and central to a supply chain worth billions of dollars annually. Over half of the world production of cocoa occurs in west Africa, on the other side of the Atlantic from where the crop originated.

The flowers of *Theobroma cacao* are white and rather unusual in form, with the anthers concealed within inflated pouches (Figure 7.5). The fruit that results from pollination of these flowers is many times the size and packed with up to sixty of the cocoa beans. Although individual trees may produce hundreds of flowers, typically only about 10% of these set fruit. Hand pollination of flowers can increase this four-fold, suggesting that pollinators are the limiting factor for larger

FIGURE 7.5 The flowers (left) and fruit (right) of the cocoa tree (*Theobroma cacao*). Photos by Chris Thorogood (flower) and Márcia Motta Maués (fruit).

crops. If this were a bee- or hoverfly-pollinated crop in Europe such as oilseed rape or apples, the advice would be to plant floral margins that are filled with nectar and pollen-rich plants to attract pollinators and increase their populations. But cocoa is different: it's pollinated principally by tiny flies, mainly midge species in the genus *Forcipomyia* of the family Ceratopogonidae (coincidentally one of the main groups of pollinators of *Ceropegia* species – see Chapters 3 and 5). As these midges rarely visit flowers, except when they are deceived into doing so, planting nectar and pollen sources would not work. However, the larvae of the flies feed on rotting organic matter, and adding old cocoa fruit husks around trees has been shown to proliferate the numbers of pollinating midges, resulting in a large increase in fruit set (Forbes and Northfield 2017). This is an excellent example of how an understanding of the detailed biology of a group of organisms gives clues as to how the local environment can be manipulated in a rather simple way for the benefit of farmers.

Despite its economic value as a crop, surprisingly little research has been conducted on the pollination biology of cocoa. A recent review of the literature found that over half of the insects found visiting flowers were not ceratopogonid midges, but belonged to groups such as ants, bees, beetles, other flies, and wasps. However, the effectiveness of these other groups as pollinators is not yet known (Toledo-Hernández *et al.* 2017). The story of cocoa and its pollinators may be even more complex than we presently know. If that's the case, then simply spreading rotting plant material around the cocoa trees may not be enough to maximise fruit set, and other habitat modifications may be required. This represents a serious knowledge gap, and there is an urgency to closing it: cocoa bean yields per farm have dropped significantly in recent years, affecting the livelihoods of many thousands of subsistence farmers across Africa and Asia for whom this is an important cash crop. The prediction is that climate change will reduce these yields even further by affecting both the growing conditions of the crop and the abundance of its pollinators (see Chapter 6). Increasing our understanding of the pollination ecology of *Theobroma cacao*, and responding appropriately at a farm level, may reduce these impacts (Toledo-Hernández *et al.* 2017).

Which insects and vertebrates are the main pollinators of crops?

As the cocoa example illustrates, crop pollinators belong to a diversity of different groups, not just bees. But not surprisingly, given what was said in Chapters 1 and 2, it is the bees that dominate agricultural pollination (Figure 7.6). Honey bees (*Apis* spp.) act as pollinators in about 40% of the crops for which we have good data, with the other, mainly unmanaged bees accounting for another 46%. The remaining pollinators belong to a range of taxa, though vertebrates (e.g. bat pollination of durian fruit) are uncommon.

Presenting such statistics broken down by taxonomy might be slightly misleading, however. A recent review by Rader *et al.* (2016) of pollinator

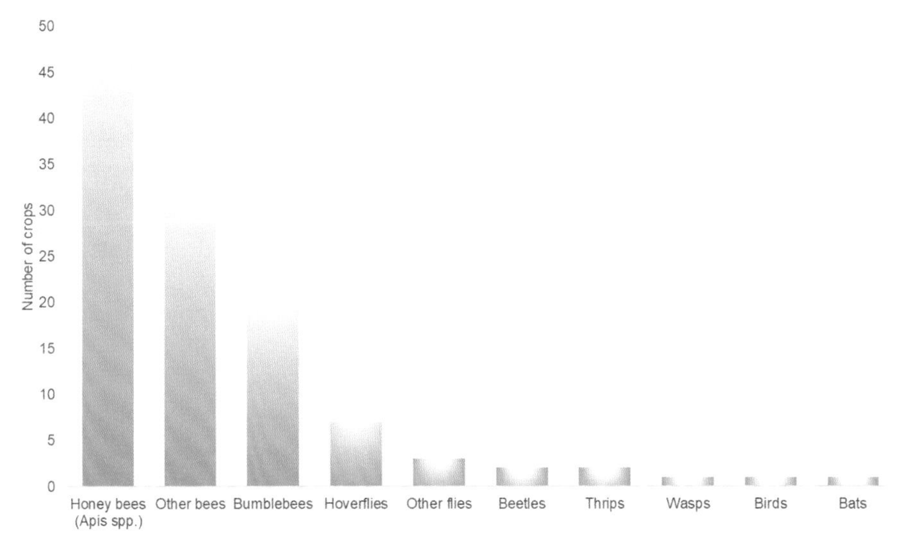

FIGURE 7.6 The frequency with which different groups of pollinators are effective pollinators of crop plants around the world. Data are mainly from Klein *et al.* (2007) supplemented with more recent studies. Note that many crops are pollinated by a mixture of different taxa, for example honey bees plus bumblebees, or bumblebees plus hoverflies plus other bees.

effectiveness in 39 field studies showed that non-bees (mainly flies, beetles, moths, butterflies, wasps, ants, birds and bats) performed between 25% and 50% of the flower visits, depending on the crop. Not only that, but these non-bees increased fruit set of crops independently of the bees, showing that they add a value to crops beyond what the bees provide. Likewise, Phillips *et al.* (2018b) have shown that shared traits such as hairiness, size and behaviour mean that bees and flies are similarly effective at delivering pollen to rapeseed (canola) stigmas. It's surprising, though, how little we still know about the effectiveness of different types of pollinator (not just bees, and certainly not just honey bees) on crops in a variety of geographical settings and in different years. It's an area where much more research needs to be done.

The value of having a diverse set of crop pollinators is partly that it acts to ensure pollination, and therefore stable crop production, in periods when some groups are in low abundance. But a further benefit comes from the way in which the animals interact with one another and force pollinator movements between flowers. There are many crops that have been shown to have an increase in yield and/or a better quality of crop when pollinators are diverse and abundant. As we have seen, coffee is one example (Klein *et al.* 2003, Classen *et al.* 2014), but the same applies to cherries (Holzschuh *et al.* 2012) and apples (Garratt *et al.* 2014, Blitzer *et al.* 2016). Diversity of pollinators is also important on small farms, which sustain the livelihoods of about two billion people across the world, as demonstrated in research by Lucas Garibaldi and colleagues (2016). Studying

33 pollinator-dependent crops on farms in Africa, Asia and Latin America, they showed that increasing the diversity and abundance of pollinators could reduce the difference between crops on low- and high-yielding farms by an average of 24%. The largest such study published to date is an analysis of 1,475 locations around the globe by Dainese *et al.* (2019), which looked at species richness of both pollinators and insects that provide pest control in agricultural crops. They found that the number of species of both types of insect (flower visitors and predators) had a positive effect on crop yield that was reduced when surrounding landscapes were simplified by the removal of wild vegetation. In the USA, a recently published study has shown that five out of seven major crops have yields limited by lack of pollinators (Reilly *et al.* 2020), a point I will return to in Chapter 10.

In Britain, wild pollinators such as bumblebees, solitary bees, hoverflies and butterflies are thought to be responsible for at least two-thirds of agricultural pollination; managed honey bees account for (at most) the other third (Breeze *et al.* 2011). However, intensive agriculture has fundamentally changed pollinator communities. For example, Deepa Senapathi and colleagues (2015) analysed historical data on the diversity and abundance of pollinating bees and wasps at fourteen sites across England in order to assess how land-use change has affected these ecological communities over the past 80 years. In three-quarters of the study sites, the diversity of pollinators has been reduced, and the greatest effect was on those sites that were surrounded by agricultural land. In contrast, there was less severe change in sites where the surroundings included urban areas, supporting a growing body of evidence that heterogeneous landscapes, including urban habitats and gardens (as we'll see in the next two chapters), support greater diversity and abundance of pollinators than relatively homogeneous, intensively farmed agricultural land.

Globalisation and the biogeography of crop pollination

Globalisation of commerce is often thought of as a modern phenomenon. But there is a deep history of movements of food, plants, animals, materials, people and other resources across the world, beginning long before there were written accounts. This includes the spread of animal-pollinated crops, many of which were (and still are) serviced by very different pollinators in their natural range compared to where they are currently grown. We've already seen the examples of coffee and cocoa, which are global products pollinated by the same types of insects, but different species, within and outside of their native ranges. But there are many other examples. Where would Indian and Italian cuisine be without chillies and tomatoes, or Hallowe'en in Europe without pumpkins? All three crops are American imports that originated in the decades following Columbus's explorations of the New World, and all are bee-pollinated to some degree, though the identity of the bees concerned varies hugely according to where they are grown.

Displaced plants are thus effectively pollinated by resident pollen vectors, in the main, though there are exceptions: in the nineteenth century, for instance, New Zealand farmers had to import bumblebees as well as honey bees to pollinate some European crops. A recent study of 'Global-scale drivers of crop visitor diversity and the historical development of agriculture' (Brown and Cunningham 2019) is one of the first to seriously address this topic. The authors show that crops grown in the region in which they first evolved and were developed by early agriculturalists tend to be pollinated by a greater diversity of bee genera than when they are grown outside of that original region. Thus, the number of pollinators of crop plants is determined by the history of human agriculture over timescales of thousands to hundreds of years, interacting with the evolutionary history of both the plants and the pollinators over the course of millions of years.

The movement of crops between the Old and New Worlds was not one-way, however. As we have seen, one of the most globally significant crops is coffee. This plant originated in east Africa and became initially popular in the Middle East and North Africa, before spreading to Europe in the seventeenth century. It was not long before the forerunners of Costa and Starbucks were opened, and there are some interesting early survivors of that period, such as the Queen's Lane Coffee House in Oxford that was established in 1654. Indeed it was the coffee houses of London that were the precursors to scientific institutions such as the Royal Society, so we can (arguably) thank bees for bringing British science into the modern era.

Coffee arrived in the New World in the eighteenth century and was well established as an agricultural crop in South America by the time Darwin visited Brazil in February 1832, when he wrote to his father that 'Nobody but a person fond of natural history can imagine the pleasure of strolling under cocoa-nuts, in a thicket of bananas and coffee plants, and an endless number of wild flowers' (Darwin 1887). Later, in April of that year, Darwin had his eye on more practical considerations. He was staying on a small farm or fazenda, '[consisting] of a piece of cleared ground cut out of the almost boundless forest'. He went on to write the words I quote above in the coffee section. Reading these accounts, I've often wished that Darwin had gone on to say 'and we have the bees to thank for this bountiful crop' – but he never did, and, as far as we are aware, he took little interest in the functional role of pollinators at this time, being far more concerned with collecting insects, seeds, fossils, mammals and birds to send back to Britain. The fact is that in the early nineteenth century there was limited interest in or understanding of the importance of pollinators for agricultural production among Western scientists. The extent to which the link between flower visitation and fruit and seed yield had been made by other cultures is currently unclear. Indigenous peoples in many parts of the world possessed a wealth of knowledge about bees and other pollinators, but how much did they know about the ecology of animal interactions with flowers? Flower visits by pollinators were certainly portrayed in the art of some cultures (see Figure 1.3), but so much indigenous knowledge has been lost (Lyver *et al.* 2015) that we will never know the full extent of their understanding.

Today, in developed countries the role of pollinators in visiting flowers and initiating seed and fruit production of agricultural crops is rather taken for granted. It's the kind of thing that we learn in primary school or see in children's television shows. But that's certainly not the case everywhere in the world. In recent years I've heard agricultural pollination researchers at conferences talk about their experiences in developing countries, and about discussions with farmers who clearly do not always appreciate the role of bees and other flower visitors. Johanna Yourstone from Lund in Sweden spent time interviewing farmers in Andhra Pradesh and encountered several who told her that bees eat or destroy flowers (personal communication 2019). They assumed that any insects on plants must be pests, and even that flower visitors play a negative role by sucking the 'goodness' out of flowers. At least one farmer sprayed pesticides to get rid of bees and burned their nests.

I do wonder if these poorly educated farmers, who are often growing crops at a subsistence level, are getting their information from pesticide merchants keen to sell more chemicals. Other researchers have encountered similar attitudes in parts of Africa, where they have been told that 'God makes the fruit' (Jane Stout, personal communication 2020). Clearly what's needed are local education programmes and demonstration sites showing the importance of pollinators to crop production – and I know of several that have been set up. Some of the South American countries are leading the way in this regard; for example, there is now a freely available guide to insect pollination of crops grown in Brazil, aimed particularly at farmers and gardeners, as well as conservationists and politicians (Klein *et al.* 2020).

However, it's not only in the developing world where we find a need for education about pollinators. In the UK the National Farmers' Union has insisted, despite mounting evidence, that neonicotinoid pesticides have played no role in pollinator decline (see Chapter 10). There's also a curious and sometimes misplaced faith in the ability of hi-tech approaches to solve agricultural problems. For example, the last few years have seen a few media stories doing the rounds about robot pollinators and how they are going to replace insects. It's nonsense, of course, and quite a number of ecologists have written blog posts and commentaries arguing that robotic bees are not the future of crop pollination (e.g. Goulson 2017a, Prendergast 2017, Potts *et al.* 2018). Perhaps the state of the robo-bee field is best summed up by Yang and Miyako (2020):

> Operability of the technology itself was unfortunately impractical, and flowers were seriously damaged. Bearing fruit was thus not achieved.

Unfortunately the authors then use this as a justification for yet another techno-fix idea, this time involving soap bubbles that could deliver the pollen to flowers after they are blown out of a drone-mounted bubble gun. Sigh...

What the technologists who are promoting these ideas don't seem to appreciate is that pollination of flowers is a complex process that has been optimised over millions of years of evolution, and robotic pollinators are unnecessary given the existence of animals that can do it perfectly effectively. All of this technology has environmental costs associated with it: the resources and pollution costs of making it, the energy costs for using it, and the disposal and pollution costs when it reaches the end of its life. Applying a greenwash of 'let's use mini-drones or bubble-blowing machines for pollinating flowers because the bees are dying out' doesn't make the technology any more environmentally sustainable – quite the opposite.

In this chapter, I hope I have shown that pollinators make a tangible and significant contribution to local, regional and national economies via their role in agriculture. But it doesn't end there; in addition, they facilitate the reproduction of the large majority of the world's flowering plants (see Chapter 1). Without pollinators, populations of most of these plants would die out over time as they would cease to produce seeds. Even within agricultural landscapes these wild plants provide us with a huge range of 'ecosystem services', which is a broad term for the benefits derived from the natural world by society. They include 'supporting' services such as soil formation, nutrient cycling and water cycling; 'provisioning' services such as wild-collected food and fuel, and natural medicines; 'regulating' services such as effects on climate, soil erosion, flood amelioration, pest control and crop pollination; and 'cultural' services, for example spiritual enrichment, recreation and aesthetic experiences. Thus we can see that without the pollinators, we lose the wild plants. Lose the plants, and we lose much of the rest of our wildlife, plus a host of ecosystem services. The 'value' of pollinators and biotic pollination goes far beyond the food that we put on our plates. Conservation of pollinators is therefore of extreme importance if we are to maintain agricultural productivity, and this is a topic we will explore in Chapters 10 and 12. But before then I want to discuss some habitats that are becoming increasingly important for pollinators: urban and brownfield areas, and gardens.

Chapter 8

Urban environments

When we think of 'nature' and 'wildlife' we often imagine rolling grasslands or shady woods, slow-flowing rivers, and beaches devoid of people. Places where the natural world can exist with little or no influence from humanity, a contrast to the roads and buildings and human domination of towns and cities. Perhaps this view is best summed up in these lines by eighteenth-century poet William Cowper:

> God made the country, and man made the town.
> What wonder then that health and virtue, gifts
> That can alone make sweet the bitter draught
> That life holds out to all, should most abound
> And least be threaten'd in the fields and groves?

Of course this is a delusion: even in the eighteenth century, the British countryside was every bit as 'man-made' as the towns and cities, the landscape having been repeatedly deforested, cultivated and grazed by livestock for thousands of years. This is also true across much of the world, where habitats have been changed by local societies. The first European explorers of Central and South America thought that they were seeing pristine wilderness, untrampled by human feet, in the rainforests, savannahs, and dry woodlands. We now know from satellite and LIDAR surveys that large parts of these biomes contain the remains of pre-Columbian human settlements; indeed, some parts were less densely forested prior to the arrival of Europeans (Loughlin *et al.* 2018). Britain has little natural habitat as such, and even our wildest and most remote areas consist of 'semi-natural' plant communities that have been created following deforestation hundreds or thousands of years ago. Thus the wild, rural landscapes that inspired so many naturalists in their formative years are 'natural' only within the bounds of human influence. In any case this distinction between 'wild' and 'anthropogenic' is an artificial one because, as many of us city dwellers see every day, nature finds a home, a habitat, a place to thrive, wherever it will. Some ecologists are beginning to acknowledge this, and there have been proposals to apply the word 'natural' to any spontaneously growing vegetation, regardless of the origin of the plants involved (Peterken 2019).

In this chapter we will explore how urban and peri-urban habitats support pollinators and the plants that they need. The process of urbanisation is frequently cited as one of the leading causes of biodiversity loss – I see this stated time and again, especially by students and conservation groups. But urbanisation itself uses up a relatively small fraction of land: in the UK it's between 6% and 10% of the land area, while globally it is thought to be less than 2% of habitable land surface, depending on how one defines 'urban'. However, the majority of the world's human population is 'urbanised' and this fraction is growing year on year. So in fact it's the land use that goes to support these urban populations that has the most impact on biodiversity, particularly agriculture, which accounts for around 70% of the UK's area and about 40% worldwide.

Urbanisation itself can have both negative and positive impacts on biodiversity, as we'll see in this chapter when we consider different kinds of anthropogenic habitats. Although they are diverse in form and structure, one of the things that these habitats have in common is that they tend to be nutrient-poor. Their thin, dry, sometimes skeletal soils contain little organic material and are deficient in the essential mineral nutrients that plants need to grow, particularly nitrates, which are water-soluble and quickly flush out of the soil. This means that the aggressive species which can dominate nutrient-rich agricultural soils – for example nettles, docks, and various grasses, all of which (coincidentally or not) are wind-pollinated – fail to take over. With the growth of these thuggish, highly competitive plants limited, other species can gain a foothold in the community and not be out-competed. It is a truism that if we want high plant diversity in an area we should reduce the level of soil nutrients, and that is something which is certainly true of many urban habitats. I explore this idea in more detail in Chapter 12, including the correlation with greater pollinator diversity (see Figure 12.3).

I've long been fascinated by the way nature inveigles its way into the urban fabric, and it's a fascination that goes right back to my childhood. So if you will indulge me, I'm going to begin with a little personal history.

Evolving a naturalist

One of the great things about the internet and social media is that you can make unexpected discoveries on a weekly basis. Around the time of my fiftieth birthday, in early 2015, I discovered that a family, the Scraftons, who lived in the same street as me when I was growing up in the 1960s and '70s, had digitised some old home movies and made them available on YouTube. In our digital age, in which every phone and camera can capture and share events as they happen, it's sometimes easy to forget that owning a movie camera in the 1960s was quite a rarity, and that the majority of kids living at that time were never filmed. These movies were taken in and around the area where I grew up in Sunderland in the northeast of England, a huge town that was, at the time, renowned for its shipbuilding and coal mining. The community into which I was born, Southwick on the north banks of the River Wear, was in the heart of this industry. My family were mainly coal miners on my dad's side and shipyard workers on my mam's. In one of the

FIGURE 8.1 The author (on the left) in about 1970 with friends on a post-demolition grassland in Southwick, Sunderland. This is a screen-shot taken of the original super-8 film footage. My thanks to the Scrafton family for capturing this footage and making it available: www.youtube.com/watch?v=rJGSruYzApw#t=236.

Scraftons' home movies there's footage of me aged about five. In Figure 8.1 I'm the youngest of the three boys, in a blue shirt, helping to build a tee-pee which I think belonged to me (I have dim memories of unwrapping a bundle of bamboo canes and bright-coloured fabric).

What's so fascinating about this footage is not just that it shows me and my friends at such a young age. The main interest is that it documents, in colour and moving pictures, one of the reasons why I became a professional naturalist with a deep fascination for biodiversity. The grassland in which we are erecting the tee-pee is not some country meadow, the kind of rural landscape cited by so many other scientists and writers as inspiring their childhood fascination with natural history. These grasslands had arisen spontaneously on cleared demolition sites, following the removal of Victorian terraced housing and tenement blocks, some of which had been slums while others had suffered bomb damage in the Second World War. Until the 1950s this area had been very built-up, with the houses, shops and pubs serving the families who were employed mainly in the local industries.

Following demolition, the sites were left to their own devices, and were colonised by plants, insects, birds and mammals from patches of habitat closer to the River Wear that had either been cleared of buildings earlier in the century, or which had never been built upon. There are some nice areas of magnesian limestone grassland nearby along the higher banks of the river valley, and typical calcicole (lime- or chalk-loving) plants such as greater knapweed (*Centaurea scabiosa*) could be found in these post-demolition grasslands. Habitats such as this, that have been spontaneously recolonised by plants, are arguably more 'natural' than much of the British countryside where intensively farmed landscapes dominate and most of the wildlife occurs in highly managed nature reserves, some of which have been restored and replanted by people. A similar point is made with respect to the neglected fringes of our conurbations by the poets Paul Farley and Michael Roberts in their book *Edgelands: Journeys into England's True Wilderness* (2011).

In other footage in the same series one can see in the background a rich flora of plants, with butterflies hopping between flowers. One of the first plants I identified was the bee-pollinated white dead-nettle (*Lamium album*), the flowers of which we would pick in order to suck out the sweet nectar. The pollinator that I could first put a name to was the small tortoiseshell butterfly (*Aglais urticae*), feeding on the nectar-rich flower heads of thistles. The first bird species that I can remember identifying, and being fascinated by its bright colours, was the goldfinch (*Carduelis carduelis*) feeding on the seeds of those same thistles later in the season.

So it's not necessary to have had a rural upbringing to appreciate and benefit from nature, and to later influence your profession and passions: any piece of land can inspire interest in children, regardless of its origin, if nature is left to colonise. And that's what nature does, it colonises regardless, making no distinction between 'natural' and 'unnatural' (Figure 8.2). These unmanaged, wild green spaces within towns and cities have huge value for flowering plants and their pollinators, for other wildlife, and for the culture of childhood. They need to be protected just as much as rural nature reserves.

Although I rarely visit the northeast now, I do know that some of the riverside grasslands still remain, and I hope that they are fascinating new generations of kids with their colour and diversity and flouncing butterflies. But the post-industrial grasslands on which I played and looked for bugs and flowers are all gone; they were cleared and built upon in a flurry of housing and retail development in the 1980s. Perhaps in the future they may return if those buildings are themselves demolished and the land allowed to lie undisturbed for a while. That is what nature does: it ebbs and flows across our landscapes in response to human, and natural, interventions, endlessly changing and endlessly fascinating to curious minds, no matter how old they are.

FIGURE 8.2 In the town of Candelaria, Tenerife, native and introduced plants have dispersed from the surrounding habitats and colonised flat roofs and old walls.

Brownfield sites

The post-industrial grasslands described above fall into a general class of land that is termed 'brownfield', and these places can be quite biodiverse, containing a high number of species, especially flowering plants and pollinators (Figure 8.3). Some of the best habitats for bees in the UK fall into this category, for example Canvey Wick in Essex (Land Trust 2020). This former coastal marsh was initially developed as a facility to refine oil in the 1960s and 1970s, but was never used as such, following which it was used to dump dredgings from the River Thames. The patchwork of wetter areas (marshland, ponds), areas of bramble and tall herbs, skeletal gravels and sand banks, both dry and wet grasslands, and the remains of industrial steel and concrete structures, is home to more than 1,400 invertebrate species. This includes important populations of rare bumblebees (Goulson 2017b). The charity Buglife has produced guidance on how best to manage such areas, and this and other useful case studies can be found on their Brownfield Hub (www.buglife.org.uk/brownfield-hub).

One of Buglife's motivations for setting up this website was that brownfield sites are frequently disparaged by local authorities and residents, and are regularly built upon, as happened in Sunderland in the 1980s. There's a difficult balance to be made here: people need homes and society requires infrastructure such as factories and industrial complexes. Brownfield sites are an obvious target for development, especially as part of urban regeneration strategies. However, these sites need to be assessed for their value in supporting wildlife before decisions are made to build. A common complaint against developing farmland is that it 'destroys the countryside'. But building houses with gardens can provide habitat for insects and birds that is far more wildlife-friendly than the original intensively

FIGURE 8.3 A plant community growing on a post-industrial site in Northampton, UK.

managed agricultural land (see Chapter 10). There's no single right or wrong answer. In principle it is neither wrong to build on brownfields nor inappropriate to develop greenbelt land, and each development has to be judged on its own merits. Indeed, there is surprising pollinator and plant diversity to be found in the most unlikely places, including in the centres of cities, to which we now turn our attention.

Urban pollinators

In summer 2018 the Royal Botanic Gardens, Kew reported that it had recorded the one-hundredth species of bee in the gardens, which is a great achievement. This announcement was made in the context of how diverse urban pollinators can be. However, Kew, while certainly surrounded on most sides by urban development, is hardly typical of most city settings: it's large (121 hectares), contains more than 30,000 species of plants, and has always been open greenspace of one sort or another. But that illustrates an important point when it comes to studying urban pollinators, or indeed city ecology of any kind: urban areas vary hugely in their size, history and form. In particular the way in which a town or city has evolved can have a huge influence on what wildlife is found there.

I was thinking about these questions in relation to urban bees when I recruited Muzafar Hussain Sirohi as a PhD researcher in 2010. Muzafar was funded by the university at which he works in Pakistan to complete a doctorate and, although his background was mainly in botany, I convinced him that urban bees were worth studying. After much discussion we focused on solitary and 'primitively eusocial' bees, i.e. excluding the honey bees and bumblebees. We did this because the eusocial bees had been studied in a lot of depth in Britain whereas the urban ecologies of the other groups were much less well known. In addition, those generally larger social bees can travel long distances, and so foraging bumblebees in the centre of town might well have come in from the surrounding countryside, whereas this was not so likely for the other, usually much smaller bees (see Figure 2.5).

It made sense to choose Northampton as a case study and focus on this one area in depth, and so together with my colleague Janet Jackson, the three of us developed a sampling strategy that encompassed a detailed survey of the centre of the town. In a circular area with a radius of 500 metres, centred on the seventeenth-century All Saints Church, Muzafar surveyed all types of habitats from road verges to churchyards, roundabouts and small gardens, and weedy patches around electrical substations. He also compared these sites to nature reserves around the periphery of the town. Muzafar's first year of field work in 2011 revealed an unexpectedly high diversity of bee species that really surprised us, and he continued his research the following year, surveying from March until November. Some of those bees were nesting in stone and brick walls that dated from the seventeenth to the nineteenth centuries, a historical legacy of how the town has evolved. There were species exploiting wall plants as pollen and nectar sources (see below), and others using small patches of weeds or garden plants. Dry areas of bare soil offered opportunities for nesting to mining bees (*Andrena* spp.). Surprisingly, busy roads seemed not to be a barrier to most species, with individuals able to fly over them and even nest in the planted areas running down the middle of a dual carriageway.

Muzafar finally completed his doctorate in 2016, but before that we published the first paper from his research in 2015 (Sirohi *et al.* 2015). It was the product of a lot of hard work to systematically sample and identify the bees, but the results were exciting. Muzafar showed that the centre of Northampton is home to at least 50 species of bees, significantly more than were found in the nature reserves at the edge of the town. One important find from the urban sites was the nationally rare Red Data Book species the grooved sharp-tail bee (*Coelioxys quadridentata*) that has declined enormously and is currently known from rather few sites. It is cleptoparasitic on several solitary bee species (see Figure 2.4), and one of those hosts, the four-banded flower bee (*Anthophora quadrimaculata*) was also found in the surveys.

The estimate of about 50 species of bees is conservative, because Muzafar focused on the more neglected groups of bees and didn't include the honey bee or bumblebees, and almost certainly missed some of the rarer species. The true figure for the centre of town is likely to be over 60 species, a remarkable number given the small area surveyed. The full diversity of bees across the whole of Northampton will of course be higher still (I have recorded bees in my urban Northampton garden that Muzafar did not find) and may approach 100 species, the number found at Kew.

Which brings us back to the idea that urban areas vary according to the history of their development. The new town of Milton Keynes lies a little over 30 kilometres to the south of Northampton. Its grid network of roads and broad green spaces for cycling and walking, created from the late 1960s onwards, is an interesting regional contrast to the dense urban centre of an ancient town like Northampton. In 2013 another of my postgraduate researchers, Hilary Erenler, was commissioned by the Milton Keynes Parks Trust to survey the bees in two of their sites. Elfield Nature Park is a narrow wedge of land between the busy A5 dual carriageway and the Milton Keynes to London mainline railway (the embankment of which has a spectacular floral display in the summer – Figure 8.4). The site's maximum width is 200 metres, narrowing to 50 metres over its 400-metre length. It comprises mounds of earth which were excavated during the building of that railway, as well as patches of trees and some ponds and temporary ditches. This tiny urban site yielded 38 species of bees, including 9 bumblebees. Stonepit Field is almost twice as large and more of a peri-urban site as it lies of the edge of Milton Keynes adjacent to farmland. But despite its greater area it is home to fewer bee species: 28 in total, 8 of them bumblebees. Clearly the degree of urbanisation does not limit species diversity for these bees, and the greater species richness at

FIGURE 8.4 The abundance of flowers growing on the railway embankment opposite platform 6 of Milton Keynes Railway Station.

Elfield Nature Park was probably a result of the old soil mounds which make very suitable habitat for ground-nesting bees. These are a contrast to the areas surveyed in both Northampton and Kew, and more similar to Canvey Wick (described above), emphasising the point that the history of an urban area is one of the most important determinants of its ability to support biodiversity.

At around the same time that Muzafar was surveying bees in Northampton, a large consortium of pollination ecologists (the Urban Pollinators Project) was doing similar work across twelve large cities and towns in England, Scotland and Wales. These urban centres were compared with surrounding farmland and nearby nature reserves. They had opted for a different strategy to us, doing broader, less in-depth surveys across more urban areas, effectively replicating their observations in space, and sampling all flower visitors, not just bees. The results, however, were equally fascinating; in summary, they found no differences in the average number of flower-visiting insects in the three types of landscape (Baldock *et al.* 2015). However, there was a greater number of bee species per unit area in urban sites compared to farmland sites, though hoverfly abundance was lower in urban areas than in the others. The design of the sampling in that study means that it's not possible to directly compare the findings with the work in Northampton or Milton Keynes, but all of this research points to the importance of British towns and cities in hosting pollinator populations.

Britain is not unique in this respect, however, and urban surveys of pollinators are showing similar results across the world. This was well illustrated by a short article entitled 'The city as a refuge for insect pollinators', written by 23 urban pollinator researchers from across the world, including our team at Northampton (Hall *et al.* 2017). Among the towns and cities where diverse bee communities had been documented we listed Berlin, Melbourne, Guanacaste Province in Costa Rica, Vancouver, Berkeley, Chicago, New York, Phoenix, San Francisco and St Louis, as well as Birmingham, Bristol, Cardiff, Dundee, Edinburgh, Glasgow, Hull, Leeds, Leicester, London, Northampton, Reading, Sheffield, Southampton and Swindon in the UK.

Since we wrote that essay, studies of other urban centres have been published and, although details of the findings vary, many show that urbanisation does not mean the total loss of pollinator diversity, and may in fact enhance it. A comparison of published research in which urban pollinator diversity was compared to that in rural areas shows that in seventeen cases urban diversity was lower, while in the same number of instances urban diversity was equal to or higher than rural diversity. Interestingly, all of the studies showing that urban diversity was higher involved just bees (Figure 8.5), possibly because the warmer microenvironment provided by urbanisation favours bees (see Chapter 6). However, as always in ecology, there are some complications to this story. Katherine Baldock has shown in a recent review of the urban pollinator literature that as well as benefiting wild pollinators, urban habitats present threats to their conservation in the form of pesticides, pollution and competition with honey bees (Baldock 2020). Likewise, a systematic review by Arne Wenzel and colleagues concluded that the responses of pollinators and pollination services was highly dependent on the degree of urbanisation, measured by things such as the proportion of built-up area and sealed surfaces such as roads and pavements (Wenzel *et al.* 2020). However, urban crops are generally well serviced by the available pollinators, something we will come on to in the next section.

The identity of the bees being supported in urban areas also needs to be assessed. A recent study by Fitch *et al.* (2019) has shown that in Michigan it

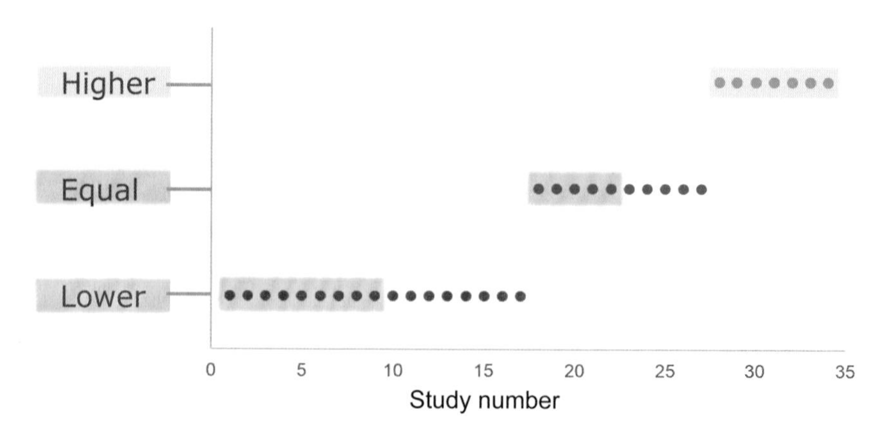

FIGURE 8.5 A comparison of 34 published studies where urban pollinator diversity was compared with rural diversity. Those studies involving just bees are highlighted by blocks of colour.

was non-native bees that mainly benefited from urbanisation, rather than native species. This type of study needs to be repeated elsewhere in the world to see if it's part of a general pattern.

The focus of the essay by Hall *et al.* (2017) was the notion of conservation *for* the city – an idea originally conceived, I believe, by Steward Pickett of the Cary Institute of Ecosystem Studies in Millbrook, New York, who has written extensively on the topic of urban ecology. In essence the term 'conservation *for* the city' refers to the fact that encouraging and managing for wildlife in our urban centres is good not just for the conservation of those species, but also for the human population. Those plants, animals and other organisms play a role in supporting the city itself through the provision of ecosystem services such as waste disposal and decomposition, flood alleviation, and, of course, crop pollination. Which brings us to the next section of this chapter and a consideration of just why we should be trying to support urban pollinators through appropriate planting and management.

Urban pollinators for urban agriculture

As I've tried to stress in this book, pollinators such as bees and flies are hugely important both ecologically (most plants require them for reproduction) and economically (a big chunk of our food production relies directly or indirectly on pollination by animals). Understanding how these pollinators are distributed across the landscape, including urban areas, is crucial to their conservation in a rapidly changing world. The research described in the previous section therefore has implications not only for conservation of biodiversity, but also for food security. Globally the United Nations Food and Agriculture Organization (FAO) estimates that 800 million people practise urban and peri-urban agriculture, accounting for up to one-fifth of world agricultural output (www.fao.org/urban-agriculture/en).

It's unclear whether the FAO figures include the kind of home gardening and allotment cultivation that's commonly practised across Europe, but I suspect not. It's very difficult to include these small patches of land in large surveys, as shown by a study by Thebo *et al.* (2014) which provided some of the first comprehensive figures on the extent of agriculture in and around the world's large towns and cities. The main message of that study is that urban agriculture is more extensive and important than previously assumed, and there are significant implications for food security and water resources. However, when discussing the limitations of their study the authors state that 'the scale and methods used … are not structured to capture very small, spatially dispersed areas of urban croplands.' In other words, urban gardens and allotments are not included in the assessment. In the UK at least this is a significant limitation, because we know that urban fruit and vegetable growing is widespread, though as I mentioned in Chapter 7, as far as I am aware there are no published figures on the volume and value of this local horticulture of food crops. We also know, from the work of the Urban Pollinators

Project mentioned above, that British urban home gardens and allotments are hotspots of pollinator diversity and abundance (Baldock *et al.* 2019), which is important for those of us who grow at least some of our own food.

Many urban gardeners and allotment holders in the UK grow food crops, and so are reliant on the ability of many pollinators (especially large bees) to provide a pollination service. In our own small urban garden (see Chapter 9) my wife Karin and I try to cultivate as much food as we can, given the constraints of space and (mainly) time. Over the last couple of years we have grown the following crops that require animal pollination to a greater or lesser degree: strawberries, apples (Figure 8.6), greengages, cherries, plums, pears, squashes, courgettes, blackberries, runner and French beans, passion fruit, tomatoes, raspberries, chillies, sweet peppers, aubergines, fennel seeds, and cucamelons. The latter are not widely known, but they belong to the melon and squash family, the Cucurbitaceae; the small fruit taste like cucumbers but look like small watermelons, hence the name. Oh, and we grow radishes mainly for the immature pods, which are delicious in a salad, and bees love the flowers.

Urban agriculture is not a new thing; for instance, much of London's food was grown in and around the city right into the twentieth century (Atkins 2003). The Allotments Act of 1925 made it a requirement for community gardens to be featured in all plans for new towns such as Milton Keynes. This came about following concerns over national food security in the First World War (see Chapter 10), a fear that was amplified in the 'Dig for Victory' campaign of World War Two. The result was that much urban 'waste' land was turned over to food production, with the number of allotments doubling by 1943. Although many of

FIGURE 8.6 Flowers of apple (*Malus domesticus*) being pollinated by the orange-tailed mining bee (*Andrena haemorrhoa*) in our garden in Northampton. This early-emerging bee is one of several *Andrena* species that are especially important for apple pollination.

these allotments were rural, by 1944 it was calculated that urban agriculture in England and Wales was producing 10% of those countries' food (Atkins 2003).

Increased mechanisation and intensification of rural farmland, and possibly stricter planning regulations, has reduced the significance of urban food production. However, many large cities are beginning to experiment with growing more and more food in vacant spaces and on rooftops. Examples include the Agripolis project in France (www.agripolis.eu) and Keep Growing Detroit in the USA (detroitagriculture.net).

As well as these large schemes, in the last decade growing food in towns and cities has taken on a new dimension and become extremely fashionable for people with gardens, or with access to community gardens. This has been driven partly by economics (it's cheaper to grow your own tomatoes and other comestibles than it is to buy them), and also by concerns about use of pesticides and other chemicals. But also it's possible to achieve very high levels of productivity even in a relatively small space (Nicholls *et al.* 2020), which is not only hugely satisfying but also fun. Some authors have dismissed this with comments such as 'contemporary urban gardening most closely resembles a middle-class pursuit for personal enjoyment' (Hallsworth and Wong 2015). I think this is overly simplistic and fails to appreciate how important this 'pursuit' is, globally, for physical and mental health, for social interactions, and for helping to support low-income households (Algert *et al.* 2016, Wood *et al.* 2016, Soga *et al.* 2017, Dyg *et al.* 2019, Nicholls *et al.* 2020). These are complex social and economic issues far beyond the scope of this book, but it does show how considerations of pollinator conservation link through to wider aspects of society.

The wider importance of urban nature

Urban pollinators are not only important for their utilitarian value in pollinating crops, however. Urban areas are home to important populations of plants, and some species can be more abundant in towns and cities than elsewhere. This is true of wall plants, an incredible variety of which grow on this apparently inhospitable stone and brick habitat. Old stone walls have held a fascination for me which goes back to the origins of my interest in natural history, as described earlier in this chapter. Growing up in Sunderland I'd see substantial walls made of the local magnesian limestone, rough-cut blocks often patterned with impressions and ridges that to my child's mind looked like exotic corals or the fossils of weird animals. When I understood more about the intriguing geology of that part of England I discovered that these patterned rocks were of chemical rather than biological origin, but no less interesting for that. Walls became a continuing backdrop to my life. As an undergraduate, my final-year research project involved clambering around on the seventeenth-century walls of Oxford University's Botanic Garden, surveying the plants that had naturally colonised them. This included the iconic Oxford ragwort (*Senecio squalidus*), which in the eighteenth century spread from the Garden onto its walls and those of the surrounding colleges. Later in the nineteenth century it travelled along the newly built rail network to colonise towns and cities across Britain.

The Garden's mural flora was an odd mix of exotics and natives, many of which had no obvious means of dispersing onto the walls. I joked at the time that perhaps the dispersal was by gardeners and ecologists working on the walls. That may have been close to the truth. Stone walls (and, less obviously, brick walls) provide a particular habitat for many plants and animals as they mimic rocky outcrops and cliff faces (Figure 8.7). There have been no published assessments of the value of wall plants for pollinators that I can locate, but I am sure that they can offer significant nectar and pollen resources.

In Britain, typical wall plants that will provide nectar and/or pollen for insects include common ivy (*Hedera helix*), various native and introduced ragworts (*Senecio* spp.), dandelions (*Taraxacum officinale* agg.), daisies (*Bellis perennis*), a range of brassicas and wallflowers (*Brassica* spp. and *Erysimum* vars.), stonecrops (*Sedum* spp.), clovers (*Trifolium* spp.) and willowherbs (*Chamerion* and *Epilobium* spp.). The daisy family (Asteraceae) is especially abundant on old walls – as Lorna Macrae, one of my undergraduate dissertation students, discovered when she surveyed Northampton town centre in 2011. Asteraceae species comprised over a quarter of the individual plants recorded, from a total diversity of more than 60 vascular plant species. This is significant, because Asteraceae are often highly generalist and provide resources for a wide range of insects (see Chapter 4). This is true not just in the UK: Asteraceae (sometimes the same genera or even species as we found in Northampton) are frequently recorded as among the most common wall plants in places as far apart as Bulgaria (Nedelcheva 2011), India (Singh 2011), the USA

FIGURE 8.7 Where there are walls, plants will grow. These images of urban wall vegetation were taken in the UK, France, Spain, Nepal and Brazil.

FIGURE 8.8 Urban road verges are often over-managed resulting in dull, close-cropped grassland (left). But careful planting of wild flowers (top right) or more thoughtful management (bottom right) results in habitats that are better for pollinators and are visually appealing.

(Zomlefer and Giannasi 2005) and China (Long 2007, cited in Qiu *et al.* 2016). Walls also provide nesting opportunities for some bee species, in soft mortar, or excavated tunnels, or large internal cavities. Walls are, I suspect, more important for urban plants and their pollinators than we realise, but are hugely under-studied. There is no doubt that more work could be done on their contribution to urban ecology.

In contrast, the urban landscape elements that are most frequently studied as important for plant populations are road verges and city parks (Figure 8.8). Traditionally, at least in the UK, these are grasslands that have been mown to within an inch or less of the soil, all in the cause of 'tidiness'. Some of this is driven by local people expecting that public land should look pristine. But I suspect that local councils are also trying to show that residents are getting value for money from their council tax. Recently, however, there has been much discussion in the printed and social media about cutting such areas less frequently in order to let plants flower, providing more resources for pollinators and allowing the plants to set seed. I'm all in favour of this, clearly. However, as I mentioned in Chapter 6, some contradictory advice has been presented, with plant conservationists suggesting that local councils should cut the verges from mid-July onwards because most plants will have set seed by then, and early mowing would reduce the vigour of competing species, especially grasses. While this might be an appropriate option as far as the plants are concerned, such an early cut would severely impact local pollinator populations, which are still in full swing at this time (see Figure 6.4). Even into late autumn, in a warm year, there will be an abundance of flower-visiting insects that require floral resources for their late-season nesting and egg-laying activities, or to build up reserves of energy to allow them to hibernate. The latter include newly mated queen bumblebees and hornets (see Figure 2.6), several species of butterflies, and a number of hoverflies.

Another factor to consider here is of course climate change. Longer, warmer autumns mean that flower-visiting insects are now active in Britain for a more extended period than was previously the case, with some species being seen for twelve months of the year in the south of the country. Advice that suggests cutting floral resources at a key time of the year for these insects is simply misguided. A cut between October and December would be much more appropriate, though this in turn could have an impact on insects and small vertebrates that are hibernating in the taller grassland. For these reasons, mowing of grasslands should be done in a piecemeal fashion, cutting sections at different times in different years, rather than all at the same time (Figure 8.8). Grassland management for ecological outcomes is a complex topic and there's no single right answer for all, or even most, situations. Expert advice from ecologists should be sought before decisions are made regarding when to manage any kind of habitat.

As I noted above, local people sometimes react negatively when road verges are left unmown, complaining that they look scruffy and unkempt. One solution is to cut 'frames' and paths around and through the longer vegetation. Not only does this make the grassland more presentable, and clearly cared for, but the shorter vegetation provides more nesting opportunities for mining bees such as *Andrena* species.

What's the future for urban nature?

In early 2016 I took part in a 'knowledge exchange' workshop entitled 'Urban grassland management and creating space for pollinators', funded by the Department for Environment, Food and Rural Affairs (Defra) and the Natural

Environment Research Council (NERC), and organised by Katherine Baldock from the University of Bristol. The workshop was well attended, with some fifty delegates from a wide range of organisations, including local and national authorities, businesses and NGOs, as well as scientists from universities. After the event, feedback from those delegates was generally positive, and most people learned something about managing urban settings for pollinators, as well as making some useful connections. I certainly learned a lot: it's good to get out of academia and talk with practitioners.

The importance of urban environments for supporting pollinator populations and the plants on which they rely is clearly a subject that's generating a lot of interest at the moment, at all levels of society from government to individuals, and there's some really exciting research being published and conservation practice taking place. However, there's clearly a lot to do if we are really to understand where pollinators are distributed across our townscapes, where they are nesting, egg laying and finding resources, and how we can best manage urban habitats to support this diversity and increase their numbers. There's no simple answer to all of this, however, and each urban landscape needs to be assessed on its own merits rather than using off-the-shelf solutions. Simply planting 'wild-flower meadows' across cities, for example, can do more harm than good if they result in the extinction of rare plants that already exist on sites, or if they contain flowers that benefit only a limited set of pollinators (see A.L. Johnson *et al.* 2017 for a nice discussion of these issues).

All human activities can potentially have an impact on the biodiversity of the local environment in which they occur. That impact can be positive or negative, depending on how the activity is managed, how impact is mitigated, and the metrics that we use to measure the effects that are occurring. This is particularly true of large infrastructure developments such as big buildings, housing developments, roads and – a category close to home for me over the past few years – new university campuses. In 2012 it was announced that the University of Northampton would be relocating to a new, purpose-built, £330 million campus to the south of the town. To be called Waterside Campus, owing to its proximity to the River Nene, this would be a redevelopment of a brownfield site that had been abandoned for decades. Initial ecological surveys in 2012–2013 determined that the site had limited ecological value, and construction began.

From the outset, the ecologists at the university were determined to influence the campus development and track its impact on the local environment. The new campus would sit at the southern edge of the Upper Nene Valley Special Protection Area, also designated as a Ramsar Wetland, and adjacent to a couple of Wildlife Trust nature reserves. Thus, it was important that the university minimised long-term disturbance and maximised opportunities for wildlife. A series of meetings between the landscape architects, the Wildlife Trust and the ecologists took place to determine what could be done. It was a fascinating process as initial disagreements were negotiated towards compromises and additions that everyone was happy with, balancing budgetary, functional and space restrictions with habitat creation and landscape enhancements such as planting of mature native trees. The campus has been zoned horticulturally and ecologically. The centre comprises formal lawns and borders of nectar- and pollen-rich perennial herbs and shrubs, with more natural planting and species-rich turf beyond that, and then retained brownfield substrate towards the boundaries of the site. In total about 10 hectares of habitat was created on the 20-hectare site (Figure 8.9).

Over the period that the campus was built we involved students and staff in surveys to track how the bird life in and around the campus and the River Nene was being affected by the development. We continued these surveys into early 2020, two years after Waterside Campus opened to students. However, the lockdown caused by the COVID-19 pandemic meant that we could not carry out the annual spring surveys. The latest analysis of the data was carried out by Kirsty Richards, one of our undergraduate students, as her final-year honours project. This showed that, while the bird communities were impacted negatively during construction, they have now responded positively, with both diversity and abundance of birds increasing since the campus opened. We also involved local naturalists and students in surveys of the flowering plants and pollinators, and the campus is currently supporting significant numbers of both: at least 25 bee species and over 150 native plants, for example. The latter includes the small-flowered catchfly (*Silene gallica* – Caryophyllaceae), a plant that had last been recorded in the county in 1843. 'Nature' is very much a part of Waterside Campus and is

FIGURE 8.9 These views of different parts of the University of Northampton's new Waterside Campus were taken in late summer 2018 when it first opened to staff. They show the mix of formal planted borders and lawns, mature trees, wilder areas and reconstructed brownfield that, together with the adjacent River Nene, makes the site so attractive to birds, bees, native plants and other wildlife.

accessible to students, staff, official visitors and the public, who are welcome to walk across the site and use some of the facilities.

There's a broader perspective to all of this, however. If you take a walk through any large conurbation in Britain, London for instance, then nature (if we define it as 'non-human life') is everywhere, if we just take the time to see it. Plants grow in the most inhospitable of places: climbers such as ivy and wisteria cover concrete cliffs; people proudly tend potted plants on the smallest of balconies; large gulls wheel overhead; house sparrows chirrup in gardens; 'weeds' pop up in the most unlikely spots. And bees and other pollinators exploit the resources and spaces that cities offer. Yes, it's commonplace stuff, and yes, much of the planting involves non-native species, and it's all highly anthropogenic. But that doesn't make it any less 'nature' or lessen our connection with it. The real question for me is about how many people actually perceive this nature, either positively or negatively, and whether they do so consciously or subliminally. I suspect there's far more of the latter than the former, but that if the non-human elements of nature were removed from even the most built-up parts of large cities like London, people would notice and respond negatively to its removal. That has certainly happened in other parts of the world. In 2013 a protest in Turkey that was initially about the destruction of one of Istanbul's few large green spaces, Taksim Gezi Park, escalated into mass demonstrations that involved an estimated 13.5 million people (17% of the Turkish populations). The demands of these protesters eventually went far beyond the initial concerns about the park, but this should serve as a warning to governments at all levels: green spaces matter to people.

Perhaps, rather than trying to reconnect people with some idealised view of 'nature' that is remote from their usual existence, we should actually be encouraging them to think about the non-human life that they encounter in their daily lives. This is a process that can start at an early age, with parents encouraging their children to look with curiosity at bees and wasps rather than panicking and shooing them away. Urban planners can help by building green space into developments, an idea that has gained traction in recent years as architects and developers realise that planting trees and flower borders for pollinators can both improve the desirability of an area and help to cool it. Even here, however, we see some important and contradictory societal implications. One of the unintended (and perhaps sometimes intended) consequences of greening our cities may be 'eco-gentrification', as property prices increase and low-income families are displaced (Haffner 2015). Such families may often be the ones who would most benefit from urban green space, and a connection with urban nature, and from growing their own fruit and vegetables. Even in our towns and cities, the implications of conservation strategies for pollinators can have far-reaching societal consequences. Nowhere is this more apparent than in gardens, which are the subject of the next chapter.

Chapter 9

The significance of gardens

Gardens and gardening represent different, and often contradictory, things to people. They can be a source of huge joy and desperate frustration, of hard work and carefree pottering, of pride and embarrassment. 'Wildlife gardening' has become immensely popular over the past few decades, promoted by television celebrities and books by conservationists. And where once the focus was on just attracting birds, frogs and hedgehogs to the garden, pollinators are now a key target for many gardeners. However, as we'll see, all planted patches can support at least some bees, hoverflies, butterflies, or even hummingbirds if you are fortunate enough to live where they occur, and you don't have to specifically 'garden for pollinators' to attract them to the garden. However, with a bit of knowledge about what flower visitors need, and some changes to gardening practices, it is possible to hugely increase the diversity and abundance of pollinators in even a small area.

I've always enjoyed gardening; it's a pleasure I inherited from my father and his father before him, both of whom kept allotments in Sunderland. But in our urban Northampton garden Karin and I don't particularly garden for wildlife, even though wildlife seems to enjoy our patch. One of the things we do try to do, however, is to avoid plants with over-bred, fussy double flowers that provide little or no floral resources. Flowers that conform to their original *Bauplan* (Chapter 3) are usually preferred by pollinators. We're not too fussy as to whether the plants are native or introduced to Britain, because our own observations suggest that the origin of the plant is less important than the type and ease of access to rewards. This is backed up at least one recent study (Rollings and Goulson 2019), though others have suggested that native plants may attract more individuals of at least some groups of species, such as hoverflies (Salisbury *et al.* 2015a).

Pollinators can be seen in every park and garden, and every municipal floral display, no matter how small or temporary. The most highly managed and artificial of flower patches will have some kind of pollinator presence, even if it's only a wide-ranging honey bee far from its hive, or a hoverfly passing through on its way to somewhere else and investigating a colourful plastic knick-knack or artificial bloom. Replace that plastic with some real flowers and the pollinators will hang around for much longer. In 2017 I attended the 19th International Botanical Congress (IBC) in China, a huge conference with around 7,000 delegates. The IBC was hosted by the government of Shenzhen, a massive city of over 12 million people that

FIGURE 9.1 The front entrance of the venue for the 19th International Botanical Congress in Shenzhen, China, July 2017.

has grown rapidly since the 1980s. The organisers of the conference had commissioned some fabulous displays of living plants (Figure 9.1), and within 10 minutes of arriving at the venue I saw lots of honey bees, one butterfly, at least two species of wasps, and a large carpenter bee (*Xylocopa* sp.) visiting the flowers. A week later the display was dismantled and the insects no doubt moved elsewhere, for Shenzhen has no shortage of municipal parks and gardens, and even the most congested of shopping malls has green spaces built into its fabric.

Temporary displays are fine in the short term, but a large and continuous floral display in gardens is the only way to maximise pollinator abundance and diversity, with a presence most months of the year. If you allow some areas to become unmanaged, provide other suitable nesting sites or areas for food plants, and other resources that they need (see Chapter 12), a thriving oasis for pollinators can be created in any plot.

In this chapter I will consider the diversity of pollinators that can be found in gardens, and suggest some reasons why gardens may be so significant to pollinator conservation in the twenty-first century – but also why they are not the main solution to pollinator decline. To begin with I'd like to focus on the small urban patch that we've created in Northampton and introduce some of the pollinators that we are supporting.

Our small urban garden

When Karin and I moved into our Victorian terraced house in January 2012 the 10 × 20 metre garden was almost wholly laid to lawn (Figure 9.2) with some

FIGURE 9.2 Our Northampton garden as it was in early 2012 (top image) and how it has subsequently developed.

narrow borders to one side, a mature greengage tree (*Prunus domestica* var.) and a large New Zealand flax (*Phormium tenax* var.), and not much else. Since then we've developed larger flower borders, put in a vegetable patch and planted fruit trees, but kept some lawn (though this reduces every year as we widen borders). We don't water or fertilise the lawn, just keep it regularly mown, except for an area that has a patch of ragwort (*Jacobaea vulgaris*) growing, which in turn supports a small population of cinnabar moths (*Tyria jacobaeae* – see below). In addition to the grasses (some native, some not) the lawn contains another ten or so native plants, including taxa which are popular nectar sources for pollinators, such as clovers, dandelions and buttercups.

The addition of a wide range of (mainly non-native) flowers that bloom during twelve months of the year means that an impressive diversity of insect pollinators has been recorded, including about 25 bee species (almost 10% of the UK total), 15 species of butterflies and day-flying moths, and at least 20 species of hoverfly and other flower-visiting Diptera (and probably a lot more, for my fly identification skills lag far behind). Some of these garden pollinators are described and illustrated in the following sections, starting with one of my favourite groups of bees – the leafcutters of the genus *Megachile*.

The UK has only nine *Megachile* species recorded, several of which are quite frequently found in gardens. The only one we've seen so far is the patchwork leafcutter (*M. centuncularis* – Figure 9.3a). It's quite distinctive, with a brush of orange hairs that extends right to the tip of the abdomen, though the colour of this can fade with age so it's not always so apparent. The brush is used by the females for collecting pollen from flowers to take back to provision the nest, which is constructed from leaf segments lining a tubular cavity in old walls, wood or occasionally soil (hence 'leafcutter' bees). The leafcutters (as with 90% of bee species) are 'solitary' in the sense that they don't have a social structure with a communal nest, a queen, and workers (see Chapter 2). It's the female bees that are solely responsible for nest building; the role of the males is simply to mate.

FIGURE 9.3 A selection of pollinators from our garden. (a) A patchwork leafcutter bee (*Megachile centuncularis*) visiting flowers of lamb's-ear (*Stachys byzantina*). (b) A worker buff-tailed bumblebee (*Bombus terrestris*) visiting the flowers of blue passion fruit (*Passiflora caerulea*). Note the large mass of pollen in the pollen basket on her rear leg. (c) A marmalade hoverfly (*Episyrphus balteatus*) visiting a flower of lemon balm (*Melissa officinalis*), a beautiful herb for cooking, but boy does it seed itself everywhere! (d) A hornet hoverfly (*Volucella zonaria*) visiting flowers of a cultivated *Hebe* variety.

I've seen this species visiting our runner beans in the garden and, given their size, they probably pollinate that crop, though not as effectively as the much hairier and more abundant bumblebees. In Figure 9.3a you can clearly see pollen on the back of the bee. This has the potential to be involved in pollination, unlike much of the pollen that's been collected by this bee under its abdomen, which is mainly destined for the nest, as food for the larvae.

In our garden the patchwork leafcutter is very fond of lamb's-ear (*Stachys byzantina*), a non-native species in the mint family (Lamiaceae), as is another solitary species, the wool carder bee (*Anthidium manicatum*). This species strips the fine hairs from the lamb's-ear, as well as leaves of other species, and uses them to line its nest. The males are aggressive, fast-flying bees that are territorial and chase away both other males and individuals of other species that visit the flowers, including much larger bumblebees. I've even see them attack and kill honey bees. Lamb's-ear attracts a lot of other species of bee, though it is a bit invasive and self-seeds everywhere, so is probably best contained in a pot. It's an exception to the rule (if it even exists) that native plants are always best for pollinators.

Another important solitary bee as far as edible produce is concerned is the orange-tailed mining bee (*Andrena haemorrhoa*) that pollinates the blossom on our apple tree (an unnamed variety which Karin and I rescued from the bargain area of a local garden centre). This species is also referred to as the 'early' mining bee due to its habit of emerging from overwintered nests as early in the year as March. In truth, however, many *Andrena* species put in an early appearance, making them important pollinators of orchard fruit, which you can see from the photograph in Chapter 8 (Figure 8.6). So 'orange-tailed' is a more descriptive name. The epithet 'mining bees' refers to the fact that species of the genus *Andrena* usually make their nests in soil, excavating deep tunnels in which to construct individual cells. It's another generalist, taking pollen and nectar from a wide variety of garden and wild flowers. Dandelions are particularly important in the first few months of the spring so it's important not to over-manage or remove them in lawns, and to allow some to flower.

Another early-emerging species is the buff-tailed bumblebee (*Bombus terrestris*) which, as I mentioned in Chapter 6, is becoming more active during winter in Britain. This is a truly social species with a (usually) annual nest comprising workers and a queen. Nests are founded by queens that have mated the previous year and hibernated. They typically choose old rodent nests in which to begin their colonies, which is why they are sometimes found in garden compost bins. An interesting question that I've not seen answered is whether the queens actively displace mice or voles from such nests. In our garden the buff-tailed bumblebee pollinates a range of crops including strawberries, squashes, courgettes, blackberries, runner and French beans, tomatoes, apples, greengages and raspberries. As Figure 9.3b shows, they also visit the flowers of passion fruit (*Passiflora caerulea*), where they seem to be more effective than the smaller honey bees and solitary bees, bridging the gap between the anthers and the stigmas (see also Chapter 14). It's a fruit that not everyone likes, but I enjoy it. The tree bumblebee (*B. hypnorum*) is another common species in our garden, and I'm going to discuss this bee in detail in Chapter 11 when we consider recent arrivals to the shores of the British Isles.

Beyond the bees, the beautifully named marmalade hoverfly (*Episyrphus balteatus*) is a frequently encountered hoverfly species in gardens (Figure 9.3c). This insect is a 'true fly' of the order Diptera that is sometimes confused with superficially similar-looking social wasps (order Hymenoptera). As the common name suggests, however, the marmalade hoverfly is translucent orange and black in colour rather than waspish yellow and black. It also has a very flat abdomen, whereas *Vespula* wasps are more rounded; and marmalade hoverflies certainly don't sting. Individually, these insects are relatively ineffective as pollinators – they are small and not very hairy, so carry little pollen compared to bumblebees for instance. However, they can be extremely abundant, and that abundance can make up for any individual ineffectiveness. These hoverflies are real generalists, visiting lots of different types of flowers in our garden, including crops such as radishes (which, as I mentioned, we grow for the succulent, mustardy pods) and raspberries. I often see individuals patrolling runner beans, not visiting the flowers but laying eggs on leaves and stems: the larvae of the marmalade hoverfly are carnivorous and feed on aphids, and so the species plays an interesting dual role as both pollinator and pest controller.

Another hoverfly that we've only seen a couple of times within the garden is the spectacular hornet hoverfly (*Volucella zonaria* – Figure 9.3d), Britain's largest species and, as the name suggests, a close mimic of the European hornet (*Vespa crabro* – see Figure 2.6).

Of the butterflies and moths (Lepidoptera), the gatekeeper (*Pyronia tithonus*) is one of the most interesting species, and one which we see every year (Figure 9.4). According to various accounts I've read, the gatekeeper is typically found in areas that have patches of shrubs near to rough grassland, or in woodland rides, and until recently it was rare for them to appear in urban settings. It's very much a species of the countryside. So what is it doing in our garden? Clearly, in order to exist in an urban setting the gatekeeper must have its basic requirements met by the habitat in which it finds itself. As I've mentioned, the lawn in our garden is quite diverse and contains a number of native species, including a range of grasses that could be used as food plants by the caterpillars, though we do keep it quite short. It's more likely that the caterpillars are feeding in some of the neighbouring gardens, which are rarely troubled by a lawnmower. This raises the question of whether neglected gardens host more biodiversity than highly managed gardens. I suppose it depends on the type of management – it's certainly possible to over-manage any habitat, including a garden, with too frequent mowing and disturbance. But in a small area where large mammals cannot graze and open up the vegetation to provide opportunities for more plants to grow (as per 'rewilding' – see Chapter 12), humans have to play that role. Otherwise a garden would very soon become overgrown with aggressive shrubs and trees.

FIGURE 9.4 A gatekeeper (*Pyronia tithonus*) taking nectar from flowers of a dark form of buddleia (*Buddleja davidii* var.).

As well as the larval food plants required by gatekeepers, there's a range of nectar sources available in a mixed native/introduced hedge along the northwest boundary of our garden. These include bramble (*Rubus fruticosus* agg.) and oval-leafed privet (*Ligustrum ovalifolium*), and the butterflies particularly love the dark, heavily scented inflorescences of the buddleia (*Buddleja davidii* var.) seen in Figure 9.4. Only one of these (the bramble) can be considered native to Britain; both the privet and the buddleia originated in Asia.

The final species I want to mention is the beautiful cinnabar moth (*Tyria jacobaeae*), the larvae of which feed on ragworts (*Senecio* spp. and *Jacobaea* spp. – Asteraceae). We allow a small patch of *J. vulgaris* (which most people know by its old name of *S. jacobaea*) to grow and flower in the lawn. Most years there are cinnabar larvae feeding on them, vivid yellow and black caterpillars whose colours warn predators of their toxicity. The adults are even more spectacular (Figure 9.5) and an interesting example of convergent evolution with the similarly striking six-spot burnet moth (*Zygaena filipendulae* – see Figure 2.2).

Although common, cinnabar moth numbers have declined in recent years and it is now a UK Biodiversity Action Plan Priority Species. Ragwort is a food plant for more than 75 other species of insect, and its flowers are visited by hundreds of others, so it's ironic that, under the Weeds Act (1959) and the Ragwort Control Act (2003), it is considered a noxious plant that's poisonous to livestock and which landowners should eradicate. Ragwort seems to be a bit of a scapegoat, as four other species are also named in the 1959 Act but do not generate the same level of hostility. The others include spear thistle (*Cirsium vulgare*) and creeping thistle (*C. arvense*), both of them important nectar and pollen sources that attract a lot of visitors. These 'weeds' provide just another example of the tensions between agriculture and the conservation of pollinators, as noted in Chapter 7. Some environmental NGOs are running campaigns aimed at having ragwort removed

FIGURE 9.5 A newly emerged adult cinnabar moth (*Tyria jacobaeae*) (top – photo by David James) and its caterpillars (bottom) feeding on common ragwort (*Senecio jacobaea* syn. *Jacobaea vulgaris*) under the rotary washing line in our garden.

from that list of noxious weeds and dispelling some of the misunderstandings about it (e.g. Friends of the Earth 2017).

Other urban gardens

The pollinator species richness in our garden is far from unusual, and in fact British gardens on the whole support surprising levels of biodiversity. The thirty-year study of wildlife in a suburban garden in Leicester by ecologist Jennifer Owen is perhaps the best-known example (Owen 2010).[2] Her garden is fairly ordinary, and photographs show some nice borders, shrubs, mature trees, a bit of lawn, and a rotary washing line much like our own. But between 1972 and 2001 Jennifer recorded an astonishing 2,673 different species, almost 2,000 of them insects, including one-fifth of the UK's solitary bees (at least as the diversity stood at the time – see Chapter 11) and more than half of British bumblebees.

Other garden surveys have also found high diversities of pollinators, including the Biodiversity in Urban Gardens (BUGS) study of Sheffield gardens (www.bugs.group.shef.ac.uk). More recently Katherine Baldock and colleagues reported on surveys of gardens and allotments in four large British cities (Baldock *et al.* 2019), and as well as finding a large number of pollinator species, showed that household income can affect their abundance. Termed the 'luxury effect' (Leong *et al.* 2018), this correlation has been found in other studies of urban biodiversity. For pollinators at least, one of the causal factors is fairly straightforward: wealthier households can afford to spend more money on planting flowers in their gardens. There are other factors involved, of course, including the higher cost of housing in 'environmentally desirable' areas, but floral diversity always correlates with pollinator diversity regardless of the origin of the plants (see next section).

Britain, however, is not unique: studies from all over the world have shown the importance of gardens for supporting pollinators and providing a wide range of other ecosystem services (e.g. in Barcelona – Camps-Calveta *et al.* 2016). But when I think about our own Northampton garden I do wonder whether some of the species in these gardens, for example the gatekeeper butterflies, have persisted since the houses were constructed in the 1880s, on land that was mainly grassy meadows and orchards, connected to nearby patches of scrubby woodland and farmland with hedgerows. But we'll never know, for the records of what occurred on our patch before the land was developed simply don't exist. One of the things that I do know enhances our garden's appeal to wildlife, however, is the landscape context in which it sits. It's just one of more than thirty contiguous gardens that collectively provide a much wider range of floral resources and nesting and hibernating opportunities (see Figure 11.2). Some of the other gardens are just lawn (good for mining bees), others are neglected with lots of brambles (great for many different flower visitors), and there are those that are more managed for flowers and food, such as our own. Diversity of structure and plants, as well as overall management, is important at all scales, as we'll see in Chapter 12.

The changing size of British gardens

The gardens in the Victorian street in which we live constitute about 50% of the footprint of the area. Having such a large proportion of the land devoted to gardens is not uncommon for late-nineteenth-century housing, though it did vary a lot. If we take the area as a whole, different streets vary in the proportion of garden from around 23% to 63%, which may have reflected the social classes of the original residents, or possibly just the whims of the builders.

But what about more recent British house-building practices? One often hears that gardens in modern houses have become ever smaller as land prices have increased, and anecdotally that seems to be the case. However, I've seen little published information on just how it has changed over time. No doubt the pattern of changes in garden size is complex and not a simple linear reduction in size over time. In fact, as I noted above, both small and large garden areas were a feature of urban building at the turn of the nineteenth century. In the period just after the Second World War, as more social housing was built by local authorities, more land was provided for growing fruit and vegetables as a response to food shortages. As a child I can remember visiting my maternal grandmother's 1950s Sunderland council house and playing in a garden that was more than twice the area of the house, and that was quite normal for the area.

The general trend at the moment is for modern gardens to be a relatively small part of the total build area, and this shows no sign of changing soon. I wonder how this will affect urban wildlife in the future. There are a lot of caveats and complications to this, however, and the area of private garden is only one factor. For example, modern housing developments often include more green space outside of the footprint of the house and garden than was traditionally the case. In addition, these developments are frequently built on the periphery of the town, adjacent to rural areas. However, this public green space tends to be over-managed by landscaping contractors who mow and strim the small areas of lawn almost to bare earth, at inappropriate times of the year, and overuse herbicides. Some local authorities in the UK are now starting to allow wild flowers to grow in these areas, or are creating 'wild-flower meadows' – though the management of the latter is not at all straightforward, as we will see in Chapter 12. All of this needs to be taken into account to get a fuller understanding of how pollinators and other animals use the small gardens that are set within a wider landscape.

Large country house gardens

At the other end of the scale from modest urban gardens are the gardens surrounding large country houses, which Northamptonshire has in abundance. It's known as the County of Spires and Squires, a nod to both the large number of churches and the historical pattern of land ownership. However, until recently we've not appreciated just how important these types of gardens are for supporting pollinator diversity within intensively farmed landscapes. In 2010 I recruited Hilary 'Hils' Erenler to investigate this question as a postgraduate researcher funded by the Finnis Scott Foundation, a local horticultural charity. Hils spent a couple of years surveying the pollinating insects found in the gardens of seventeen large country houses around Northamptonshire and into adjacent counties. Some of these are open to the public but others are privately owned. Many of them retained their walled gardens, fascinating relics of a time when such large households of families and their staff relied on these sheltered, productive patches to provide food twelve months of the year (Figure 9.6). One private garden I visited with Hils had an avenue of some of the oldest espalier apple trees I've ever seen. Thick and gnarled and festooned with epiphytic lichens and mosses, they must have been planted at least a hundred years ago. Solitary bees such as the red mason bee (*Osmia bicornis*) were nesting in cavities in the surrounding walls and pollinating these apples, as well as pears, cherries, nectarines, beans, squashes and other insect-reliant crops. No doubt this has been going on for hundreds of years.

During two years of sampling, Hils recorded an astonishing diversity of pollinators. Across the seventeen properties surveyed she found 27% of British hoverflies (74 species), 68% of bumblebees (15 species) and 22% of solitary bees (55 species) (Erenler 2013). One could interpret this as an extreme example of the 'luxury effect' that I described above in action, but regardless of the cause, it is

FIGURE 9.6 The walled garden of a Northamptonshire country house, now growing ornamental flowers rather than fruit and vegetables. Photo by Hilary Erenler.

highly significant. These gardens have been there for a long time, centuries in most cases, and are acting as oases of pollinator diversity within what might otherwise be considered relatively depauperate, quite intensively managed, agricultural land.

Another interesting finding from Hils's study was that, despite 90% of the plants being non-native to the UK, the flower–pollinator network had identical overall structure to natural networks in terms of modularity, nestedness and levels of specialisation (Chapter 4). I think that this tells us a lot about the flexibility of pollinator species, that they can adapt to exploit suitable resources, no matter their origin. This includes artificial feeders (see below) and nesting sites, though the latter can be a little hit and miss. The success of artificial 'bee hotels', 'bee bricks' and bumblebee nests depends very much on where they are sited and the alternatives that are present in an area.

A Brazilian hummingbird garden

In November 2013 I was fortunate to be invited to Brazil for a month of teaching, giving talks at universities and symposia, and field work, hosted by my Brazilian collaborators Marlies Sazima and André Rodrigo Rech. One of the constant delights of that trip was seeing hummingbirds visiting flowers, both in wild habitats and in planted areas. I kept a special ear and eye out for their whirring wingbeats and rapid, darting movements, because they are significant pollinators in Neotropical plant communities, and at various times members of my research group have studied their ecology. Stella Watts, for example, worked on hummingbird–flower interactions in Peru for her PhD (Watts *et al.* 2012, 2016), and a Danish colleague Bo Dalsgaard spent a year in Northampton during his PhD research on Caribbean hummingbirds, since when we have collaborated on some macroecological questions about hummingbird specialisation in relation to current and past climates (see Chapter 6). Before all of that I did some work on their role as flower visitors and (probable) pollinators of some forest Apocynaceae (see Chapters 3, 4 and 5) in Guyana during field work in the late 1990s. So, it's always special to see them.

The bird guide I used for the trip to Brazil listed more than 80 hummingbird species that occur in the country, many of which were found within the Atlantic Forest region where I was travelling with a group of postgraduate researchers. During most days of field work we might see two or three species, but on one memorable day we saw eleven species in just an hour. We were visiting a private garden belonging to a retired gentleman named Jonas d'Abronzo, who had been feeding the hummingbirds in and around his property constantly for about twelve years, using home-made hummingbird feeders (Figure 9.7). The day we visited, Jonas had thirteen bottles of sugar solution hung up around his house and we estimated that over a hundred individual birds were using them. It's hard to be more accurate, as these birds move so fast, disappearing and reappearing without

FIGURE 9.7 Hummingbirds visiting artificial feeders in a private garden in Brazil. These fast-flying pollinators are difficult to successfully photograph without very good camera equipment.

warning, like hyperactive kids on an outing to a chocolate factory. It was a quite stunning sight.

The eleven species we observed are about half of the total number Jonas has recorded since he began feeding the birds. It's an incredibly species-rich area for hummingbirds – though the density and richness of birds in this one small property is clearly artificial, and we saw nothing like it out in the forest. There's a regular annual rhythm to their appearance, presumably in response to temperature and plant flowering in other parts of the country; as I mentioned in Chapter 6, it's a myth that the tropics have no seasons. Jonas was concerned that by feeding the birds so frequently (he was using five kilograms of sugar a day and replenishing each feeder several times) he might be negatively affecting plant pollination in the surrounding forest. We reassured him that this was probably not the case and that his efforts were likely to be positive, certainly compared to some of the other activities that go on around the area, such as building, forest clearance and agriculture. Assuming that food availability limits the population size of these birds (which may or may not be the case), then feeding the hummingbirds should result in a population increase in that area which will spill out into the wider forest. Similar arguments apply to feeding garden birds in the UK, particularly in the winter. However, there have been concerns about bird diseases being spread by feeders, so they need to be regularly cleaned. There is also the possibility that a higher density of more aggressive bird species might locally displace those that are less aggressive as they compete for food and nesting sites. These issues certainly require rigorous study.

As we watched the birds crowd and jostle around the sugar-water feeders, frequently erupting into conflict and chase, we discussed testing our assumptions, and the next year some of the postgraduates set up an experiment in the garden which was subsequently published (Sonne *et al.* 2016). Sure enough, as predicted, there was no effect of the high hummingbird density on pollination of a species of hummingbird-pollinated *Psychotria* (Rubiaceae) further out in the surrounding forest. Although these feeders have a significant local effect on hummingbird abundance, there's no evidence that they affect plant reproduction in the vicinity. I'd like to see this study repeated with more plant species to assess how general this finding is, though a similar study of artificial bat feeders in Ecuador also found no effect on reproduction of a bat-pollinated plant (Maguiña and Muchhala 2017).

The collaborations with researchers in Brazil and Denmark in which I've been involved over the years have been very productive: we've published fifteen research papers and book chapters to date and we've all learned a huge amount. The research has focused particularly on networks of plant–pollinator relationships (see Chapter 4), especially hummingbirds and the flowers that they visit. One of the things that it's revealed is just how flexible hummingbird–flower interactions can be. Not only do these birds use artificial feeders, but they will visit non-native plant species from which they can access nectar, either planted in gardens or in naturalised populations (Maruyama *et al.* 2016). In this respect they are no different from many of the pollinators I see in my garden or which Hils observed in her country house gardens: flexibility is the rule, not the exception. In a sense, gardens are experiments that test some of the ideas that I discussed in Chapter 4.

Scientists and gardens and pollinators

It should be clear from this chapter that I love gardens of all kinds: I love being in the space that a planted area encompasses, the coming together of nature and mind, but I also love the act of moving through that space, of smelling, touching, working with soil and leaves and flowers. 'Pottering' is a great word for it, though it doesn't quite do justice to acts as physically enjoyable and psychologically enriching as digging soil, dead-heading a plant, or tying in tomatoes to their canes. The scent of tomato foliage always transports me back to the 1970s, to my father's allotment in Sunderland. There, in a greenhouse constructed from old window panes, he grew luscious, sweet tomatoes, fed and watered by 'filtered beer'. It was some years before we realised that he was filtering the beer through his kidneys, which didn't impress my mother, who stopped eating them. But as Stephen King captured it beautifully: we don't buy beer, we only rent it,[3] and feeding tomato plants rather than flushing it down the toilet is certainly the environmentally savvy solution. Clearly my dad was an environmentalist before his time.

These childhood allotment memories represent my first exposure to horticulture, an interest and a practice that has remained with me ever since. I've always gardened and, even when I didn't own or rent a house with a garden, I grew house plants. This link between scientists and their gardens is a persistent one, certainly not unique to me. For example, in *The Invention of Nature*, Andrea Wulf's great biography of the explorer and biogeographer Alexander von Humboldt, gardens feature several times as places of calm and inspiration for both Humboldt and his mentor Goethe.

Historically, there are many other scientists who have used and been inspired by the gardens they have cultivated. Humboldt's friend and colleague Aimé Bonpland maintained a garden during his time in South America. Charles Darwin's garden at Down House certainly inspired the great man, and he carried out numerous experiments on plants and earthworms there. The University of Uppsala maintains the garden in which Carl Linnaeus, one of the founders of modern taxonomy, cultivated plants that he used in his teaching and research (I've visited this a couple of times, and it's well worth the trip if you are in that part of Sweden). Benjamin Franklin and Thomas Jefferson are two other scientists/inventors who also gardened.

More recently I can think of a number of prominent scientists in my own area of pollination ecology and plant reproductive biology who are also keen gardeners. These include John Richards, who is a specialist on alpine plants and wrote the book *Plant Breeding Systems* (1997); Spencer Barrett, whose website features a garden photo gallery that shows the location where he did some of the work on the mating costs of large floral displays, subsequently published in *Nature* (Harder and Barrett 1995); Dave Goulson, who has written eloquently about his own love of gardening in *The Garden Jungle* (2019); Simon Potts, who had long-term experimental plots set up on his lawn in which he tested ways to enhance flower diversity (until he moved house); Juliet Osborne, whose house in Cornwall comes with a large patch of semi-natural grassland and woodland that is slowly being restored; and Alexandra-Maria Klein, who uses her garden to carry out experiments to determine whether the fruit and vegetables that she grows require pollinators, similar to the bagging experiment on passion fruit that I describe in Chapter 14.

The garden that Karin and I are developing in Northampton serves many functions: as a centre of quiet relaxation and a place to write, to be inspired by the pollinators and their behaviour, to enjoy physical labour, to grow food, to watch the birds on feeders in the winter, and (occasionally) to collect data (see Chapters 11 and 14). I cannot imagine being a scientist without a garden – as Francis Bacon said, 'it is the purest of human pleasures'. However, he was writing in the sixteenth century before the advent of pesticides, herbicides, inorganic fertilisers, electric mowers and other gardening modernities that, one way or another, can have a profound environmental impact. Good gardening must be tempered with a sense of how we go about those activities in a way that minimises that impact.

Gardens, urban and rural, modest and posh, temperate and tropical, are important for supporting pollinator diversity, that much is clear. But they are no substitute for conserving and enhancing the wider countryside and making agricultural landscapes more suitable for pollinating insects and vertebrates. Davies *et al.* (2009) estimated that domestic gardens in the UK cover an area of 432,964 hectares, which the authors note is greater that the county of Suffolk. However, that's less than 2% of the area of the UK; in contrast, agricultural land amounts to about 70% of the country. Parks and gardens, as important as they are, can be no panacea for changes that are happening in our wider countryside. In the next chapter we'll consider just how these changes have affected pollinator diversity and abundance, in the UK and elsewhere in the world.

Coda: gardens and scientists during the COVID-19 crisis

As I was putting the finishing touches to this book during the first quarter of 2020, the SARS-CoV-2 coronavirus pandemic was accelerating and the lives of billions of people across the world were changing fundamentally. In the UK, schools, colleges, universities and many businesses closed, and people were restricted in their movements outside of their homes. At the University of Northampton, as in other higher education institutions, all face-to-face teaching and field trips were cancelled and lectures and tutorials went online. It would be the first year since 2003 that I hadn't made a trip to Tenerife (see Chapters 3 and 6). Those of us fortunate to have gardens at least had somewhere to go outside and engage with some aspects of nature. When it became clear that, for many pollination ecologists, it would be impossible to do any field work, I sent an email out to my network of researchers in the UK suggesting that we make the most of the situation and use our gardens to collect standardised flower-visitor network data (see Chapter 4). These data could then be used by postgraduate and postdoctoral researchers whose time and funding were running out, so that a whole year of data collection would not be lost.

The response was very positive, and indeed the idea caught on internationally, with ecologists all over Europe, Australia and North and South America interested in being involved. As of early August, more than fifty ecologists have already submitted information, amounting to almost 15,000 rows of data, with more promised. At this point it's too early to say how much data we will collect and how it will be used, but given travel restrictions a number of us are certain to be surveying our gardens for the rest of the summer. I'm especially interested to see how the ornamental and wild plants are providing nectar and pollen resources for the edible crops that we grow, and how this shifts during the season. There are lots of other questions that can be addressed by the data set, and the collective efforts of these ecologists are not only providing useful and usable data to the research community, but also helping us to keep active and mentally healthy during the crisis. This is summed up by an email I received from one of those taking part in the lockdown garden surveys: 'Doing these surveys keeps the ecologist inside of me alive!'

Chapter 10

The shifting fates of pollinators

John Clare (1793–1864) is one of the nineteenth century's most celebrated English poets of rural landscapes and nature. To quote his biographer, Clare was 'the greatest labouring-class poet that England has ever produced. No one has ever written more powerfully of nature' (Bate 2003). Not only that, but he was born and lived for much of his life in my adopted county, hence his epithet 'the Northamptonshire Peasant Poet'. A number of his poems focus on, or mention, bees: it was clearly a topic that fascinated him. One of these poems, the pertinently titled *Wild Bees*, was written sometime in the period 1819–1832.

> These children of the sun which summer brings
> As pastoral minstrels in her merry train
> Pipe rustic ballads upon busy wings
> And glad the cotters' quiet toils again.
> The white-nosed bee that bores its little hole
> In mortared walls and pipes its symphonies,
> And never absent couzen, black as coal,
> That Indian-like bepaints its little thighs,
> With white and red bedight for holiday,
> Right earlily a-morn do pipe and play
> And with their legs stroke slumber from their eyes.
> And aye so fond they of their singing seem
> That in their holes abed at close of day
> They still keep piping in their honey dreams,
> And larger ones that thrum on ruder pipe
> Round the sweet smelling closen and rich woods
> Where tawny white and red flush clover buds
> Shine bonnily and bean fields blossom ripe,
> Shed dainty perfumes and give honey food
> To these sweet poets of the summer fields;
> Me much delighting as I stroll along
> The narrow path that hay laid meadow yields,
> Catching the windings of their wandering song.
> The black and yellow bumble first on wing

> To buzz among the sallow's early flowers,
> Hiding its nest in holes from fickle spring
> Who stints his rambles with her frequent showers;
> And one that may for wiser piper pass,
> In livery dress half sables and half red,
> Who laps a moss ball in the meadow grass
> And hoards her stores when April showers have fled;
> And russet commoner who knows the face
> Of every blossom that the meadow brings,
> Starting the traveller to a quicker pace
> By threatening round his head in many rings:
> These sweeten summer in their happy glee
> By giving for her honey melody.

This piece is a stunning example of Clare's ability to make detailed observations of the natural world and to translate them into poetry. So good are those observations that it's possible to identify Clare's bees from the descriptions he gives of their colour, their nesting behaviour, and the flowers they visit. Furthermore, his words may give us an early insight into the topic of this chapter: the ways in which pollinator abundance and diversity have changed over time.

When Clare writes of 'the white-nosed bee that bores its little hole / In mortared walls and pipes its symphonies', he is referring to the male hairy-footed flower bee (*Anthophora plumipes*). Male bees of many species quite often have 'white noses', or at least white or yellow faces, but the real giveaway is in the next two lines: the 'never absent couzen, black as coal, / That Indian-like bepaints its little thighs' has to be the female of this species, which is all black except for the orange pollen brushes on its rear legs. The males are rarely far away ('never absent') from the nests of the females, which are often in old walls, as they patrol looking for mates.

'The black and yellow bumble first on wing, / To buzz among the sallow's early flowers, / Hiding its nest in holes from fickle spring' refers, I think, to the buff-tailed bumblebee (*Bombus terrestris*). The queens tend to emerge earlier than other, similar species, hence 'first on wing'. It usually nests in rodent holes, particularly in hedge banks or beneath garden sheds and compost heaps.

'In livery dress half sables and half red, / Who laps a moss ball in the meadow grass' – that must be the red-shanked carder bee (*Bombus ruderarius*), the only red and black bee in Britain that makes a mossy nest above ground. This species is related to the 'russet commoner who knows the face / Of every blossom that the meadow brings', which refers to one of my favourite bumblebees, the all-brown common carder bee (*Bombus pascuorum*). I am fond of this species because it is as common as the name suggests, and is renowned for foraging on a wider range of flowers than most others, so that it 'knows the face of every blossom'. When I'm doing field work in Britain it is always with a sense of familiar greeting that I encounter the common carder bee, a frequent and important pollinator of wild flowers.

The Northamptonshire landscape that Clare was describing in his poetry and prose was a very different one from that we see today, and indeed the same is true for most of Britain. Hay meadows and other species-rich grasslands were much more common, though (in Northamptonshire at least) woodland was no more extensive than it is currently. Importantly, farming was less intensive, with no use of pesticides, herbicides or inorganic fertilisers, and limited mechanisation. Clare may have been familiar with species that no longer occur in the county, such the great yellow bumblebee (*Bombus distinguendus*) and the short-haired bumblebee (*B. subterraneus*), or which have only recently re-established themselves, including birds such as red kite (*Milvus milvus*) and raven (*Corvus corax*).

Of the four species of bee that Clare describes, and which we can identify with some certainty, three are still common in Northamptonshire and in fact across much of Britain. The exception is the red-shanked carder bee, which has seen a huge decline throughout its range, particularly in the southwest and the north. It still occurs in Northamptonshire but is far from common. The mixed fates of Clare's bees, with some species still common and others greatly declined, is indicative of the state of Britain's pollinators as whole, in that there have been winners and losers. It's very far from the 'pollinator crisis' that the media (and some conservation organisations) like to portray (see below and Chapter 13). Nonetheless, there is a worry that we are losing a significant fraction of the pollinator biodiversity of the British Isles, both locally and nationally. Internationally the situation is also concerning, as I will show.

Texts such as John Clare's poem provide some hints as to how rare or common species may have been almost 200 years ago. In truth there is precious little hard data for that period, even for a nation whose natural history is as well documented as Britain's. *Monographia Apum Angliae* (1802) by Suffolk clergyman William Kirby was the first book to deal exclusively with the wild bees of Britain. However, Kirby has only a few comments to make about commonness and rarity of individual species, and some of what he writes needs to be taken with a pinch of salt. For example, he discusses the potter flower bee (*Anthophora retusa* – now one of our rarest and most declined bees) and talks of finding it 'in great abundance frequenting the walls built with Kettering stone at Wansford and Ilford in [Northamptonshire]'. The potter flower bee nests in the ground, and Kirby is clearly mixing it up with the superficially similar hairy-footed flower bee (*A. plumipes*) that we have already discussed.

By the middle of the nineteenth century, however, naturalists were starting to become a little more systematic in how they recorded information and preserved specimens. It is these more systematic data that have allowed us to reconstruct the patterns of decline and extinction of the major groups of British pollinators over the past 150 years. Not only that, but we can relate these patterns to large-scale social changes in our country, wrought by industrial advances, government policies, and warfare.

Pollinator extinctions in Britain

Extinction of species is perhaps the most fundamental assault that humans can inflict on the rest of the natural world, and it takes a range of forms. At its most tragic level extinction refers to the loss of an entire species. The dodo (*Raphus cucullatus*) is the most famous anthropogenic species extinction, but there are many others. However, extinctions can also occur at a national or local level (sometimes referred to as the 'extirpation' of a population). In the British Isles, country-level extinctions are highly significant, as the surrounding seas limit opportunities for species to recolonise areas where they previously lived (though, as we shall see, this does sometimes occur).

Since the nineteenth century the UK has definitely lost two of its 26 original bumblebees – the short-haired bumblebee (*Bombus subterraneus*) and Cullum's bumblebee (*B. cullumanus*). The status of a third species, the apple bumblebee (*B. pomorum*), is debatable: it may not have been established in Britain. In addition, eleven solitary bees from other genera have gone. These losses are mainly due to destruction of habitat and changes in land management associated with intensification of agriculture. This includes more severe (and inappropriately timed) hedgerow management, mechanisation, and greater use of chemical fertilisers and pesticides. The short-haired bumblebee has been the subject of a reintroduction programme; at the time of writing it appears that this has failed, but other rare bee species have benefited from the habitat restoration work that was undertaken (Bumblebee Conservation Trust 2020).

Other British pollinator extinctions include ten species of flower-visiting wasps and numerous moths and butterflies. Indeed, the most (in)famous pollinator extinctions are butterflies such as the large blue (*Phengaris arion*), now reintroduced and thriving, and the large copper (*Lycaena dispar*), which has never been successfully reintroduced despite several attempts. There are no substantiated hoverfly extinctions, though it's possible that early extinctions were missed as flies were traditionally not as well studied as other groups.

At least 10 beetles have also gone extinct in historical times, though as far as we know none of them were regular flower visitors. However, we do know that Britain has lost some beetles that were potential pollinators in prehistoric times. The evidence comes from the preserved remains of oak trees from former marshland in East Anglia. Within these bog oaks occur the cavities formed by larvae of wood-boring longhorn beetles (Cerambycidae), the adults of which are often found on flowers in the summer (see Chapter 12). At the Natural History Museum in London some examples are preserved that can be identified to species level (Figure 10.1). One such is the great capricorn beetle (*Cerambyx cerdo*), which probably went extinct in Britain many centuries ago, though it occasionally turns up when imported with infested wood (Harding and Plant 1978, Salisbury *et al.* 2015b).

Reconstructing the history of how Britain lost these insects gives us an insight into the ways in which changes in agriculture and land use have affected the wildlife of our country (Ollerton *et al.* 2014). Early deforestation was clearly

FIGURE 10.1 The great capricorn beetle (*Cerambyx cerdo*) from 4,000-year-old bog oak in Cambridgeshire, now preserved in the Natural History Museum, London.

responsible for the loss of the great capricorn beetle before anyone bothered to record it. But historical records can give us much more information. In particular, changes in the rate of extinction, such as the number of species lost per decade, tell us about specific periods in time. We can interpret any patterns in relation to broader changes in society, for example in agricultural practices and policies, and conservation strategies.

This has been done using over half a million records of the presence of bees and flower-visiting aculeate wasps at particular localities and times held by the Bees, Wasps and Ants Recording Society (BWARS). This is probably the most extensive data set on these insects available for any country in the world, and therefore an important resource for understanding changing patterns of pollinator occurrence and distribution. As of 2020, the number of records has increased to about 1 million (Stuart Roberts, personal communication).

The BWARS data reveal 23 extinctions of bees and flower-visiting wasps (Ollerton *et al.* 2014). Dates of the last records of these insects range in time from the crabronid wasp *Lestica clypeata* (which has no common name and was last

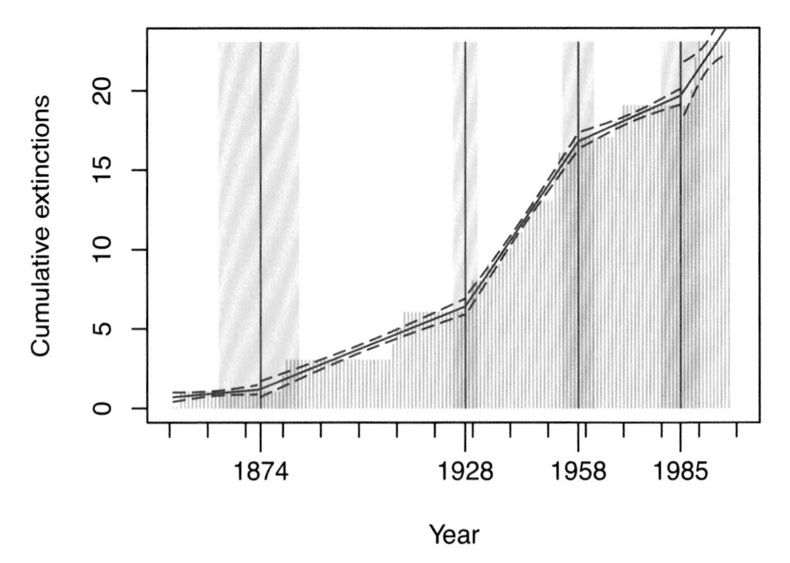

FIGURE 10.2 The cumulative number of British bee and wasp extinctions that occurred between the 1850s and the 1990s. 'Extinction' here is defined as at least 20 years since the last recorded occurrence of the species in Britain, which is why the data stop in the 1990s. Single early records of species that cannot be verified as representing stable breeding populations have been excluded from the analysis. This is updated from Ollerton *et al.* (2014), with thanks to Robin Crockett for the analyses.

observed in 1853) to the solitary Burbage mining bee (*Andrena lathyri*) not seen since 1990. All of these species still occur on mainland Europe, so these were country-level extirpations, not species extinctions.

The rate of extinction is highly variable over time, with more species lost during some periods than others, as shown by the steepness of the regression lines in Figure 10.2. One of the most interesting findings from this analysis of the BWARS data was that the main period of species loss followed changes to agricultural policy and practice just after the First World War. This is much earlier than previously believed: it has usually been the Second World War and the subsequent Common Agricultural Policy which have been seen as the main drivers of pollinator loss. The four years marked in red on the figure are the points where we estimate the rate of extinction changed (with 99% confidence intervals shown in pink). The most rapid rate of extinction (shown by the solid blue lines and dashed 99% confidence intervals) occurred between the late 1920s and the late 1950s. This may be the cumulative effects of agricultural changes precipitated and then augmented by the First and then the Second World Wars, respectively.

The earlier period of extinction, from the late nineteenth into the early twentieth century, was probably caused by greater imports of South American guano as soil fertiliser. This increased grass productivity at the expense of wild-flower diversity. It also reduced reliance on strict rotational cropping, including fallow periods with nectar- and pollen-rich weeds, and nitrogen-fixing legume years. However, it was the invention of the Haber process in 1909, allowing industrial manufacture of inorganic nitrogen fertilisers for the first time, that fundamentally affected British agriculture.

The slow-down of the rate of extinction from the early 1960s to the mid-1980s is not easily explained, given the continued intensification of farming, encouraged by Common Agricultural Policy subsidies. It could be due to the most sensitive species having been already lost, leaving behind more robust species. Alternatively, the reason might be that early conservation initiatives were successful, including the establishment of more nature reserves by organisations such as the Wildlife Trusts and the RSPB, habitat restoration and management by groups such as the British Trust for Conservation Volunteers, and more farmers switching to organic methods. Or it could be a combination of these, plus other factors that we have not considered.

There is a final period that shows an increased rate of extinctions, from 1985 onwards. This could be seen as evidence against the slowing in the rate of decline of pollinators in northwest Europe that was found in a study by Luisa Carvalheiro and colleagues (2013). However, we need to be cautious here as there's a large degree of statistical uncertainty around the calculated extinction rate: the four extinctions between 1988 and 1990 could be an isolated cluster, or the start of a further period of relatively high extinction rate. Only time will tell.

As I've shown in previous chapters, bees, wasps and other pollinating insects are absolutely vital to the functioning of our natural ecosystems. We've known for some time that many of these pollinators are declining in Britain, but now we can see how historical agricultural changes may have caused species to become extinct. The big question is whether these extinctions have stopped or whether they will continue in the future. The species that have been lost to Britain still survive on the European continent, and there is the possibility of natural recolonisation (which happened for one species while we were doing our analyses – see Chapter 11) or artificial reintroduction. However, in order for this to be successful we must restore as much natural habitat as possible within our farmland, which after all covers most of the British land surface. The irony of finding that agricultural changes are the main reason for pollinator extinctions is, of course, that pollinators are also vital for agriculture, as the various National Pollinator Strategies recognise (see Chapters 7 and 13).

Assessments of decline and extinctions of species such as these illustrate the importance of maintaining the year-on-year effort of recording natural history data: this level of understanding simply wouldn't have been possible without the BWARS records, which were mainly collected by amateur naturalists. The Joint Nature Conservation Committee (JNCC) has in recent years started to use the data from BWARS and the Hoverfly Recording Scheme to produce an index that measures change in the number of one-kilometre grid squares across the UK in which bees and hoverflies were recorded in any given year. This is used to indicate changes in distribution of those species for which enough records exist for robust analysis. The most recent (Figure 10.3) included records for 365 species (137 bees and 228 hoverflies). The decline in some species is striking, but for more than 50% there has been no change or even an increase in their range (Figure 10.4). Overall, however, by 2016 the index had declined by 31% compared to its value in 1980, with the long-term trend for the 'average' pollinator species being assessed by JNCC as declining.

In 2019 a more detailed assessment of the British data was published by Powney *et al.* (2019b), and this again showed a mixture of winners and losers among pollinators. The main losers were species associated with upland habitats (a signal of climate change?) and species already considered rare, while a group of relatively widespread bee species that are considered to be important pollinators of crops had increased in distribution by 12%, possibly due to the instigation of

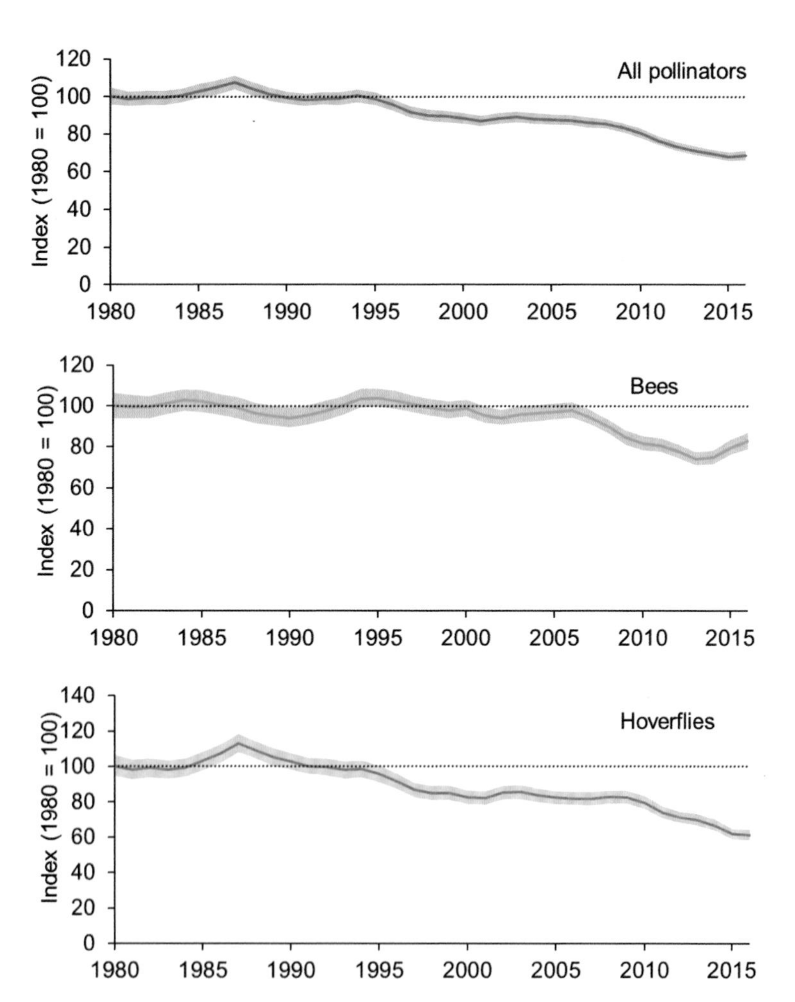

FIGURE 10.3 Changes in the distribution of all bee and hoverfly pollinators (top graph), bees ($n = 137$ species, middle graph) and hoverflies ($n = 228$ species, bottom graph) in the UK between 1980 and 2016. Data from Powney *et al.* (2019a)

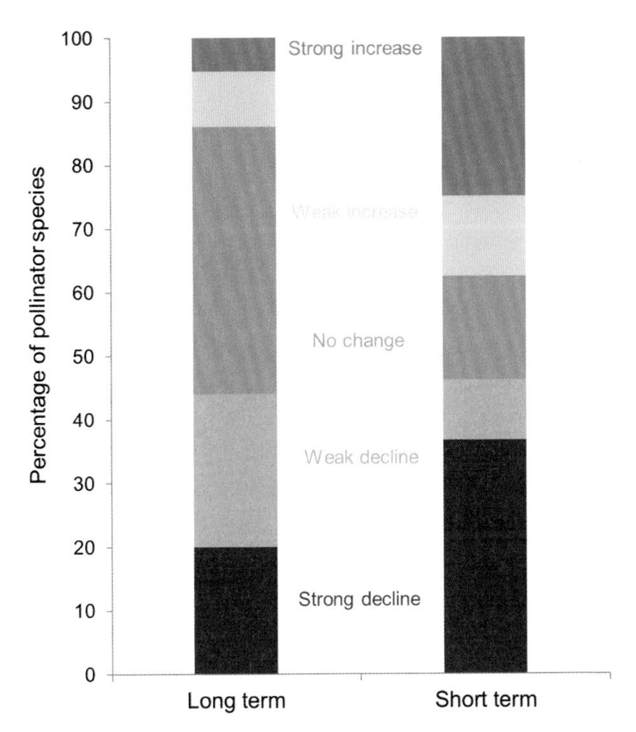

FIGURE 10.4 The percentage of UK bee and hoverfly species in each of five trend categories, calculated using the mean annual change over (a) the long term (1980–2016) and (b) the short term (2011–2016). Data from Powney *et al.* (2019a).

agri-environment schemes. Importantly, while these analyses tell us about how the areas in which species have been recorded have changed, they do not provide information about changes in the abundance of the insects. To do this we require repeated observations of insect numbers in the same places over time, which is one of the main objectives of the Pollinator Monitoring Scheme (see below).

Future bee extinctions?

Although the causes of extinctions have historical roots right back into the nineteenth century, some of the bees described above were lost in living memory. The second year of my PhD (1990) was the final year that the Burbage mining bee was observed in Britain. That was the last bee extinction to occur in this country, though there had been one a couple of years previously when I was still an undergraduate: the short-haired bumblebee was last seen in 1988. However, other bees could follow them if we are not vigilant to the issue. There are a number of species in Britain that are causing concern among conservation organisations because only a few populations are known. These include the six-banded nomad bee (*Nomada sexfasciata*) and the fringe-horned mason bee (*Osmia pilicornis*).

There are also others that are now gone from most of their British and Irish range, such as the great yellow bumblebee (*Bombus distinguendus*), which is extinct in Wales and England but still has Scottish and Irish populations.

All of this sounds pessimistic, but there are two things that give me some hope for the future, one social, one ecological. The first is that there is increasing public awareness that we need to conserve Britain's biodiversity. That awareness can be (and is being) turned into action by organisations such as the RSPB, the Wildlife Trusts, Buglife and the Bumblebee Conservation Trust, in collaboration with farmers and land owners, citizen scientists (see Chapter 14) and wildlife recorders. The second is that most of the British extinct species (though not necessarily endemic subspecies) still occur on the near-continent. These could naturally recolonise, or even be reintroduced if appropriate habitats were found and extended by ecological restoration. But small nature reserves are not enough: the speed of climate change is such that we must have larger, better-connected protected areas to allow species to move at their own pace (if they can). Buglife's B-Lines project (www.buglife.org.uk/our-work/b-lines) is a great example in that regard.

Providing enough nectar- and pollen-rich plants at a landscape level is part of the solution, although clearly not the whole of it, as pollinators need other resources too. But it is a good starting point. In an article in the journal *Nature* a few years ago entitled 'Historical nectar assessment reveals the fall and rise of floral resources in Britain', Mathilde Baude and colleagues (2016) showed how the main nectar-producing plants have changed in extent and composition over the twentieth century. There have been huge fluctuations, and that's the probable cause of at least some of the losses of pollinators. Restoring those nectar- and pollen-rich flowers back into our agricultural landscapes should be a priority, for wild pollinators and for the use of managed honey bees, to which we now turn our attention.

The honey bee situation

Honey bees are seen by many as the archetypal pollinators; they are certainly the ones that are most familiar to the general public, and they receive a disproportionate share of media attention (see Chapter 13). This is partly due to the publicity generated by such issues as colony collapse disorder (CCD) and the varroa mite, as well as more recent concerns in relation to pesticides. To what extent is there a crisis in the abundance of honey bees? Declines in numbers of managed hives have certainly been documented in some parts of Europe, although in southern countries they have increased. Likewise there have been both reductions and increases in different parts of North America, and globally the trend in numbers is upwards. But there is evidence that world demand for honey bee pollination services is outstripping supply, making the role of wild, unmanaged bees more important than ever (Aizen and Harder 2009).

In Britain the situation is far from clear. One would have thought that we would be able to cite robust statistics for long-term changes in numbers of hives, but that's not the case. Indeed, official statistics seem to be rather unreliable, probably because most beekeeping is hobbyist rather than professional, and registration of hives is voluntary. The earliest continuous data available are those in Bailey and Perry (1982) that span 1946–1982. This should be fairly accurate for England and Wales. However, their estimate of 800,000 hives in the 1920s needs to be treated with caution; they make a number of assumptions in their regression-based analysis that may be incorrect, and a report from 1919 cites an official government estimate of 32,500 hives. It seems impossible that this would have grown to 800,000 hives within a decade (Anon. 1919), and a more probably trajectory is shown in Figure 10.5. There has been some speculation that during and just after the Second World War the number of hives being kept was exaggerated by some unscrupulous beekeepers in order to claim a larger ration of sugar to sustain the bees during the winter months. I don't know of any evidence for this, and certainly beekeeping became more popular during this period, so we can be reasonably confident that numbers of hives in England and Wales have fallen steeply from their post-World-War-Two level. Since the late 1970s hive numbers, while they have fluctuated, have remained fairly constant overall. Indeed at the moment numbers are higher than they have ever been in the last 50 years (Figure 10.5). Unfortunately the UK stopped returning official numbers of hives to the United Nations Food and Agriculture Organization (FAO) in 1977, and their data up to 1987 are an unofficial estimate. From 2003 the UK had to report bee-hive numbers to the European Union (EU) to claim money for the National Apiculture Programme (European Commission 2020), but the data were suspiciously constant

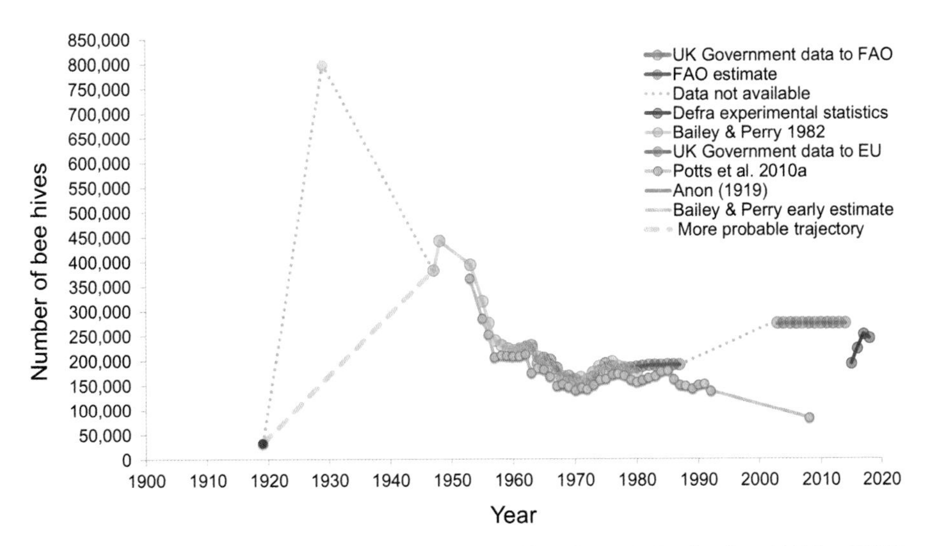

FIGURE 10.5 Changes over time for the number of bee hives in England and Wales (1918 to 1980) and Great Britain (1980 onwards) Data are from various sources (see text). I am grateful to Defra for providing the hive data from 2015 to 2018, to Simon Potts for sending me the data from Potts *et al.* (2010a) and to Andrew Hubbard for drawing my attention to Anon (2019).

between years. More recently beekeepers have been encouraged to update their hive record annually on BeeBase (www.nationalbeeunit.com), and hopefully these new estimates are more realistic and more responsive to changes over time.

Hive data from a range of sources were plotted by Potts *et al.* (2010a), and their curve follows the data that I've graphed very closely until we get to the mid-1980s. At this point the lines diverge considerably. There are two possibilities: either numbers of hives crashed to fewer than 100,000 by 2008 (Potts *et al.* 2010a), or they increased hugely to more than 250,000 (my data). Both scenarios cannot be correct, and indeed both may be wrong! We simply don't know. Given the wide range of the low and high estimates, the fact that beekeeping has become more popular over the past decade, and that the recent data sit more or less within this range (at least initially), I wonder whether honey bee numbers have actually remained quite stable over the past 25 years, and indeed have hovered around 150,000 hives or so since the 1970s.

There ought to be a correlation between numbers of hives and amount of honey being produced, but again the UK numbers are confusing (Figure 10.6, top panel). In fact, FAO data show an exponential increase in honey production from the early 2000s that is not matched by hive numbers in Figure 10.5. Annoyingly, from 2004 onwards the FAO statistics are estimates that increase year-on-year, and could well be far too high. EU data (hopefully more accurate) state that about 4,500 tonnes of honey was produced by the UK in 2014, while FAO data estimates that it was more than double that amount. Once again, they can't both be right. Just using the official UK/EU statistics, and Defra experimental statistics for hive figures since 2015, there is indeed a correlation between number of hives and amount of honey produced (Figure 10.6, bottom panel), though there's a lot of variability in that trend – which is to be expected, as honey production also relies on factors such as the weather.

Putting all of this together, it appears that there are as many bee hives, and as much honey being produced, now as 50 years ago. Though there have been considerable fluctuations over that period there is no indication of a honey bee 'crisis' occurring, despite what some maintain (see below and Chapter 13). It also demonstrates just how problematic it is to assess long-term trends in insects, even (supposedly) well-documented ones such as honey bees.

It is worth noting that in some parts of the world where the honey bee is not native, or is native but maintained at very high hive densities, honey bees themselves have been shown to have a negative impact on other pollinators, dominating the floral resources available to wild pollinators, and on the pollination of endemic plants (Mallinger *et al.* 2017, Hung *et al.* 2019, Valido *et al.* 2019, Herrera 2020). While honey bees are undoubtedly a good thing for agriculture and for food production, as the saying goes, it's sometimes possible to have too much of a good thing.

Having considered what's happening in Britain, let's now turn to the situation in the world as a whole.

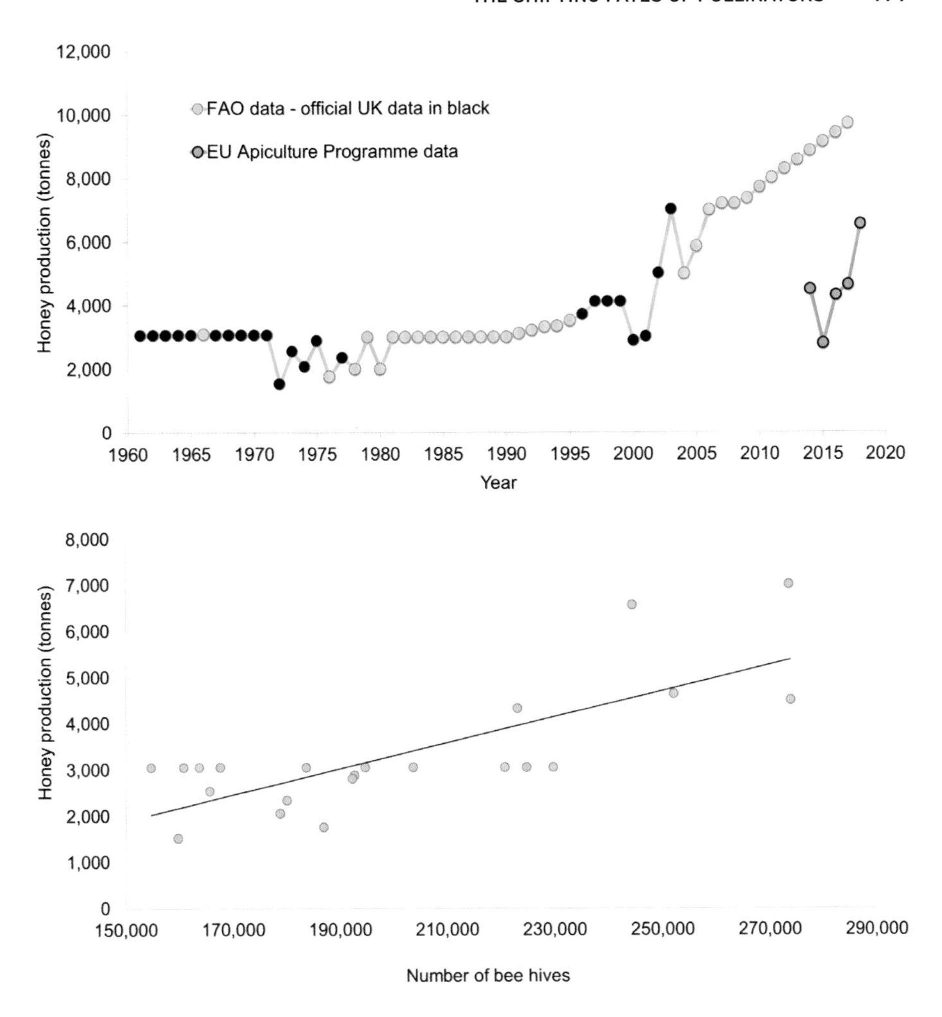

FIGURE 10.6 UK honey production (top) and the correlation between official/experimental statistics for hives and honey production (bottom). Data are from various sources (see text and Figure 10.5).

A global perspective on pollinator loss

Not all scientists and naturalists who study pollinators are comfortable with the idea that we are currently experiencing a global 'pollinator crisis'. Crisis suggests that we are in danger of losing most of our pollinator diversity and abundance, the apocalyptic (but unlikely) scenario modelled by Smith *et al.* (2015), who assessed the health benefits of micronutrients and minerals provided by animal-pollinated crops (see Chapter 7). Differences of opinion are often voiced in meetings or in private conversations, and occasionally spill out into the scientific literature. Among a minority of scientists there are suggestions that the evidence for declines in pollinator abundance and diversity are mainly issues relating to honey bees and bumblebees in developed countries in Europe and North America (Ghazoul 2015). Others argue that this is not the case, and that the evidence for wild pollinator declines is considerably more robust (though the honey bee situation, as I've described above, is rather more complex). In my opinion, while it may not be a 'crisis', it is a concern.

Wild bees (including bumblebees, and solitary and primitively eusocial bees) have shown significant reductions of abundance and diversity. This has been documented at local, regional and country levels in Britain, and in parts of continental Europe, North America, South America, Africa and Asia (Matsumura *et al.* 2004, Kosior *et al.* 2007, Benton 2008, Inoue *et al.* 2008, Xie *et al.* 2008, Grixti 2009, Williams and Osborne 2009, Cameron *et al.* 2011, Bartomeus *et al.* 2013, Nieto *et al.* 2014, Schmid-Hempel *et al.* 2014). There is at least one example where we know that the local extinction of a bee has negatively affected the reproduction of plants with specialised pollination systems (Pauw 2007).

Some of these declines are probably due to climate change, with further losses of diversity predicted in the future, particularly for bumblebees (see Chapter 6). As I mentioned in Chapter 4, there's been some interesting work in the USA reassessing Charles Robertson's Illinois site and looking at how the bee fauna has changed. A repeat survey in the 1970s revealed a bee community very similar to that observed by Robertson in the late nineteenth and early twentieth centuries (Marlin and LaBerge 2001). However, surveys in 2009–2010 found a significant loss of bee species, with 50% of those on Robertson's list not re-recorded (Burkle *et al.* 2013). Again, degradation of land, especially for agriculture, is implicated as a major cause of these losses. Not only that, but through detailed observations of the flower visits made by individual insects, the authors of the study found evidence that pollination services to wild plants had suffered from the changes to the pollination network: for one early-spring-flowering species, a vital floral resource for many pollinators, the amount of pollen being carried by bees had declined by more than half since the nineteenth century.

A lack of such historical data on pollen movement and seed set means that few studies have addressed this question of how the pollination of plants has changed over time, though it's a fundamental one if we are to go beyond monitoring the loss of species and consider how ecosystem functions are also changing. One interesting approach is to use preserved herbarium specimens to document changes in seed set over time. Duan *et al.* (2019) did this for 109 Chinese species in the pea and bean family (Fabaceae) and found that, over the past century, only four species showed a decrease in number of seeds per pod. Nine species had increasing levels of seed set, and the majority showed no change at all. This is far from the pollinator or pollination 'crisis' that we have come to expect, but many more such studies are needed before we can say how generalisable this finding might be.

Hoverfly diversity declines at a local level have been documented in the Netherlands and Britain (Biesmeijer *et al.* 2006), but the situation elsewhere in the world is unclear because of a lack of monitoring records. The same is true of flower-visiting wasps, of which (as shown above) there has been a reduction in country-level diversity in Britain, but we know very little about other countries. In contrast we have much more information for butterflies and moths and it is clear that the diversity and abundance of Lepidoptera has declined dramatically in the UK (e.g. Gonzalez-Megias *et al.* 2008, Fox 2013) while in North America some fifty species are Red Listed by IUCN criteria – and there is particular concern about the iconic monarch butterfly (*Danaus plexippus* – Agrawal 2017). Likewise, a significant fraction of butterflies in other parts of the world, including southern Africa, Australia and Europe, are of conservation concern.

Bird and mammal pollinators were recently assessed at a global level by Eugenie Regan and colleagues (2015), using IUCN Red List criteria. They concluded that, as a general trend, the conservation status of pollinating birds and mammals was getting worse over time: more species are moving nearer to extinction than away from it. Birds are often considered 'bio-indicator' species whose fortunes reflect those of other, less well-studied wildlife, so this is particularly worrying.

Of course there are studies that have shown that many pollinator species are doing well and have remained stable or are even increasing in abundance (including honey bees in some parts of the world – see above). But at the moment the evidence points towards declines in pollinator abundance and diversity for a wide range of groups. This is occurring at a range of scales in all areas that have so far been assessed with any rigour. A few years ago I co-supervised the MSc thesis of Ceri Green in Ireland, who looked at as many Red Lists of pollinator groups compiled using IUCN criteria as she could find. The summary graph from that work is sobering: in all groups, a significant fraction of the assessed species are declining and of conservation concern (Figure 10.7). Just as worrying is the fact that for most pollinators we are 'data deficient', in other words, we don't know how their populations are performing. They could be doing well, but they may not be.

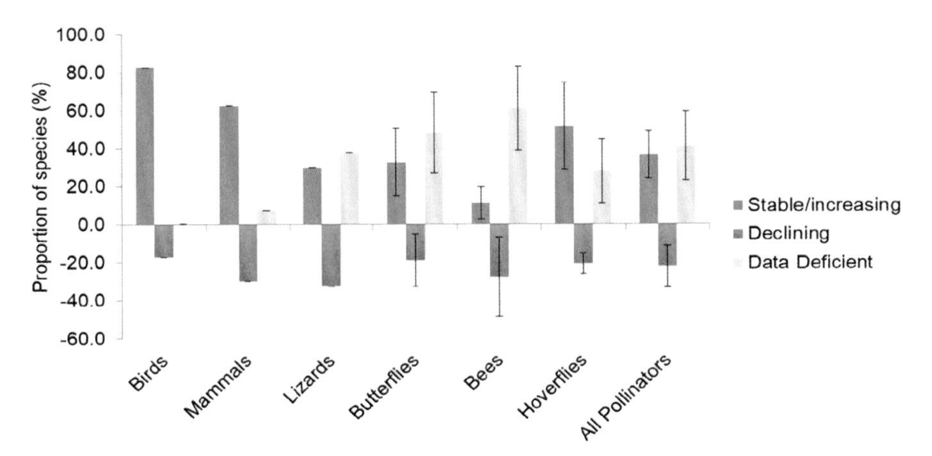

FIGURE 10.7 The mean percentage (with standard deviations) of pollinators in IUCN Red List assessments that are stable, declining or data deficient. Data are split by taxonomic group and then combined to give an assessment for all pollinators. Reproduced with permission from Green (2017).

The historical causes of loss of pollinator diversity and abundance relate to changes in agriculture and the destruction of natural vegetation for farmland and development, as we have seen. But more recently these have been augmented by other causes that are giving us reasons to be concerned (Potts *et al.* 2010b).

The role of pesticides and diseases in pollinator decline

Changes to agricultural land use and management of the sort I have described above encompass many different farming activities. These wax and wane in their initial importance; for example, inorganic fertilisers are commonplace now and the initial 'shock' to the ecological system has already happened, in terms of reducing wild-flower diversity, though their continuing use is still problematical. Nutrient build-up in soils is a major reason why grassland nature reserves lose species: high levels of nitrates and phosphates, in particular, encourage the growth of grasses, docks and nettles at the expense of less competitive species. However, it's not only farming practices that are responsible: in small nature reserves problems can also be caused by dog owners who take their pets for a walk to empty them, and don't bag-and-bin the results.

In recent years, however, some other factors have received increased attention as we begin to realise that they constitute the latest assaults on the natural world. Two of these are the use of pesticides, in particular neonicotinoids, and the emergence of new bee diseases.

Neonicotinoids (or 'neonics') are chemically similar to naturally occurring nicotine. These pesticides were introduced to Britain in the mid-1990s as a less harmful alternative to chemicals such as organophosphates. However, they were quickly implicated as one of the causes of colony collapse disorder in honey bees, and the evidence has since grown that they are having a major effect on Britain's wildlife, including pollinators (Godfray *et al.* 2015, Goulson *et al.* 2015). Some of these claims are disputed, with the main naysayers being, not surprisingly, the agrochemical companies that have invested a lot of money in developing and marketing these pesticides. Here, for example, is a quote from Nick von Westenholz, CEO of the Crop Protection Association, which represents the firms that produce neonicotinoid pesticides: 'the latest studies … must be seen in the context of ongoing campaign to discredit neonicotinoid pesticides, regardless of what the real evidence shows' (quoted by Briggs 2015). The studies he was referring to were published in *Nature*, the world's most prestigious scientific journal, and his comment demonstrates a bizarre lack of understanding of how science works. Apparently, he believes all of us scientists are ganging together to discredit things. It just ain't so.

The question of whether or not neonicotinoid pesticides are negatively impacting agricultural pollinator abundance, diversity and behaviour has certainly focused the minds of researchers and conservationists. It's an issue that has been almost constantly in the news since the earliest suggestions that they were harming pollinators. Those concerns led to temporary EU restrictions on the use of these chemicals, a decision that was partially overturned in the UK (see Chapter 13).

Clearly insecticides kill insects – that, after all, is what they're designed to do. So it should come as no surprise to find that, at certain doses, neonicotinoids kill bees and other pollinators. These are doses that can be found in farmers' fields, particularly on crops such as oilseed rape. For example, Mickaël Henry and colleagues in a study entitled 'Reconciling laboratory and field assessments of neonicotinoid toxicity to honeybees' showed that although the chemicals are lethal to individual honey bees, the overall impact of the loss of the bees is buffered by the fact that the colonies can simply produce more worker bees to compensate for the losses (Henry *et al.* 2015). This is interesting but needs to be judged in the context of the fact that honey bees are very unusual and atypical compared to most other pollinators, and indeed most other bees. They produce very, very large colonies with a unique social structure (see Chapter 2), and so this compensation might be expected. Clearly, they are not a good model on which to test the general impact of pesticides, though until recently they were the only pollinator to be so assessed.

Less well appreciated is the fact that even at much lower doses these pesticides have negative effects on bees. That's because they change the behaviour of the insects, specifically their ability to learn and remember, making bees less effective at foraging for nectar and pollen, and thereby also less effective as pollinators (Wright *et al.* 2015). Subtle behavioural changes such as these are not looked at in standard toxicological safety assessments for pesticides, which are mainly focused on whether or not the chemicals kill non-target animals, and at what dosage. But for plant–pollinator interactions (including agricultural pollination) such changes in pollinator behaviour could be crucial. As Irish ecologist Dara Stanley and colleagues showed, we now have evidence that the sublethal effects on pollinator behaviour could translate into an effect on pollination of apple crops (Stanley *et al.* 2015a, 2015b). Using artificial bumblebee colonies and caged apple trees, Dara and colleagues carried out an experiment in which they tested the effect of two different levels of exposure to a neonicotinoid pesticide on pollinator behaviour and subsequent pollination services to the fruit trees. They found a clear effect of exposure to the higher level of pesticide, resulting in a change in bee behaviour and a reduction in apple quality. This in turn could translate into economic losses for farmers.

All of this focus on pesticides and bees has overshadowed, to some extent, the fact that other pollinating insects might also be affected by them. One of the few studies to address this is by Andre Gilburn and colleagues (2015), who asked 'Are neonicotinoid pesticides driving declines of widespread butterflies?' The study showed that between 2000 and 2009 there was a 58% decline in butterfly abundance on farmed land in the UK despite a doubling of spending on conservation over the same period. Much of this spending was on agri-environment schemes on that very same farmed land. Using a statistical modelling approach, the authors concluded that the introduction of neonicotinoid pesticides in the mid-1990s is strongly implicated as a likely driver of those declines.

Other studies have also implicated neonicotinoid pesticides in pollinator declines in Britain, for example Woodcock *et al.* (2016), and as of mid-2020 the EU partial ban on neonicotinoids is still in place. Worryingly, however, a new generation of pesticides (sulfoximine-based chemicals) that are seen as a replacement for neonicotinoids have recently been shown to be just as harmful to bumblebees as the neonicotinoids (Siviter *et al.* 2018). This does not seem like a step forward for pollinator conservation in agricultural systems.

The other issue that has come to the fore is that of bee diseases. It's long been known that honey bees are affected by various viruses and bacteria, but until recently these were not a cause for wide-scale concern. Like most diseases, they could be dealt with at a local level using isolation of hives and appropriate treatments. However, the large-scale trade and movement of honey bee colonies for pollination of mass-flowering crops has now been implicated in the spread of diseases such as deformed wing virus (DWV), which is carried by varroa mites (Wilfert *et al.* 2016). This emphasises what many have been saying, that loss of honey bee colonies in some parts of the world is largely a veterinary and husbandry problem, rather than a conservation issue *per se* (Ollerton *et al.* 2012).

DWV is only one of more than twenty viruses known to be carried by honey bees. Some of these viruses have been shown to jump the 'species gap' and be passed on to bees that were not previously known to be affected by them. For example, DWV and slow bee paralysis virus (SBPV – originally thought to be restricted to honey bees) are both known to be affecting bumblebee populations (Manley *et al.* 2017, Yañez *et al.* 2020). Indeed, bumblebees foraging near to honey bee apiaries are more likely to be infected with a range of bee diseases, presumably picked up from the honey bees. The route of transmission is currently unclear, but it could be from flowers that both types of bees have visited. Clearly the increasing reliance on honey bees for crop pollination in parts of the world where they are not native is a cause for concern. Likewise, the commercial introduction of non-native bumblebees and solitary bees for crop pollination has been implicated in the spread of diseases to native species (Ravoet *et al.* 2014).

In conclusion

In this chapter I have summarised our current understanding of how pollinators are faring in Britain and further afield, though you would be forgiven for still being confused as to whether there really is a 'pollinator crisis'. Don't worry, all of us are: the situation is complex and the information required for understanding changes to pollinator abundance and diversity is not as robust as we would like. Hopefully the recently launched Pollinator Monitoring Scheme in the UK, which aims to undertake repeated, systematic sampling at the same sites over time, will provide better data in the future, as will those proposed for the EU and for other countries (see Chapter 13). What is certain, however, is that we are not (yet) at the point where we need to be concerned about an 'insect apocalypse', as it's been termed in social and some print media. As Saunders *et al.* (2020) have pointed out, this 'narrative is fuelled by a limited number of studies that are restricted geographically ... and taxonomically'. Indeed, in some relatively pristine environments, abundance and diversity of insect pollinators has not changed over 20 years (Herrera 2019). Most long-term studies of insect decline have been conducted in habitats that have been highly modified by humans, and there is a discrepancy between terrestrial species, which have declined in abundance since the 1960s, and freshwater species, which have increased (van Klink *et al.* 2020).

A question that I frequently get asked is whether loss of pollinator diversity and abundance is having a measurable impact on the pollination of wild and crop plants. Certainly there are examples of wild plants which have reduced reproduction because they have lost their pollinators, as I discuss in the next chapter. But in agriculture, up until a decade ago there was little evidence that availability of pollinators was limiting agricultural yields (Aizen *et al.* 2008). Since that review by Marcelo Aizen and colleagues, some new studies have addressed this question and the results are both worrying and show us how we can improve the situation. It is thought that apple production in the UK could be pollinator-limited and that more sensitive orchard management, such as planting wild flowers that provide nectar and pollen before and after the apples have flowered, could improve yields (Garratt *et al.* 2014). More recently Reilly *et al.* (2020) have shown that of the seven crops that they studied in the USA, five have their yields limited by lack of pollinators. Significantly, even in areas of highly intensive farming, wild bees provided as much pollination service as honey bees. So again, habitat restoration and creation could certainly help to support farmers. More research like this is needed to fully understand how changes in pollinator communities are affecting agriculture, not to mention the almost 90% of wild plant species that need their services. But as a general precautionary principle, maintaining pollinator abundance and diversity should be a priority: we would be foolish to wait until things get worse.

One of the other (many) unanswered questions in my mind is: why are some pollinator species doing so much better than others? Are some species winners because we have forced others to become losers? For example, consider the six or seven British bumblebee species that are doing relatively well: have they spread and become more successful because we have extirpated populations of other bumblebees that would have competed with them? Or could we have predicted that these would become successful from a knowledge of their natural history? Having such 'back-up pollinators' emphasises the importance of maintaining pollinator biodiversity as well as abundance: if we can't predict what we're likely to lose in the future, a diversity of species can act as an ecological insurance.

With this in mind it's interesting that, over recent years, some new pollinators have arrived in the British Isles, and are changing their distribution in other parts of the world. We're going to discover them, and consider why they have been so successful, in the next chapter.

Chapter 11

New bees on the block

With alliterative apologies to the other types of flower visitors discussed in the chapter.

The previous chapter has a rather pessimistic tone, for which I make no apologies: pollinators are important and their conservation and support should be a priority both for agricultural and ecological reasons. The first part of this chapter will be more optimistic, because it deals with species' successes, and the (arguably) natural enrichment of regional biodiversity. I'm going to focus on Britain because it's the country that I know best, and because it has some impressive long-term data sets with which to track such changes. But the general principles of what I have to say apply to all areas of the planet – we just don't have the data to show the changes that have occurred in most other places.

We often think of species as being fixed in their distribution, always tending to occur in the same places. When a plant or animal is found in an area where it did not previously occur, we are likely to assume that this has happened because of human agency, either by deliberate or by accidental movement of that species. This is patently incorrect. Organisms must be capable of shifting their distributions, often dramatically – otherwise, how would species such as the ancestors of Darwin's finches have arrived on remote oceanic islands long before people came on the scene?

Birds provide a number of other striking examples of less dramatic (but nonetheless important) range shifts in what are now common British species. For example, the collared dove (*Streptopelia decaocto*) was hardly known in the UK prior to the 1950s (first breeding record was in 1955) but is now one of our commonest garden birds. Likewise, species such as Cetti's warbler (*Cettia cetti* – first breeding record 1972) and little egret (*Egretta garzetta* – first breeding in 1996) demonstrate that this is what wildlife does: over different time scales (seasonal to decades to hundreds or thousands of years) it shifts and changes its distribution. The earliest record of a little egret in Britain was almost 200 years ago, in East Yorkshire in 1826. However, since breeding was first recorded in Dorset in the 1990s it has gone from being an uncommon bird whose rare arrival would have generated a flurry of local twitching, to a species that now hardly gets a mention on birding websites, we are so familiar with it. Not so the great white egret (*Ardea alba*), which still raises some excitement when it appears, though less than it did a decade ago – it too has become a British breeding species. Although it was also recorded as early as 1821 in Britain, great white egrets only began to breed in Britain in 2012, and there is considerable anticipation that it will follow the little egret in expanding its UK population. Other bird species will undoubtedly follow in the future to take up permanent residence on our shores.

But why am I focusing on (non-pollinating) birds in a book about pollinators and pollination? The simple reason is that, because there has traditionally been greater public interest in the natural history of birds, we have long and detailed data sets with which to make comparisons, giving us a better understanding of how species expand and contract in their distributions. Although we know much less about pollinating insects, there's no reason to suppose that the ranges of such species cannot change in just the same way as those of birds. Indeed, some pollinators are extremely good long-distance fliers with seasonal migrations similar to birds. Examples in Britain and continental Europe include the painted lady butterfly (*Vanessa cardui*) and the marmalade hoverfly (*Episyrphus balteatus*), which fly north from the Mediterranean and North Africa, while in the Americas the monarch butterfly (*Danaus plexippus*) is well known for its journey from Mexico to Canada and back (though multiple generations are involved).

If species broaden their distribution naturally, then the same must be true of the opposite: some range contractions and local extinctions must be natural. Indeed extinctions of whole species occur naturally; it's just that, at the moment, humans have considerably increased the rate of those extinctions. Some (though certainly not all) of the species that went extinct in Britain since the mid-nineteenth century probably did so naturally, without any human influence. As a mirror image of this, we now have some new pollinators on the British list, and the rate at which they are arriving seems to be increasing.

Recent additions to the pollinating fauna of the British Isles

Historically we know that insects new to the British Isles have long been arriving, but until recently our records have been patchy. For example, the bilberry bumblebee (*Bombus monticola*) arrived in Ireland only in the 1970s, but we don't know when it first came to Britain. There are certainly records from the mid-nineteenth century, but how long before that did it fly over the English Channel or the North Sea? In contrast, we do know that the hornet hoverfly (*Volucella zonaria*) arrived in Britain in the 1940s as part of a regional spread that saw it moving eastwards and northwards. It's our largest hoverfly and very distinctive (see Chapter 9), so we can be sure that the species was not present at least as far back as the late eighteenth century when serious interest in British insects was just beginning. But was it here earlier than that? In medieval or prehistoric times? We will probably never know, though huge advances have been made in recovering environmental DNA (eDNA) from ancient sediments (Thomsen and Willersley 2015) so I wouldn't rule it out completely. But what we do know is that, at the moment, we are in the midst of a very interesting period of change in British pollinators.

More intensive monitoring, especially by interested (and often highly skilled) amateur entomologists, has shown just how dynamic a regional invertebrate fauna can be. Since the turn of the millennium the British Isles has seen the addition of at least seventeen flower-visiting bees, flies and wasps to its list of species. Table 11.1 shows the species concerned and their status as currently understood. At least some of these seem to be natural range shifts rather than artificial introductions (though anthropogenic activities such as urbanisation and climate change may certainly have played a role). Others are range shifts by species that are introduced in continental Europe and are finding their way here, such as the Asian hornet and the grass-carrying wasp.

One of these species (the tree bumblebee) is shown in Figure 11.1 and I'm going to discuss its fascinating life history in more detail later in the chapter. However, all have interesting ecological stories to tell, though the history of their discovery is also often fascinating. The anthracite bee-fly is a parasite of solitary bees in the genera *Osmia* and *Megachile* (BWARS 2019). It's especially

TABLE 11.1 New species of bees, wasps and flies that have turned up in Britain since 2000.

Species	Date of first British record
Tree bumblebee (*Bombus hypnorum*)	2001
Ivy bee (*Colletes hederae*)	2001
Fringed furrow bee (*Lasioglossum sexstrigatum*)	2006
Small-headed resin bee (*Heriades rubicola*)	2006
European orchard bee (*Osmia cornuta*)	2014
Grey-backed mining bee† (*Andrena vaga*)	2014
Asian hornet* (*Vespa velutina*)	2016
Viper's bugloss mason bee (*Hoplitis adunca*)	2016
Grass-carrying wasp* (*Isodontia mexicana*)	2016
Species of spider wasp (*Agenioideus apicalis*)	2016
Large bear-clawed nomad bee (*Nomada alboguttata*)	2016
Variable nomad bee (*Nomada zonata*)	2016
Anthracite bee-fly (*Anthrax anthrax*)	2017
Hawk's-beard nomad bee (*Nomada facilis*)	2017 (1802)
Species of parasitoid wasp (*Leucospis* sp.)	2018
Dusky-horned nomad bee (*Nomada bifasciata*)	2018
Species of paper wasp (*Polistes nimpha*)	2019

* Indicates species that are not native to Europe and have undoubtedly been transported by human agency. † Formerly extinct in Britain (see text). Sources: Goulson and Williams (2001), Kirby-Lambert (2016), Notton (2016, 2018), Cross and Notton (2017), Notton and Norman (2017), Falk and Earwaker (2019), www.bwars.com, social media, BBC News website.

FIGURE 11.1 A tree bumblebee (*Bombus hypnorum*) resting on the author's hand in his Northampton garden not long after it was recorded in that county for the first time.

associated with the European orchard bee, which itself may have been introduced, as cocoons, deliberately for its pollination services; the parasite could therefore have come with it (Stuart Roberts, personal communication 2020). Likewise, the *Leucospis* wasp is a parasitoid of *Megachile* bees and was first spotted by accident when some footage was uploaded to the Bees, Wasps and Ants Recording Society (BWARS) Facebook group in summer 2018. This is not just a new species but a whole new *family* of Hymenoptera (Leucospidae) that was not previously known from Britain. The small-headed resin bee, as the name suggests, collects plant resin to construct its nest; it may have been accidentally imported with thatching straw from Hungary (Stuart Roberts, personal communication 2020). The history of another species is older but more complex: the hawk's-beard nomad bee was first recognised as a British species in 2017, but specimens had actually been collected in the nineteenth century, though misidentified.

The grey-backed mining bee is a species that we included in our analysis of extinct bees and wasps (Ollerton *et al.* 2014 – see Chapter 10), and it was discovered to have (we assume) recolonised Britain while we were revising the paper following referees' comments. We had to hastily add a note to the appendix pointing out that the species had returned after an absence of almost seventy years. However, it's a great example of the natural flux of biodiversity that needs to be appreciated if we are to disentangle anthropogenic from natural range expansions and species declines.

In the following sections I'd like to explore two of the new bee species in more detail and explain a little about their ecology and their significance as 'new bees on the block'. I'll begin with the tree bumblebee and then move on to the ivy bee. I've chosen these two species because they have been intensively monitored by BWARS, and the timing of their arrival coincides with the ability of entomologists to record and share their observations via the internet. In addition, they have generated a lot of attention among the public and the media, plus (in the case of the tree bumblebee) I've made some observations that I think pose interesting questions about why this species has been so successful.

The rise and rise of the tree bumblebee

In July 2001 Dave Goulson was in the village of Landford in Wiltshire, to the north of the New Forest, surveying bumblebees, when he collected a male foraging on bramble. It was one of those moments that every naturalist encounters, the feeling of 'here's something a bit different, something I don't recognise, I need to find out what this is'. Dave's instincts turned out to be correct – comparison with material in the Natural History Museum in London showed that he had collected the first British specimen of the tree bumblebee (*Bombus hypnorum* – Figure 11.1). This is a species that is common on the continent, in Scandinavia, and across Asia, but which had previously never been found in these islands. Dave subsequently wrote up the findings with Paul Williams of the Natural History Museum, and it was published in the *British Journal of Entomology and Natural History* later that year (Goulson and Williams 2001).

In their report Dave and Paul speculated that the fresh appearance of this male suggested that it was from a colony that was already established in the area, rather than having flown across the Channel. However, they also wondered whether the species might have originated from a nest deliberately imported for pollination of glasshouse crops, though in retrospect this is unlikely as it's not a species that is commercially bred. At the end of their short paper Dave and Paul noted that 'they would be very interested to learn of any further sightings of this species'. Little did they realise how prophetic their words were to be.

Records of tree bumblebees were subsequently collated by BWARS, who set up a specific mapping scheme to try to understand the spread of this species. For the first few years of the new millennium there were rather few reports, just a scattering of observations in the southeast of the country. Then suddenly, from 2006 onwards, the records started to flood in as the species moved north and eastwards, while at the same time consolidating its position in the southeast of the country. In 2007 the first record from the borders of Scotland was noted. In 2008 it was recorded in East Anglia and the Welsh borders. By 2010 it had spread westwards in earnest and was consolidated in Wales and the West Country, down into Cornwall (van der Wal *et al.* 2015).

So it went on, with more and more records appearing on the BWARS maps; in 2013 it was found well into Scotland for the first time, in 2015 it had travelled over to the Isle of Man, and by 2017 it was in Ireland. As you can see in the

National Biodiversity Network (NBN) account for the species (species.nbnatlas.org/species/NHMSYS0000875564), tree bumblebees are now found across most of the British Isles except for upland areas and northern and western Scottish islands. This seems to be part of a general westwards range expansion for the species because of increased urbanisation in Europe and Scandinavia from the 1980s onwards (Rasmont 1989), rather than as a result of climate change as some have suggested. The species even appeared in Iceland in around 2008, probably as an accidental introduction (Kratochwil 2016, Prŷs-Jones *et al.* 2016).

Despite considerable interest in the species and its spread, there's been little published to date on the local natural history of tree bumblebees and their ecological fit into existing pollinator assemblages in the British Isles. One exception to this is the work of Liam Crowther and colleagues at the University of East Anglia. Liam studied the tree bumblebee for his PhD and has published a couple of really interesting papers (Crowther *et al.* 2014, 2019), with more in the pipeline. Liam's work suggests that the species has spread fast across Britain because queens of *B. hypnorum* can potentially disperse up to 39 kilometres from their place of origin, though distances of less than 5 kilometres are much more frequent (Liam Crowther, personal communication). This is not unusual for bumblebees; I've seen them follow ferries between mainland and islands across distances of up to 8 kilometres, while Mikkola (1984) reported similar behaviour observed from the Estonia to Helsinki ferry – a stretch of water 80 kilometres wide. But even this is nothing compared to what young ecologist Will Hawkes documented in 2016, that queen bumblebees may engage in mass migrations (involving thousands of bees) across the North Sea from England to the Netherlands, a distance of 165 kilometres (Hawkes 2016).

As well as Liam, a few other biologists have published research on the tree bumblebee that may give an insight into why it has been so successful. Dave Goulson and his team looked at nest survival and found that in comparison with other species, *B. hypnorum* had the highest proportion (96%) of nests that went on to produce reproductive females (gynes) (Goulson *et al.* 2018). This is even though queens from the British population suffer a greater incidence than other bumblebee species of the high-impact parasites *Sphaerularia bombi* (a nematode worm) and *Apicystis bombi* (a protozoan) (Jones and Brown 2014).

The lack of published information on tree bumblebees to some extent reflects the global situation for bumblebees more generally, even in Europe with its long tradition of natural history observations. Despite the abundance of some species, and public and scientific interest in them, there are still many aspects of the natural history of *Bombus* species that are unknown or poorly understood, particularly in relation to their nesting behaviour. It turns out that the tree bumblebee has some very interesting peculiarities in this regard, which may at least partly explain why it has become so successful, as I'll outline in the next section.

The tree bumblebee as an irruptive species

Lying in bed early one sunny Sunday morning in June 2012, I was intrigued to see a steady stream of large bees flying past the bedroom window, heading up towards the eaves of our house in Northampton. Sticking my head out of the window I was pleased to find that they were heading into and out of a hole just below the gutter: we had a tree bumblebee nest in the house! As I showed in Chapter 9, we live in an area of Victorian terraced housing. Most of the houses date from the 1880s, with some later in-filling, and are typically of brick construction with original slate or later ceramic tile roofs. The traditional mode of roof construction leaves a small gap below the eaves to permit air flow through the roof space. This gap allows a range of animals to colonise such houses, including various species of bats, common swift (*Apus apus*) and house sparrow (*Passer domesticus*), as well as the tree bumblebee. As the name suggests, tree bumblebees naturally nest in holes in trees. So it's not surprising that they favour aerial nesting sites such as cavities in the eaves of houses and between the rows of bricks, garden nest boxes, and so forth. They will nest at ground level, but it's more unusual.

Over the next few weeks, whenever I was sitting in the garden and taking a break from planting and cultivating, I would time the bees as they passed in and out of the colony. This was a busy nest: at their peak, bees were entering and leaving at a rate of three bees per minute, or one every twenty seconds. This reflects the size of tree bumblebee nests, which can be quite large, housing more than 150 workers at a time.

The nest that I observed in the eaves of the house in 2012 diminished over the next few weeks, and by mid-July was abandoned as the resident queen ceased her reproductive activities and the new batch of queens and males dispersed. To my surprise this nest site was subsequently reused in 2013 and 2014 (though not in 2015 or any years since). The reuse of nest sites by bumblebees has been previously observed by others but has not been formally studied, so it is unclear whether the sites are recolonised by mated queens produced by the colony in the previous year, or by unrelated queens attracted by the scent of an old nest.

These initial observations got me interested in how the bee was performing in my local area, and I started keeping records of when I spotted their nests. Summer 2014 proved to be a quite extraordinary year for the tree bumblebee, with lots of people reporting that they were nesting in their roofs and in bird boxes (Morelle 2014). I was able to document this in my local patch because of an oddity of tree bumblebee behaviour. Locating the nests of most bumblebees can be frustratingly difficult, as they are often hard to spot unless one happens to be in the right place at the right time to see the bees coming and going from the colony. In comparison with most species, however, nests of the tree bumblebee are relatively easy to locate because, later in the season, the entrance is typically patrolled by large aggregations of male bees waiting for virgin queens to emerge so that they can mate with them. This behaviour, coupled with the synanthropic (i.e. living close to human habitation) habit, means that it's not uncommon to find pairs of mating bees in gardens, something that is rarely observed in most other species.

During the summer of 2014 I saw a lot of these male aggregations in and around the streets where we live. In total I spotted 24 tree bumblebee nests in an area 300 × 400 metres, or 12 hectares (Figure 11.2).

This must be considered a minimum number in the area, as it was usually impossible to observe the sides of the houses that do not face the street, other than those I could see from our back garden, or the garden bird boxes which are also often used by this species. If we conservatively estimate that 50% of the area available for bee nesting was not surveyed, this suggests that the nest density for this species at this site in 2014 was between 2 and 4 nests per hectare. How does this compare with nest densities recorded for other species in Britain? In Figure 11.3 I have collated all of the available published information that I could locate, and it turns out that even my lower estimate of nest density is greater than the highest value ever recorded for any other bumblebee species in the UK. At the same time Liam Crowther was also analysing nest densities in Norwich using

FIGURE 11.2 Aerial view of the part of Northampton in which we live. The blue lines mark the boundary of the area of Victorian housing that I've surveyed for tree bumblebee nests. Yellow markers indicate nests observed in the eaves of houses in 2014. The red dot is the nest site that was used in 2012, 2013 and 2014.

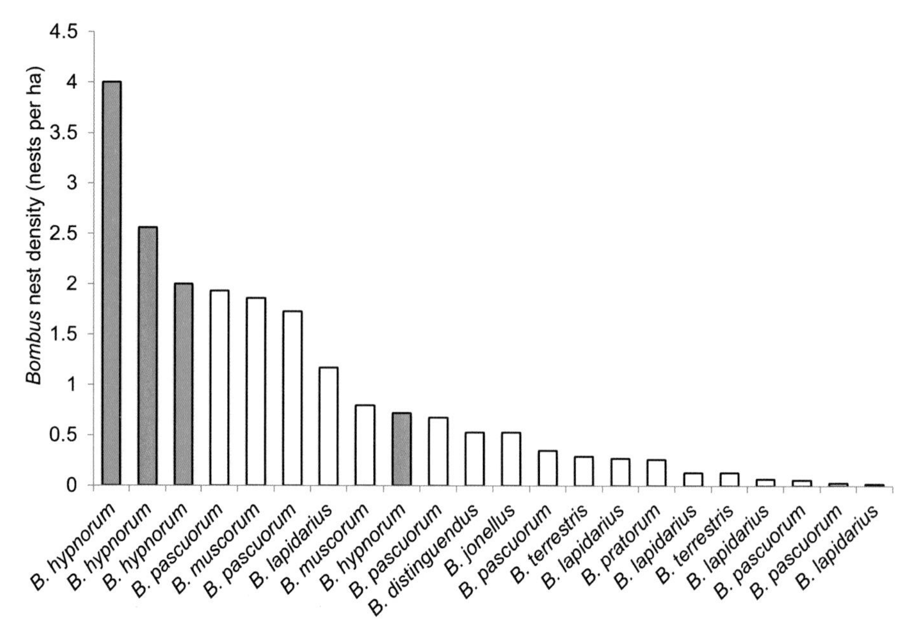

FIGURE 11.3 Nest densities for some common species of British bumblebees taken from the published literature and compared with my upper and lower estimates of tree bumblebee nest density in Northampton in 2014 (green bars). The purple bars are Liam Crowther's estimates for the same species in Norwich in 2014 (higher figure) and 2015 (lower figure).

molecular genetic methods, and his estimate for 2014 is between my upper and lower estimate for Northampton.

It appears that 2014 was an anomalous year. In 2012 and 2013 I noticed only the nests in the roof of our house, and between 2015 and 2017 full surveys of the whole area failed to locate any nests. I'm sure they were there, as I was seeing a lot of tree bumblebees in our garden, but the nests were just not as numerous, and bees may have been coming in from outside of my survey area. In 2018 I spotted one nest; in 2019 I was out of the country during the main period of male tree bumblebee flight activity and so I can't say for certain what the nest densities were like. However, it's possible that there were unusually high nest densities in some parts of Britain that year. I say that because a blog post that I had written in 2016 called 'How to deal with bumblebees in your roof' suddenly became extremely popular. In 2018 that post had 6,398 views, up from 717 in 2017. In 2019 it received over 15,000 views. Comments on social media also suggested that wildlife organisations were receiving unusually high numbers of enquiries that summer.

These observations, and Liam's research, show that high nest density in urban areas is an exceptional phenomenon that does not happen every year. I suspect that it's related to seasonal weather conditions, which can be assessed using the Met Office's summaries (www.metoffice.gov.uk/research/climate/maps-and-data/summaries/index). The winter of 2013/14 experienced temperatures well above the long-term average, with a notable absence of frosts. The spring of 2014 also had fewer air frosts than usual and produced average temperatures of 9.0°C (1.3°C above the long-term spring average). In contrast, the 2014/15 winter was very close to the long-term average for that season, while spring 2015 was 'fairly unexceptional overall' compared to long-term data. Since then Britain did not experience the 2013/14 combination of very mild weather in both winter and spring, until 2018/19, when winter was 'milder than average' and the following spring was 'slightly warmer than average'. This combination seems to be the key to irruptive breeding behaviour in tree bumblebees, because it results in lower than usual mortality of overwintering queens and greater flower resources earlier in the season, leading to more colonies being founded the following spring. This is all anecdotal and correlative, however, and needs to be tested experimentally.

Irruptive species are those that in some years appear in very large numbers, compared to typical years in which they are much less abundant. There are lots of examples in the animal world, including one of my very favourite birds, the Bohemian waxwing (*Bombycilla garrulus*) as well as some species of locust (Orthoptera). However, I don't know of any published examples of bee species that show this kind of irruptive breeding behaviour. I'm sure they exist; they've just not been documented.

Mean minimum and maximum spring temperatures in Britain are broadly rising as a result of climate change, and therefore such exceptional nest densities of tree bumblebees may become more common in the future, if my weather hypothesis is correct. What we don't yet know is what effect, if any, this species will have on British plant–pollinator communities and agricultural crops.

The ecological and human impact of tree bumblebees

Not only is *Bombus hypnorum* spreading rapidly through the British Isles, but it is apparently an irruptive species capable of achieving much greater population densities than is typical for bumblebees in this country. It can also produce a small second generation in some years, which has no doubt also helped its spread. At the moment we do not know what effects tree bumblebees may have on more long-established *Bombus* species in Britain, through both competition and transmission of disease, though at the moment there is no serious concern about the latter (Jones and Brown 2014). There was some worry that the bees may exclude birds from garden nest boxes, but that seems to be rather rare (Broughton *et al.* 2015). The tree bumblebee is here to stay and certainly deserves greater study in the future, but not only for British ecology. The species is found widely across Eurasia and the far north and east of Siberia (Goulson and Williams 2001), from where it's only a matter of hopping between islands before it can arrive in Alaska and subsequently spread across North America. Climate change is bringing increasingly warm weather to that part of the world and so one could predict that this species will expand its range into the New World at some time in the future. We do not know what the implications might be for North American native bees – which is why studying the impact of its spread in the British Isles and northern Europe is so important.

As well as the potential ecological impacts of new bees on the block, there can be implications for people too. The tree bumblebee is now one of the most common bee species to be observed in gardens in the south of Britain and, certainly in our garden, contributes to urban food production by providing pollination services to rosaceous fruit such as greengages, raspberries and blackberries, among others. However, over the past few years I've had increasing numbers of enquiries from colleagues at the University of Northampton asking for advice on what to do about colonies of tree bumblebees that have set up home in their roofs. As I noted above, that blog post that I wrote on this topic in 2016 is now my all-time most viewed. Because of their association with human settlements, tree bumblebees are significant pollinators of garden produce, as I've mentioned. But having a bee nest in your home is, for many people, a real concern. The colony in our house was quite a large one, but it's unlikely that there were more than 150 bees in it, possibly far fewer. However, the behaviour of the males patrolling the front of the nest can make the colony seem larger than it actually is, and that is often why people notice them and get concerned. A friend of mine who had a colony in a nest box positioned quite low in the garden was worried about the bees stinging his young daughter. He didn't want to kill them so he waited until late at night when most of the bees would be back in the colony, and sealed up the hole in the nest box with duct tape. He then took the nest box to a woodland some distance away, nailed it to a tree, quickly removed the tape, and scarpered. The bees were not impressed, but at least the colony was not killed.

The tree bumblebees in our roof never caused any damage, and like most bumblebees they are unlikely to be aggressive unless you stick your fingers in the nest. The colony dies over the winter and the newly mated females fly off quite early in the summer; by late August or September (perhaps earlier if the weather is warm) the bees should have gone (unless they have started a second colony cycle, which happens in some years). At that time one could seal the entrance to the roof space to ensure that they don't return next year. The first response of some people would be to call a pest controller, who would kill the colony. But my advice is to let them be and take pride in your own bee colony – they are very discerning and don't nest just anywhere!

Of course, if you or your family have a particular sensitivity to bee stings you may need to think carefully about this advice, but in my experience bumblebees are only aggressive if they feel directly threatened. In over thirty years of field work focused on bees and other pollinators, I've only ever been stung a few times, and mainly by honey bees.

The ivy bee – another recent colonist

As you can see from Table 11.1, the other pollinator that appeared for the first time in 2001 was the ivy bee (*Colletes hederae* – Figure 11.4), so named because its main pollen source is common ivy (*Hedera helix*). Of course, ivy is a very late-blooming species in Britain (see Chapter 6) and consequently the ivy bee emerges to found its nests very late in the season, appearing typically in late August to early September, long after most other bees have ceased reproduction.

As a ground-nesting solitary bee it is very different to the tree bumblebee in its ecology, but its life history is no less interesting, though even less well studied. In fact the species was only described as new to science in the 1990s (Schmidt and Westrich 1993); prior to that it had been mixed up with the sea aster mining bee (*Colletes halophilus*) a coastal species (hence *halophilus* – 'salt-loving') which looks very similar but is on the wing rather earlier and provisions its nest with the pollen of sea aster (*Aster tripolium*) and other members of the daisy family. In Europe the ivy bee has a mainly southern European distribution, and its spread further north may well have been facilitated by climate change. By way of contrast, the sea aster mining bee is also likely to be affected by climate change, but in this case negatively as coastal areas are flooded by rising sea levels (Buglife 2019). As I noted in Chapter 6, there will be winners and losers as the effects of climate change play out.

Once again, BWARS was instrumental in setting up a monitoring project for the ivy bee, and from its first recording in Dorset it's been possible to track its advance across the country (species.nbnatlas.org/species/NBNSYS0100002538). The spread of this species has been nowhere near as rapid or as extensive as that of the tree bumblebee. That may be because its exact nesting requirements are less

FIGURE 11.4 The ivy bee (*Colletes hederae*) foraging on ivy flowers (*Hedera helix*). Photo by Steven Falk.

common: as well as plenty of ivy it typically requires light soil with a south-facing aspect. Nesting aggregations can be very dense, with thousands of nests in an area, but suitable areas are at a premium. In addition it has only one generation per year.

Since learning of its arrival in Britain I had been looking out for the ivy bee in likely sites in Northamptonshire, where we commonly find pockets of sandy soil derived from the underlying Jurassic sandstone. My aim of having the first record for the county was thwarted in 2016 by the arrival of an energetic young graduate ecologist, Fergus Chadwick, who worked with me for a couple of months to gain some research experience before going on to do a PhD. One morning not long after he arrived he casually informed me that he'd seen what he was sure were some ivy bees. Even more galling was that they were on a busy main road only a few minutes' walk from my office.

As discussed previously in the book, ivy is one of our most ecologically valuable and underrated plants. It's a hugely important nectar source to a wide range of overwintering bees, flies, beetles, hoverflies, wasps and other insects. In addition, its berries are a vital food source for many fruit-eating birds. As a pollinator of ivy, the ivy bee is probably contributing to the role of this plant in supporting bird populations. However, we know even less about the ecology of this species than we do about the tree bumblebee, and more research should be a priority (Bischoff *et al.* 2004).

Negative (and positive) effects of invasive species

The two case studies I have presented above, of tree bumblebee and ivy bee, are relatively benign examples of species moving into new areas and fitting into the local ecology without (apparently) any negative effects on the species that already exist in that region. If only all such examples were as benign. Unfortunately, there are many other examples of invasive species that have detrimental effects on the local flora and fauna. Some of these species are pollinators, demonstrating (once again) that popular conceptions of bees and other pollinators as kindly animals helping plants to reproduce do not reflect ecological reality.

Pollinator diversity has been artificially enhanced in many areas of the world by the establishment of introduced species. This has occurred across a wide taxonomic range from hoverflies to wasps to birds. However, it is bees that have enjoyed the greatest anthropogenic spread (Goulson 2003a, Russo 2016), either purposefully for agricultural pollination, or accidentally. Oceanic islands are especially sensitive to introduced species (Olesen *et al.* 2002), and there are examples where the introduced species outnumber the native ones. For example, in the Azores a recent checklist shows that as many as 15 out of the 19 recorded bee species were almost certainly introduced by people (Weissmann *et al.* 2017). The Azores are exceptional in this regard, but elsewhere in the world the number of introduced pollinators is certainly increasing.

The effects of such introductions on the local ecology vary from harmful to neutral to positive, and where they lie on that spectrum will very much depend on the species involved, when and where it was introduced, and its abundance in a region (Russo 2016). By far the most abundant and impactful bee to have been introduced, anywhere in the world, is the western honey bee (*Apis mellifera*). Vast numbers of hives are maintained to provide agricultural pollination services far beyond its native range. Studies have demonstrated that introduced honey bees compete for floral resources with native pollinators, though this is dependent upon hive density and landscape context, and especially whether or not an area has a large proportion of natural habitat in which pollinators can forage (Herbertsson *et al.* 2016). At low density, honey bees become integrated into local flower visitation networks (Chapter 4) without obvious negative effects, though any effects they may have are difficult to quantify (Watts *et al.* 2016). However, they can visit a very high proportion of plants within a region and clearly exploit any available floral resources they can access. Recent work by Carlos Herrera has shown that over the past fifty years, honey bees have replaced other bees as pollinators in the Mediterranean basin as hive numbers have increased in the region (Herrera 2020). Although *Apis mellifera* is native to that part of the world, it does not naturally occur in such large numbers, and the potential ecological impact of these findings is of concern. In particular it could result in smaller, less aggressive bees being locally extirpated, leading to a reduction in pollination services to wild plants and crops.

The buff-tailed bumblebee (*Bombus terrestris*) is another species that was introduced as an agricultural pollinator into areas far beyond its natural range of Europe, Scandinavia and North Africa. It is now considered to be invasive in Japan, Tasmania, New Zealand, the Middle East and South America, among other regions (Dafni *et al.* 2010). Most significantly, this species and a second, the ruderal bumblebee (*B. ruderatus*), have been implicated in the extirpation of one of the world's largest bumblebees, *B. dahlbomii*, across much of its former range in southern South America (Morales *et al.* 2013). However, as well as direct competition from introduced species, the other factor that is of concern to conservationists is that diseases may be passed on to native species, against which they have no natural defences (e.g. Arbetman *et al.* 2013).

All of this sounds very gloomy, and I had promised at the start of this chapter to be a bit more optimistic – and indeed there are some positive aspects to pollinator introductions. A number of studies have shown that introduced species are facilitating reproduction in endangered plants where the original pollinators are not doing the job because they are now rare or have gone extinct. One of the classic pieces of research in this area is from the 1980s when Paul Alan Cox showed that bird pollination of a vine (*Freycinetia arborea* – Pandanaceae) in Hawaii was being carried out by the warbling white-eye (*Zosterops japonica*) following the extinction of its native pollinators (Cox 1983). Working on the same plant genus, Janice Lord in New Zealand demonstrated that the introduced common brushtail possum (*Trichosurus vulpecula*) was now fulfilling the role of two native bats (one rare, the other probably extinct) in pollinating *F. baueriana* (Lord 1991). In more recent times Fox *et al.* (2013) discovered that the most reliable pollinator of the endangered North American western prairie fringed orchid (*Platanthera praeclara*) was a Eurasian moth (the spurge hawkmoth – *Hyles euphorbiae*) that had been introduced to control invasive plants. However, it's unclear whether the introduced moth has also out-competed the native pollinators, so this might not be such a straightforwardly positive example. Indeed, at the moment it appears that the most effective pollinators for rare plants that are facing extinction, such as the critically endangered *Brighamia insignis* (Campanulaceae) from Hawaii, are humans and their dextrous hands (Walsh *et al.* 2019).

The effects of pollinator introductions on local flora and fauna are clearly complex and depend on the specific ecologies of the species involved and where in the world they have been introduced. However, local circumstances can change quite rapidly, and what are at the moment fairly benign introductions may ultimately present us with conservation problems. For this reason it is probably wise to invoke the 'precautionary principle' and limit future introductions, especially where agricultural crop pollination can be done by native species (Hogendoorn *et al.* 2006, Hingston 2007). However, in order for this to be achievable there needs to be suitable habitat in and around farmland that can support those pollinators, which is the topic of Chapter 12.

Understanding the ebbs and flows of biodiversity, both native and introduced, over time, requires data to be collected – and we are fortunate in Britain to have a number of active monitoring schemes that regularly survey different groups of organisms, including pollinators. Some other countries also have, or are setting up, such schemes, as we'll see in Chapter 13. This activity is vital if we are to be able to monitor wildlife and to take action if we see declines or record potentially invasive species – for there are other pollinators from other parts of the world that are also changing their distribution, either naturally or with human aid. For example, the bumblebee *Bombus haematurus* was historically found in the region from northern Serbia to northern Iran. But it has increased its range by about 20% since the 1980s by expanding northwestwards into central Europe (Biella *et al.* 2020b). This is probably a response to climate change, and many other species are likely to do the same in the near future (see Chapter 6). Sustained, long-term monitoring across a wide geographical area that informs political and societal action, including habitat restoration, is required if we are to understand changes in pollinator diversity and their effects on crop and wild plant pollination. These topics are discussed further in Chapters 12 and 13.

Chapter 12

Managing, restoring and connecting habitats

The decline of pollinators, along with a decline in biodiversity more generally, has been linked to habitat destruction, fragmentation and degradation, as we saw in Chapter 10. The notion that we must manage and restore habitats in order to halt and reverse biodiversity loss is therefore ingrained within the consciousness of most conservation professionals, whether academic or hands-on. Indeed, there are journals devoted solely to habitat management and restoration, and every week thousands of volunteer conservationists across the world work with wildlife organisations to clear invasive species, plant trees, put up bird boxes, remove plastic waste and otherwise do their bit for biodiversity. Running counter to all of this is the notion of 'rewilding' landscapes, which is currently seen by many as the future for nature conservation. Rewilding 'aims to restore self-sustaining and complex ecosystems, with interlinked ecological processes that promote and support one another while minimizing or gradually reducing human interventions' (Perino *et al.* 2019). However, in truth rewilding is also management, just on a much larger scale and with only an initial human input. But most wildlife sites in Britain and elsewhere in the developed world, such as nature reserves and urban greenspaces, are (currently) too small to be truly rewilded. In the absence of large herbivores to keep areas of habitat open, and large predators to keep the herbivore populations in check and moving around the landscape, some form of management is almost inevitable.

As well as those journals, whole books have been written about habitat management, and the fine details of how to look after calcareous versus acid grassland, or the different types of deciduous woodland, are complex, variable, and hotly debated. However, from looking at some of this literature and from my own experience of talking and working with conservationists from charities such as the Wildlife Trusts, Buglife and the RSPB, as well as companies with large landholdings, it seems to me that there are some basic principles of managing, restoring and connecting habitats for biodiversity. In this chapter I want to deal with these principles as they relate to pollinators and their plants. Our starting point is straightforward: we must understand the organisms that we are trying to conserve.

First principles

The guiding principle behind any form of habitat management or restoration of a conservation area for a particular species, or group of species, is that one must have a knowledge of the natural history of those species and their ecological requirements. For example, managing an area of woodland for bats without providing adequate roosting sites such as bat boxes, in the absence of hollow trees, is not going to be successful. Likewise managing an area to increase the local population of a plant of open habitats, without ensuring that it is not overshadowed by taller trees and shrubs, will result in the loss of that species from the community. Natural history is therefore key to conservation. As an example, consider the reintroduction of the large blue butterfly (*Phengaris arion*) to Britain after it went extinct in 1979; following a series of failed attempts it was discovered that for part of its life cycle it maintains a critical symbiotic relationship with ants. In the absence of those ants, and its larval food plant *Thymus drucei* (Lamiaceae), no reintroduction was going to be successful (Thomas 1995). Once that part of the ecology of the species was understood, the reintroduction was successful.

Managing and restoring habitats for pollinators is no different to managing and restoring for other species, with an obvious exception: in the preceding chapters

FIGURE 12.1 A 'bug hotel' of the kind that has sprung up in many gardens and parks across Britain to provide overwintering and nesting habitat for invertebrates. This one is in Becket's Park adjacent to the University of Northampton's Waterside Campus, near Northampton town centre, and was built by the Buddies of Becket's, a local volunteer group.

I have emphasised the diversity of natural histories of pollinators, with different species (even within the same broad taxonomic groups) having very different requirements. Faced with this diversity, how can conservationists ever hope to manage or restore sites 'for pollinators'? The short answer is that they cannot, of course, ever fulfil the needs of all the potential pollinators within the regional pool of species. For example, within Northamptonshire, there are around 100 species of bees that we know, or suspect from historical records, live in the county. There's also probably a similar number of species of hoverflies, and about 700 species of butterflies and moths. Some of these (such as the bumblebees *Bombus terrestris* and *B. pascuorum*, or the drone fly *Eristalis tenax*) are very broad generalists that are found foraging in a wide range of habitats, from open grassland to woodland, gardens, river-edge vegetation and so forth. Conservation managers would be surprised if they did not turn up on their sites. Others are restricted to just one particular habitat type, for example deciduous woodland or limestone grassland, and without that sort of vegetation in a vicinity, such species will not occur no matter what kind of management is done or however many bee hotels and bug habitats are constructed (Figure 12.1). Clearly there must be realistic expectations of what species to expect in any conservation area; knowing what the requirements of the various species are can help to realise those expectations.

The requirements of pollinators

One way to think about the requirements of pollinators is to simplify their natural history into three broad categories: the flowers that they exploit, their basic reproductive requirements, and any supplementary resources that they need (Figure 12.2).

In this pollinator requirements triangle, 'flowers' are the most obvious necessity and are the ones that are particularly focused on in agri-environment schemes, garden planting advice and other guidance. The internet is awash with ideas for 'planting for pollinators', and books are beginning to appear that provide advice for farmers and other land owners. Marek Nowakowski and Richard Pywell's book *Habitat Creation and Management for Pollinators* (2016) is especially good in this regard and has useful lists of plant species that will do well in the UK. Clearly the type of species to be planted will vary considerably with geography, but the diversity (number of species) and abundance (how many of each) of nectar- and pollen-providing flowers are important considerations. Diversity means more species will be attracted to a site; there is a positive correlation between plant diversity and pollinator diversity, which means that every additional plant species adds at least one pollinator to a community (Figure 12.3).

A high abundance of flowers means that the habitat can sustain larger populations of those pollinator species. The other elements within this requirement (such as a long flowering phenology and the presence of specialist plants for those pollinators that need them) often go hand-in-hand with diversity: the more species

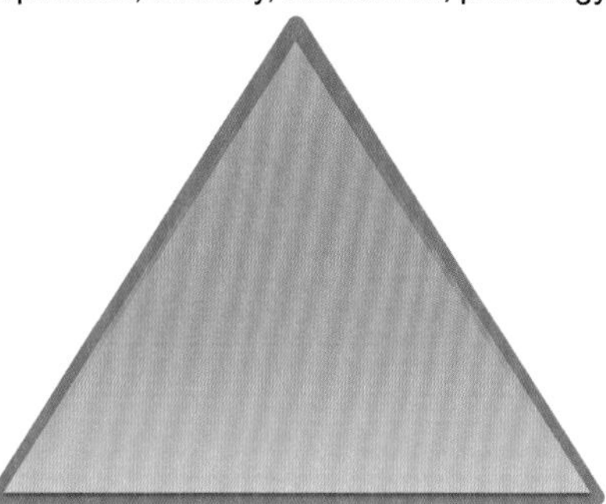

FIGURE 12.2 The *requirements of pollinators* triangle. In order to complete their life cycles, pollinators have broadly three sets of requirements: flowers on which to feed, places in which to reproduce, and supplementary resources that they need for part or all of their life cycles. Successful habitat management and creation, from the pollinator's perspective, needs to provide each of these requirements.

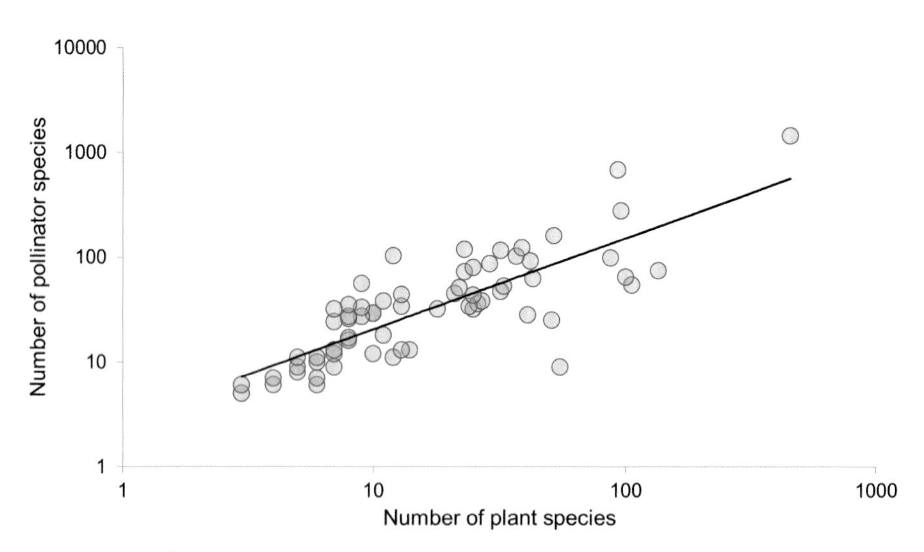

FIGURE 12.3 There is a strong positive correlation between the diversity of plant species in a community and the number of pollinator species. Data are from 65 communities across the globe. Note the log scale on each axis. Data from Ollerton (2017).

there are, the wider the range of flowering times that are represented, and the greater the chances that specialist plants are present. However, if particular species are the focus of conservation efforts, this may need to be taken into account. For example, some bees, especially in the tropics, collect oil and resin from flowers as well as nectar and pollen, and these types of plants would need to be present in a community to support those bees. Likewise, many solitary bees use pollen from just a single species, genus or family of plants – for example the ivy bee (*Colletes hederae*) that we discussed in the previous chapter.

Working clockwise around Figure 12.2, we come to 'reproduction', a category that includes nesting sites for pollinating birds and bees, host plants for pollinators with herbivorous larvae such as butterflies and moths, wet areas for some hoverflies that have aquatic larvae, specific microhabitats for other species such as decaying wood (see below), and so forth. The more heterogeneous a habitat, the more species that can be supported.

Finally, we come to the most miscellaneous and arguably problematic requirement, which I am terming 'supplements' – by which I mean resources additional to those from flowers and aspects of the habitat that may be only partly linked to the reproduction apex of the triangle. This includes a huge range of resources such as additional food, for example insect prey of omnivorous wasps or pollinating birds, or carnivorous beetles that visit flowers to catch other insects. It also includes nesting materials such as particular leaves for leafcutter bees, or mud for mason bees; symbionts such as the ants mentioned above for the large blue butterfly; and hibernation sites for many insects, including queen bumblebees and wasps, and some butterflies and hoverflies. This is just part of what could be a very long list of resources that don't fit easily into the 'flowers' or 'reproduction' categories.

Without requirements in all three of these categories being met, a locality will never be able to support a diverse pollinator community. Even a common bumblebee such as *Bombus terrestris* has a particular set of requirements: usually an old rodent nest for establishing its colony, a diversity of flowers on which to feed (including early blooming species such as willows), and an area with loose earth for hibernating queens. This species is wide-ranging and may fulfil its requirements across a very large area. Other species, however, are much more site-constrained and may not move far from their place of birth. However, anyone who tried to manage or restore a site by attempting to accommodate all the possible pollinators that might be present in an area would become extremely frustrated by the competing demands of, for example, species that require wet areas versus those requiring dry sandy soil in which to nest. The guiding principles should therefore be that, as far as possible, diversity of plant species and physical heterogeneity of the habitat must be maximised, but not at the expense of providing a sufficient abundance of plants and area of habitat to enable pollinators to complete their life cycles.

Diversity in this context means as wide a variety of plants as possible, preferably native to an area (though see Chapter 9 on the role of gardens). Heterogeneity of the habitat can correlate with plant diversity but also depends on how those plants are managed (e.g. not mowing all of a grassland to the same height in one fell swoop) and on the topography of the landscape, presence of water, south-facing slopes, etc. The scale of that heterogeneity is also important; some small solitary bees may spend their whole life feeding and nesting in an area of a hectare or less, whereas large bumblebees can travel kilometres on a single foraging trip, and nomadic butterflies may range far across the landscape.

Diversity in time also matters, which means having floral resources available throughout the season (see Chapter 6). As an example of this, consider the flowering phenologies of typical, abundant hedgerow and woodland-edge plants in Britain and northern Europe that have fairly open, easily accessible flowers providing nectar and/or pollen for a diversity of species. By planting an appropriate selection of species within a hedgerow or new woodland one could provide floral resources for pollinators for as much as twelve months of the year, as well

FIGURE 12.4 Cherry plum or myrobalan plum (*Prunus cerasifera*) flowering in an old, peri-urban hedgerow in Northampton, February 2019. This long-established introduction from southeast Europe and western Asia has now become an important early-flowering component of hedgerows. It's also the progenitor species for many garden cultivars. It is visited by a wide diversity of early-emerging insects.

as providing habitat for some of those pollinators and a host of other species. Late winter species include goat willow (*Salix caprea*) and cherry plum (*Prunus cerasifera* – Figure 12.4) that can flower from February to April; in spring there are other *Prunus* species such as blackthorn (*P. spinosa*); summer-flowering species include bramble (*Rubus fruticosus* agg.) and various roses (*Rosa* spp.); late summer and autumn bloomers include old man's beard (*Clematis vitalba*) and common ivy (*Hedera helix*). Nowakowski and Pywell (2016) have a useful list of species, including gorse (*Ulex europaeus* – Fabaceae) which can produce flowers in every month ('when gorse is in bloom, kissing's in season' goes the old saying).

In other parts of the world a different selection of plants would produce a similar continuity of flowering, but it very much depends on the environment, and particularly weather conditions. In the Canary Islands, for example, flowering time and intensity is hugely influenced by rainfall in the winter and early months of the year. If it's dry during this period, some plants fail to flower at all.

It is important to distinguish between managing for a single species and managing for 'pollinators' generally. Plant and manage appropriately and pollinators will find a site and hopefully flourish. But knowing how to support any one particular species, for example the focus of a Biodiversity Action Plan, requires an intimate knowledge of the species in question, especially where species reintroductions are concerned. For example, I'm currently working with the charity Butterfly Conservation on a project to reintroduce the chequered skipper (*Carterocephalus palaemon*) into the Rockingham Forest area of Northamptonshire, where the last English population went extinct in 1976 (the species is still hanging on in Scotland). An (undisclosed) area of the forest has been managed to provide suitable habitat, including widening the rides between woodland blocks. These rides are packed with plant species, among them at least five grasses that could be used as caterpillar food plants for the skippers, plus dozens of potential nectar sources that the adults can use. This is being monitored by a postgraduate researcher, Jamie Wildman, who as I write is just completing his third field season of the project, which was subject to pandemic-induced social isolation rules. The project has proven to be successful and has resulted in the first chequered skippers to be born in England in over forty years, and what seems to be a viable population. Although this is a project focused on a particular species, it's about more than just the chequered skipper. It's also about understanding how managing a network of sites for this flagship species can help other organisms, particularly invertebrates and plants, that will benefit from the management interventions.

As I've noted, there is an abundance of literature on how to manage, restore and create habitats for pollinators, written by experts far more knowledgeable than I am, and I don't propose to repeat in detail what is in those books and articles. In the next two sections I will look in a bit more detail at two contrasting habitat types, one of which (woodland) is, I think, rather neglected in discussions of pollinator conservation. The other (grassland) is much more widely considered but with often conflicting advice. These sections will explore the diversity of pollinators associated with those habitats when they are well managed and restored, and how inappropriate management can be detrimental to pollinator communities and the plants that require them.

Woodland habitats

Most of the native trees in Britain and Ireland are insect-pollinated, though the larger species such as oak (*Quercus* spp.) and beech (*Fagus sylvatica*) are mainly wind-pollinated. In warmer climates closer to the equator, up to 100% of the trees in a forest may be pollinated by animals (see Figure 1.5). These trees, and much of the ground flora, provide pollen and/or nectar for flower-visiting insects. However, woodland trees also supply a host of other resources for pollinators, and a well-managed woodland can fulfil the requirements of all three points of the triangle in Figure 12.2 for pollinators that are woodland specialists. There is a school of thought that believes a 'well-managed' woodland is one that is neat and tidy, with fallen branches, and dead and dying trees removed, and stands of similar-aged trees precisely spaced. All of this makes sense for a commercial woodland where the priorities are accessibility for machinery and workers, as well as reducing attacks by fungi and other 'pests' that can affect the quality and yield of timber. However, this minimalist approach is anathema to nature conservationists, who have long known that the most biodiverse woodlands are 'messy' and have a varied age structure of trees, some open gaps, and significant amounts of fallen wood. Standing dead trees are especially important for a wide range of specialist invertebrates that feed on or in decaying wood. These saproxylic organisms include some species that are known to be pollinators.

In the early 2000s Keith Alexander of the Ancient Tree Forum compiled a preliminary checklist of the insects that are associated with living and decaying timber in the British Isles (Alexander 2002). This report lists almost 1,800 species for Great Britain and a little over 600 for Ireland and, although it does not specifically mention pollinators, a proportion of these insects certainly visit flowers. For Great Britain this includes most or all of the longhorn beetles (family Cerambycidae – 47 species), nine other beetle families, many of the 246 species of Hymenoptera, and flies from at least six families, especially the 39 species of hoverflies (Syrphidae).

While most saproxylic beetles are not themselves pollinators, their wood-boring activities can directly benefit other insects that do visit flowers. However, the aculeate Hymenoptera are choosy about which holes they will nest in. A study in Sweden of nesting behaviour by solitary wasps and bees by Westerfelt *et al.* (2015) showed that less than 2% of the available holes in wood were used by bees or wasps. In comparison to these natural sites, similar artificial holes in timber poles were occupied about 30% of the time. The critical factor seems to be moisture: most of the natural holes were in rather wet wood, which is not favoured for nest construction. This fits with the idea that, by and large, bees are adapted to drier, warmer conditions (see Chapter 2).

Ten species of bees are named in the Alexander (2002) report as always or frequently nesting in dead wood. Eight of these are within the family Megachilidae, but they also include the honey bee (*Apis mellifera*), the natural European habitat of which is hollow trees. Lack of hollow trees has been suggested to be a limiting factor for feral honey bee colonies in Britain, but most of those hollows that do occur do not get colonised, so that can't be the whole story.

About 15% of Europe's 800 species of hoverflies (Syrphidae) are thought to be associated with dead and veteran trees. They use different parts of the trees for hibernating, egg laying and adult feeding, including the timber of mature trees; microhabitats such as deep bark crevices and holes that occur on old living trees; 'sap runs' in which a wounded part of the tree still bleeds; rot-holes and concavities in branches that fill with rainwater, dead leaves, and decaying material (Figure 12.5); and 'insect workings' such as the galleries caused by the saproxylic beetles described above. The detailed larval life histories of many of these species are still not known, but as adults all species visit flowers for nectar and pollen, and thus are potential pollinators of woodland and woodland-edge plants.

It's not only ancient woodlands that can host these insects. 'Unnatural' plantings comprising non-native species such as traditional apple orchards can also be important, where veteran trees with cavities and dead branches are allowed to persist (Alexander *et al.* 2014).

FIGURE 12.5 Mature trees and fallen wood provide a wide variety of resources and habitat for pollinators. Water-filled hollows are sometimes referred to as 'phytotelmata', a term that includes water bodies held in other living plants such as teasels (*Dipsacus fullonum*); they can be used as egg-laying sites by some hoverflies. Holes of various types are made use of by insects which hibernate as adults, such as the female marmalade hoverfly (*Episyrphus balteatus*). All photographs taken on the same day in early February 2019 at Rushmere Country Park, Bedfordshire.

Before we leave woodlands behind, it's worth saying something about coppicing, one of the most frequently used approaches to woodland management, certainly in Britain and northern Europe, though the practice is also known from other areas, including India. Coppicing involves cutting trees to ground level on a rotational basis and allowing them to re-sprout, such that every 5–20 years a crop of wood can be harvested to use for furniture or construction or charcoal, depending upon the size of the timber poles. By coppicing different areas at different times, a diversity of physical structure can be maintained, and with it a greater diversity of plants and animals across the woodland as a whole. Plants in recently coppiced areas tend to flower more abundantly in response to increased levels of sunlight reaching the woodland floor. This in turn attracts more pollinators. As the coppice stools grow, the shade in that part of the woodland increases, providing habitat for those insects that require more humid and closed conditions.

Coppicing may strike us as very artificial, consisting of people interfering (once more) with nature. In one respect it is, but it also mimics the damage and removal of trees that would naturally occur in such woodlands if they were rewilded and species such as beaver and large herbivores (and their predators) were allowed free rein. The ancient landscapes of Europe prior to large-scale human exploitation were not just 'wild woods' with a completely closed tree canopy (Sandom *et al.* 2014). Over the past 130,000 years or so, the proportion of open vegetation has fluctuated enormously, but it has always been an important component of the British landscape. Likewise, at any one time during the Holocene, open, species-rich grasslands accounted for 30–50% of habitats in central Europe (Feurdean *et al.* 2018). However, highly biodiverse grasslands have suffered most from habitat loss, degradation and agricultural improvement, so restoring and creating such sites plays an important role in sustaining pollinator populations.

Grassland habitats

The current explosion of local councils and businesses wanting to 'do something for the pollinators' (see Chapter 13) has led to a particular focus on the importance of grasslands. This includes both large expanses of habitat and lowly village greens, road verges and roundabouts. No doubt this is because it is in such grasslands where the effects on managing and restoring for pollinators can best be seen, where the diversity and abundance of flowers and their visitors are most visible. Most of the requests that I get for advice on how to manage habitats for pollinators relate to grasslands in one way or another, and sharing some of those examples will hopefully give a sense of what is good (and bad) practice in this regard.

In 2015 I provided some advice to an ecological consultancy who were creating a large area of habitat as part of a huge rail–road freight development in west Northamptonshire. The site would subsequently be owned and managed by the Wildlife Trust, who were the main advisors on the project, but I was asked for some specialist input regarding the pollinators. The development, like so many in the UK, is on former intensive farmland. One of the most prominent features of this project was the building of a long ridge of land created with the spoil from the nearby development, 17.5 metres in height at its maximum. It was proposed to plant this ridge with a mix of broad-leaved trees interspersed with grassland patches. The south-facing slope of the ridge had great potential as habitat for ground-nesting bees and wasps, if it was properly constructed and managed. The proposal was to seed the grassland habitats on the south slope with a wild-flower mix and to vary the management of these grassy slopes, with some areas experiencing hay cuts in late summer and others having low-intensity grazing by sheep. One of my suggestions was to sculpt the surface profile of the ridge to provide a variety of local slopes from near vertical to very shallow, to allow bees with different preferences to choose where to nest. In addition, I suggested that some of the 'management parcels' (i.e. areas with designated management protocols) should have little or no vegetation. This would include grassland maintained with a very short sward height (ideally no more than a couple of centimetres) and with bare earth being created by occasional mechanical disturbance. Such disturbance would facilitate not only bee nesting opportunities but also the establishment of early-flowering 'weeds' (not a term I like) that provide important nectar and pollen sources in the spring. Again, it's the diversity of plants and the heterogeneity of the landscape at both small and large scales that's important for such habitat creation.

The most frequent request for advice that I get is from local community groups wanting to create 'wild-flower meadows' on their village green or urban park. My advice is always the same: don't rush into it, and think carefully about whether you really want to do it. Although making a wild-flower meadow can be an important focus of community action, it will require a lot of work to create and (especially) to maintain. For many such sites, simply reducing the frequency of mowing and allowing plants that are already present to flower will provide more than enough resources to sustain a diversity of pollinators. That's a much cheaper and easier option in the long term, though it often seems not to be the advice that people want to hear: they want a Big Project with Impact.

When I suggested this to the local Friends Of an urban park in Northampton, some members raised objections such as problems of 'messiness', places to hide drugs, and human defecation. I pointed out that all of this could equally apply to a wild-flower meadow: these can be 'messy' for much of the year and provide just as many opportunities for such hypothetical antisocial behaviour. Not only that, but the site in question already had a diverse flora which was being suppressed by an over-keen mowing regime, something that was not widely appreciated either by the group or by the other users of the park. There were already significant areas of wild flowers that would create a lovely flowering grassland if the local council reduced the frequency of mowing in places and let nature take its course. Some plants we know have been present on the park since at least 1900 – there are lists from that time published in the journal of the Northamptonshire Natural History Society. Additional surveys with one of my colleagues, Duncan McCollin, had turned up over 30 species, including some which are rare in the county, and simple mechanical disturbance in places would no doubt increase this number by allowing seeds long-dormant in the soil to germinate and grow. In contrast, starting a new wild-flower meadow would have a significant cost in relation to setting it up and maintaining it in the long term, and could in this case do more harm than good. Digging up and reseeding any areas could have a detrimental impact if it destroyed existing plant populations. Big Projects are all very well, but I think that often we need to value, and look after, what we already have rather than trying to create something new.

An important message from this is that nature will often do a much better job of colonising a site if it's left alone to just get on with it. This was also one the conclusions of the work carried out by two postgraduate researchers that I helped to supervise, Sam Tarrant and Lutfor Rahman, both of whom I've mentioned in earlier chapters. Funded by the SITA Environmental Trust with money from the UK Landfill Tax Credit Scheme, Sam and Lutfor's linked research focused on landfill sites in the region that had been closed and either seeded with perennial grasses or allowed to colonise naturally. They compared the biodiversity of these sites, including plants, pollinators, molluscs, beetles and birds, with those of nearby grassland nature reserves and other sites of conservation value. Overall there were no differences in the diversity and abundance of floral resources across these three types of site, though there were differences in the sorts of plants to be found on each (Tarrant 2009, Rahman 2010, Rahman *et al.* 2013, Tarrant *et al.* 2013). Likewise, pollinator diversity was similar across all the site types. One particularly important finding was that floral resources, and pollinator diversity, were much higher in the late season on the restored landfill sites than they were on the nature conservation sites. This is because the nature reserves are routinely mowed or grazed at the end of the season in order to maximise plant diversity. This can come at the expense of pollinators that are active late into the season, especially those which hibernate as adults.

As I mentioned in Chapter 6, there is frequently a tension between the demands of grassland management for plant diversity and (especially in urban areas) neatness, and the requirements of pollinators. The only real solution to this is to have a range of different management regimes within an area, with some parts left untouched for a year or two while others are mown. This allows pollinators to move from the disturbed areas and find refuge in areas that are undisturbed. Framing the unmown areas with shorter-cut grass, as is often done in the Netherlands for instance, can convince the critical observer that such areas are not neglected weedy patches (Figure 12.6). Putting up interpretation boards explaining the ecology of pollinators and pollination, and how the site is being managed for their benefit as well as for wildlife more generally, can also help. However, even without such boards, there's no doubt that well-managed grassy strips are attractive to both people and to pollinators (Figure 12.7).

FIGURE 12.6 Urban species-rich grassland framed by short-mown grass in Wageningen, the Netherlands. See Figure 8.8 (bottom right) for a British example.

FIGURE 12.7 Two views of the same urban grassy bank in Kettering, Northamptonshire, showing a mixed management style that is likely to maximise pollinator diversity.

Making connections

Most organisms are mobile for at least part of their life cycle, even plants, the seeds of which must disperse into new areas if populations are to persist across a region. Small nature reserves and other wildlife sites should therefore ideally be connected to others to facilitate this dispersal. However, these reserves are often embedded within a matrix of unsuitable landscape that limits or entirely prevents any dispersal. How is a small hoverfly or solitary bee to travel across kilometres of intensively farmed land and find another site in which to establish a population? Wildlife corridors that connect smaller patches have long been known to be effective in allowing such dispersal. They are one of the three pillars of the Lawton Review's recommendation that British wildlife sites should be 'bigger, better, and more joined up' (Lawton *et al.* 2010 – see Chapter 13).

But establishing large corridors of connecting habitat is not possible in many areas, especially where there is intensive agriculture taking place. Smaller linear features, however, such as hedgerows and ditches, may help to link habitats together, at least for pollinators. That was the conclusion that we came to during Louise Cranmer's PhD research (Cranmer *et al.* 2012). The origins of this research lay in a study which showed that butterflies will often follow straight lines in a landscape, even if they are just red and white warning tape laid flat (Dover and Fry 2001). Louise repeated this in natural hedgerows and in artificial hedgerows made of black weed-control fabric, and looked at bumblebees and hoverflies as well as butterflies. She also assessed the consequences of insects following such linear features by studying how these movements affected pollination in experimental

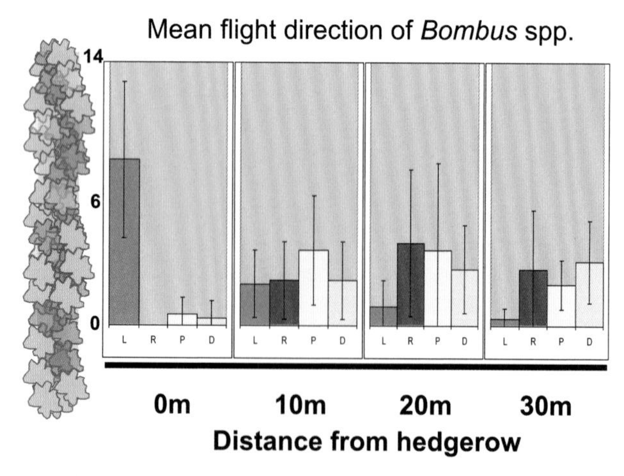

FIGURE 12.8 Bumblebees (*Bombus* spp.) respond to the presence of linear landscape features such as hedgerows by flying parallel to them. The graph shows the average number of bees per observation period flying in different directions relative to the hedgerow: L = linear flight parallel to the hedge; R = random with respect to the hedge; P = perpendicular to the hedge; D = diagonal to the hedge. Data from Cranmer *et al.* (2012). Image is based on one produced by Louise Cranmer.

patches of meadow clary (*Salvia pratensis* – Lamiaceae). Although hedgerows have long been known to provide floral resources to support pollinators, what Louise found was that insects use such features to guide them across the landscape in much the same way as a light aircraft pilot will use major roads to navigate (Figure 12.8). When I visited her on the farm she used for her experiments it was fascinating to stand adjacent to a hedgerow and watch bumblebees sweeping past parallel to the hedge. The consequence of this for the salvias was profound: plants in patches that were more connected by hedgerows had more bumblebee visits and greater seed set. In fact, plants in patches that were at the confluence of four or five hedgerows could have up to 50% greater seed set than those plants that were not connected by hedges. This suggested to us that small isolated populations of rare plants could be linked quite easily by such linear landscape features.

When is it restoration?

An open question (for me at least) is whether the interventions that farmers and other ecological practitioners make to encourage and support pollinator populations deserves the name 'restoration'. I frequently see research papers discussing 'restoring habitats for pollinators', and my reaction is usually 'restoring to what?' Ecological restoration is a widely used term that is generally defined as putting nature back to the way it was before humans interfered with it (though there's a whole discussion to be had on the extent to which humans are a 'natural' part of the ecology of any landscape). There's a useful analogy here with furniture restoration. Imagine buying an abused Victorian wooden dining table at an auction with a view to 'restoring it' and putting it to good use. We might start with stripping off successive layers of paint that previous owners had added to update the table's looks, starting with the grey and white, minimalist shades of the 1990s and 2000s, working back through the pastels of the 1980s, to the vivid psychedelic additions from the 1960s and 1970s, and finally getting to the original brown wood that sufficed for most of the table's life. As restorers we could stop at any of these layers and return the table to how it looked during a particular decade, including (ultimately) French polishing the table to return it to its original state. Alternatively, we could choose another approach and decide to live with this overpainting and 'enhance' the table for the twenty-first century with the addition of new paint and stencilled motifs, or an embossed leather surface, shorter legs, and so forth – features that have never been previously been part of the table.

In our furniture project, the options, of restoration to a particular point in time or enhancement to suit our present needs and tastes, are limited only by our imaginations. In a sense it's the same set of options that are available to ecological restorers. All landscapes have been modified over timescales of millions of years to decades. Decadal changes are frequently imposed by people, for example felling a natural forest to turn the land into agricultural fields. But earlier, long-term modifications are natural: before it was a forest that habitat may have been swamp or salt marsh, with changes occurring as rivers shift their course or sea levels fall. Faced with a modern agricultural landscape of large, open fields (currently a common sight not only in Britain, but in much of the rest of the world), what point in time should we aim for in our restoration? Back to forest, or to an earlier state? Do we add more hedgerows of the type that were removed in the 1960s and 1970s? Or remove most of the existing hedgerows to create large, open grasslands of the type that were common before the enclosures of the 1800s? This was the type of landscape that John Clare was familiar with, and perhaps his *Wild Bees* flourished better in such habitats than they might do otherwise (Chapter 10). There is no right answer to the question of what point in time to restore a landscape back to, and often we must take a utilitarian view and restore to what is feasible given the limits of time and budget.

The differences between habitat 'restoration', 'management' and 'creation' are not at all clear, and these three aspects of practical conservation merge into one another. For example, is grazing of rough grassland to promote plant diversity restoration or management? I'm not convinced that the distinction matters, in the grand scheme of things, but I know that some people have very strict views about how to define such activities. However, there is certainly a clear distinction between restoration and enhancement, of furniture or of habitats. Enhancement of a habitat is frequently the option that is meant when we talk of 'planting for pollinators', with strips of 'wild flowers' sown along the edge of arable fields or within orchards, for instance. Depending upon the mix of plants used, such strips may be more or less natural in the context of the local landscape, but frequently they contain non-native species such as blue tansy (*Phacelia tanacetifolia*) added to lengthen the flowering time of the strip. This is habitat enhancement, not restoration, and in a sense no different to wildlife gardening on a grand scale. There is, however, clear evidence that it works for at least some species: bumblebees (*Bombus* spp.), for example, benefit from sown flower patches in agri-environment schemes in England (Carvell *et al.* 2015).

Discussions about the definitions of the terms we use in conservation are not just academic musings. They reach deep into the philosophical heart of environmentalism as it relates to our treatment of habitats and landscapes: what do we want our world to look like? Personally, I am comfortable with urban settings that are highly modified and include a lot of non-native plants and animals, as I think that there is something inherent in the human psyche that impels us to surround ourselves with diversity and variety of all kinds. However, beyond the limits of human habitation I would much prefer that we endeavoured to restore and manage landscapes in a way that both supports biodiversity and maintains the ecological processes such as energy flows and nutrient cycles that go with it, in as 'natural' a way as possible. Clearly pollinators can play a vital role in such systems for most terrestrial, and some aquatic, habitats, and their needs should not be ignored. There's growing evidence that quite straightforward restoration interventions, for example removal of invasive non-native species from degraded habitats, can have positive effects on plant–pollinator interactions. For example, removing exotic shrubs from sites in the Seychelles resulted in an increase in diversity of pollinators, more visits to flowers, and greater fruit set of native species (Kaiser-Bunbury *et al.* 2017).

Management, including enhancement and restoration of habitats, needs careful consideration before being put into practice. Questions need to be asked about what we are planning to do, whether it is feasible given constraints of budget and people-power, and what we are hoping to achieve. Beyond these immediate practical ecological considerations, however, there are other dimensions that also need to be considered. These include economic drivers, public opinion and political will, which we turn to in the next chapter.

Chapter 13

The politics of pollination

During a seminar at my university a couple of years ago, the veteran American activist, action researcher and scholar Lonnie Rowell told us, 'If knowledge is power, then knowledge has a politics.' That phrase neatly encapsulated an issue that had been bothering me for some time: the relationship between ecological knowledge, societal power, and politics. I'm going to use that word – politics – in several senses throughout this chapter, including the 'political' use of corporate social responsibility (CSR) by large businesses and the political game playing of scientific and conservation organisations, as well as politics in the narrower sense of elections, governments, policies and legislation.

It is almost impossible to have an interest in pollinators and be unaware that there is a political dimension to their conservation. In the UK parliament, virtually every MP and member of the House of Lords who has ever expressed any kind of environmental sentiment has, at some point in the last few years, mentioned bees or, if they are truly well informed, pollinators more broadly. As recently as March 2019 there was a debate about pollinators in Westminster. Elected representatives in many other countries, including President Barack Obama in the USA, have done the same, raising the 'plight of the bees' or the 'pollinator crisis' as a sign of their (often genuine) concern for matters environmental. A range of commercial organisations, such as banks and supermarkets, as well as environmental NGOs, have also started campaigns focused on pollinator conservation (Figure 13.1).

Yet this is a very recent phenomenon. For most of the time that pollination ecologists of my generation have been studying these animals and their ecological interactions with plants, there has been almost no political interest in the subject. Indeed, in Britain there was precious little scientific interest, either, which is why many of us looked to Scandinavia and the Americas for our collaborations and conferences. When attending UK conferences in the early 1990s I always got the impression that those of us interested in pollinators and in plant reproduction were considered old hat, that there were no more interesting questions to answer in this field, and that the future of ecological research lay in other areas. That soon started to change, as shown by the international increase in research on pollinators and pollination that I illustrated in Figure 1.6. But it wasn't until the early 2000s that the public really started to become interested. How this political dimension has expressed itself in wider society, including governments, businesses and NGOs, is

FIGURE 13.1 A small selection of logos from 'save the pollinators' campaigns by banks, supermarkets and non-governmental organisations (NGOs).

the subject of this chapter. It can't possibly cover all facets of this topic in depth, that would be a book in itself. But hopefully this chapter will give a flavour of where we currently are, and how we got here, and what the future may hold.

The origins of political interest in pollinators

Tom Brereton, formerly Head of Monitoring at the charity Butterfly Conservation and my university's Visiting Professor in Conservation Science, once said to me that campaigns to 'Save the Bees' are to the new millennium what 'Save the Whales' was to the 1970s. I think he has a very good point, and in some respects both campaigns started in the same milieu, in the raised environmental consciousness of the 1960s.

The best-known of the earlier expressions of concern about conservation of pollinators is in Rachel Carson's influential 1962 book *Silent Spring*. In that book she notes the irony of spraying orchards with pesticides that kill pollinators, and states that humanity 'is more dependent on these wild pollinators' than we usually realise. However, there had been some earlier authors voicing such worries; for example, here's the Royal Horticultural Society's entomologist George Fox Wilson writing just after the Second World War: 'these preparations are toxic to beneficial insects – parasites, predators and pollinators ... [gardeners should] ... avoid upsetting the natural equilibrium between parasite and host on one hand and between flower pollination and fruit production' (Wilson 1946). Much earlier still, accidental poisoning of bees by pesticides had been of concern to beekeepers in the USA in the 1870s (Johansen 1977). These examples presage the more recent (and ongoing) worries about neonicotinoid pesticides, and their replacements, that I discussed in Chapter 10.

Despite the influence of Carson's book, however, during the rest of the 1960s through to the 1990s (a period, remember, that took us from Woodstock to Nirvana via the Sex Pistols and Live Aid) there was relatively little written about conserving pollinators, or why pollinators were important. Scientific research continued during this time (see Figure 1.6) but little of this focused directly on conservation. There were some exceptions, for example work by Peter Kevan in Canada, who in the 1970s was publishing articles with titles such as 'Pollination, pesticides, and environmental quality', 'Pollination and environmental conservation' and 'Blueberry crops in Nova Scotia and New Brunswick – pesticides and crop reductions' (Kevan 1974, 1975, 1977). Likewise, Anderson and Atkins (1968) had written about 'Pesticide usage in relation to beekeeping'; and Benedek (1972) discussed the 'Possible indirect effect of weed control on population changes of wild bees pollinating lucerne'. Much of the earlier literature was summarised by Martin and McGregor (1973) in their review 'Changing trends in insect pollination of commercial crops'. Although this review mainly considered honey bees, the authors did point out that pesticides which directly kill managed bees and herbicides that reduce the amount of forage available to colonies 'are equally if not more detrimental to wild pollinators'. Similar points were raised a few years later by Johansen (1977) in his review of 'Pesticides and pollinators'.

These examples give a flavour of some of the scientific literature on pollinator conservation that was emerging at this time. No doubt I've missed publications that were not written in English, but nonetheless the scientific literature in this area was relatively scant, and much of it focused on conserving managed honey bees. Certainly there was almost nothing written about how honey bees themselves might be a conservation *problem* (see Chapter 10), though an exception was Roubik (1978), who was concerned about 'Competitive interactions between neotropical pollinators and Africanized honey bees'. But this was very much an outlier at the time. It is notable that *The Pollination of Flowers* by Michael Proctor and Peter Yeo, the foremost summary of the field in the early 1970s and into the 1980s, devoted less than one page to pollinator conservation (Proctor and Yeo 1973). In the mid-1990s a revised version, retitled *The Natural History of Pollination* and with the addition of my PhD supervisor, Andrew Lack, as an author, doubled the space devoted to pollinator conservation to a full page and a half (Proctor *et al.* 1996). This is not meant as a criticism of those authors; it simply highlights the fact that pollinator conservation was not on many people's radar at the time.

It was in the mid-1990s, with the publication of one highly influential book and one foundational review paper, that concerns about pollinator conservation really began to enter the wider consciousness, of both scientists and society at large. *The Forgotten Pollinators* by Gary Paul Nabhan and Stephen Buchmann hit the shelves in 1996 and was immediately popular with green-minded readers. Gary's background is in the agricultural ecology and ethnobotany of the American Southwest, and he founded the Forgotten Pollinators Campaign and the Migratory Pollinators Conservation Initiative. Steve is a pollination ecologist whose research has largely focused on bees. Both now work at the University of Arizona, and *The Forgotten Pollinators* dealt mainly with the migratory species such as hummingbirds and bats that are a feature of that part of North America. Nonetheless, the book had a far-reaching effect on both the public and those of us already working on pollinators and their interactions with plants. Many of us read it and were inspired to focus at least some of our efforts on conserving plant–pollinator interactions.

The second publication from this period that I think was especially important was a review paper by Carol Kearns, David Inouye and Nick Waser entitled 'Endangered mutualisms: the conservation of plant–pollinator interactions' (1998). The opening lines of this review sum up what was to come:

> The pollination of flowering plants by animals represents a critical ecosystem service of great value to humanity, both monetary and otherwise. However, the need for active conservation of pollination interactions is only now being appreciated.

These publications, and others of the same vintage, brought pollinator conservation to international attention, culminating in the United Nations Convention on Biological Diversity *São Paulo Declaration on Pollinators* (Dias *et al.* 1999). This arose from a 'workshop on the conservation and sustainable use of pollinators in agriculture with emphasis on bees' held in October 1998. It was written by three scientists based in Brazil, Braulio Dias, Anthony Raw and Vera Imperatriz-Fonseca, collaborating with an international set of advisors and working groups. The focus of this report was very much on agriculture and bees, especially *Apis mellifera*, but it is nonetheless a foundational international statement that galvanised pollinator conservation efforts. At about the same time other regional meetings were taking place, such as the First Congress of the Systematics Society of South Africa in January 1999, from which emerged the African Pollinator Initiative that met for the first time in February 2002 in Kenya.

How have governments responded to the 'pollinator crisis'?

The São Paolo Declaration, the African Pollinator Initiative, and other meetings and reports were followed by a period of campaigning around issues such as colony collapse disorder (CCD) in honey bees, effects of pesticides on pollinators, and of course research into the ecological and agricultural importance of pollinators. Some of that research was funded by specific programmes, such as the UK's £10 million Insect Pollinator Initiative (nerc.ukri.org/research/funded/programmes/pollinators). In addition there have been pollinator-specific initiatives by private organisations such as the JRS Biodiversity Foundation (jrsbiodiversity.org), which specifically funds data collection and sharing, and expertise capacity building, in Africa.

The cumulative effect of all of this has been to spur many national governments to produce (or start planning) 'national pollinator strategies' or 'pollinator initiatives'. One of the earliest was the National Research Council's *Status of Pollinators in North*

America document in the USA (2007), and countries such as Brazil, Canada, Norway and France have followed, as have Britain and Ireland (see below). In addition, a number of wider regional plans have emerged – not only the African Pollinator Initiative but also others such as the Oceania Pollinator Initiative and a European Union Pollinators Initiative. As recently as February 2020 a meeting of scientists took place in Kolkata, India, to discuss setting up an Asian Pollinator Initiative.

There is an *All-Ireland Pollinator Plan* that (very sensibly) includes both Northern Ireland and the Republic of Ireland: biodiversity does not respect political boundaries. Yet for Britain there are separate documents for all of the devolved administrations of Scotland, Wales and England. Most of these plans are broadly similar (though some focus more on honey bee management than others), so it's worth considering the ten-year *National Pollinator Strategy* for England (Defra 2014) in some detail as an example. The aim of the Strategy is to deliver across five key areas, and I've quoted liberally from the text in Box 13.1.

Box 13.1 National Pollinator Strategy for England (2014): five key areas

1. **Supporting pollinators on farmland** – Working with farmers to support pollinators through the Common Agricultural Policy [CAP] and with voluntary initiatives to provide food, shelter and nesting sites. Minimising the risks for pollinators associated with the use of pesticides through best practice, including Integrated Pest Management (IPM). [Clearly Brexit has made any reference to the CAP redundant, and it remains to be seen what will now occur.]

2. **Supporting pollinators across towns, cities and the countryside** – Working with large-scale landowners, and their advisors, contractors and facility managers, to promote simple changes to land management to provide food, shelter and nest sites. Ensuring good practice to help pollinators through initiatives with a wide range of organisations and professional networks including managers of public and amenity spaces, utility and transport companies, brownfield site managers, local authorities, developers and planners. Encouraging the public to take action in their gardens, allotments, window boxes and balconies to make them pollinator-friendly or through other opportunities such as community gardening and volunteering on nature reserves.
3. **Enhancing the response to pest and disease risks** – Working to address pest and disease risks to honey bees while further improving beekeepers' husbandry and management practices to strengthen the resilience of bee colonies. Keeping under active review any evidence of pest and disease risks associated with commercially produced pollinators used for high-value crop production.
4. **Raising awareness of what pollinators need to survive and thrive** – Developing and disseminating further advice to a wide range of land owners, managers and gardeners. Improving the sharing of knowledge and evidence between scientists, conservation practitioners and non-government organisations (NGOs) to ensure that actions taken to support pollinators are based on up-to-date evidence.
5. **Improving evidence on the status of pollinators and the service they provide** – Developing a sustainable long-term monitoring programme so we better understand their status, the causes of any declines and where our actions will have most effect. Improving our understanding of the value and benefits pollinators provide, and how resilient natural and agricultural systems are to changes in their populations.

Within these five areas, the *Strategy* stated that the aim was to achieve the following outcomes:

- More, bigger, better, joined-up, diverse and high-quality flower-rich habitats (including nesting places and shelter) supporting our pollinators across the country.
- Healthy bees and other pollinators which are more resilient to climate change and severe weather events.
- No further extinctions of known threatened pollinator species.
- Enhanced awareness across a wide range of businesses, other organisations and the public of the essential needs of pollinators.
- Evidence of actions taken to support pollinators.

It was not surprising to see 'more, bigger, better, joined-up ...' used in the *National Pollinator Strategy*, as this has been the buzz phrase in British conservation since the publication in 2010 of a government report entitled *Making Space for Nature*, usually referred to as the Lawton Review. One of the outcomes of that report was the setting up of twelve flagship Nature Improvement Areas (NIAs), one of which is the Nene Valley NIA, a project on which my research group collaborated. The *Strategy* mentions the NIAs several times and talks about 'extending the monitoring and evaluation framework for Nature Improvement Areas to include pollinators' as one of its interim aims. But the funding for the NIAs finished at the end of March 2015 and the Department for Environment, Food and Rural Affairs (Defra) had indicated well in advance that there would be no additional government money. This is just one example of good intentions (on paper) that are not backed up by resources. It is also noteworthy that there is little legal protection for most pollinators, in comparison to birds and mammals, for instance. That's not just a feature of England, but is true of the environmental legal frameworks of just about all countries: invertebrates as a whole are relatively unprotected (Cardoso 2012).

Why should governments care about pollinators?

One cynical view of why politicians might be interested in pollinators links directly to a subject that is often at the forefront of their minds: food security. For it is suggested (according to a quote attributed to MI5) that the UK is never more than four hot meals from a revolution. Pollinators do indeed play a significant role in agriculture, as we saw in Chapter 7, but perhaps not significant enough for greater action: loss of pollinators is not likely to cause rioting in the streets of Northampton. But it is worth emphasising that pollinators underpin, directly or indirectly, much of terrestrial biodiversity and the ecosystem goods and services that flow from that biodiversity (Figure 13.2). These elements in turn are embodied within many of the United Nations Sustainable Development Goals (www.un.org/sustainabledevelopment/sustainable-development-goals). Income from small-scale beekeeping is contributing to the goal of 'No Poverty' via the work of charities such as Bees for Development (www.beesfordevelopment.org), for example, while the links between 'Zero Hunger', 'Good Health and Well-being' and the part played by pollinators in agriculture should be obvious in the light of Chapter 7. In fact any of the Goals that are underpinned to some extent by the ecological attributes of trees and other plants, such as 'Clean Water and Sanitation', 'Sustainable Cities and Communities' and 'Climate Action', can also be linked back to pollinators, which play a key role in the long-term persistence of plant populations (Chapter 3). Indeed there is a clear social justice aspect to the conservation of all biodiversity, including pollinators and the plants they service. For example, studies have shown that pollinator diversity is associated with richer neighbourhoods in cities, and that supporting pollinators within a locality can increase yields in the home gardens

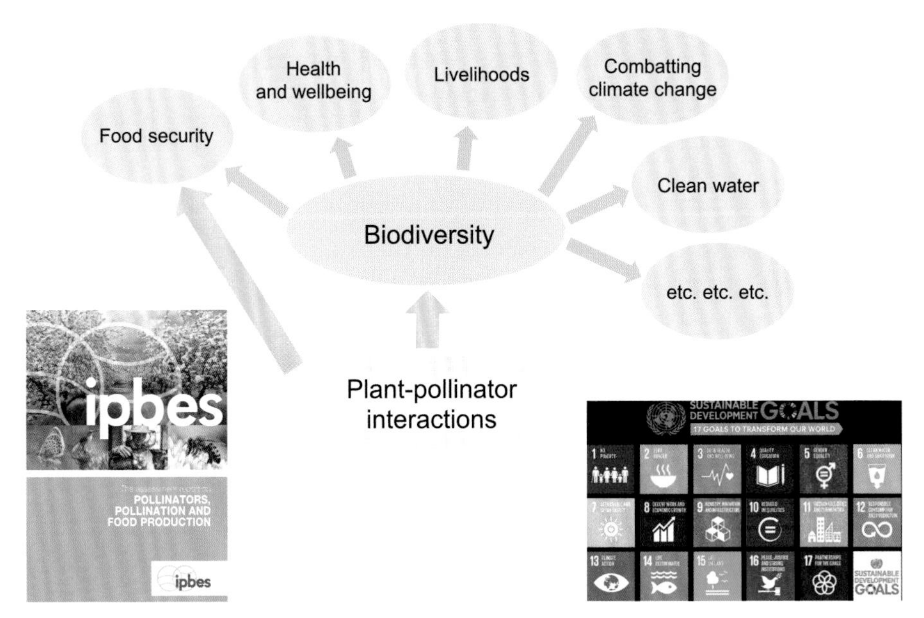

FIGURE 13.2 International agencies such as the various arms of the United Nations, including the FAO & UNEP, and IPBES, recognise that biological diversity is important for human society from a range of perspectives. Much of the terrestrial component of that biodiversity is itself underpinned by the activities of pollinators.

of poor people across the developing world (see for example Baldock *et al.* 2019, and studies cited in Chapter 8).

All of this is great, positive stuff and, as Steven Pinker has pointed out in his book *Enlightenment Now: The Case for Reason, Science, Humanism, and Progress*, many of the original Millennium Development Goals (which were superseded by the Sustainable Development Goals in 2015) have been met or even exceeded (Pinker 2018). However, the environmental cost has been, and continues to be, significant, and some of the 'knowledge' that is supporting campaigns and initiatives to conserve pollinators is shaky at best, and downright misleading at worst. This is problematic because, as Pinker and others have noted, accurate information about the natural world and our societies is a prerequisite for both human well-being and environmental protection.

Reactions to the 'pollinator crisis' shaped by social, print and broadcast media

Albert Einstein said a lot of things, and was a hugely important and influential scientist, so it's saddening to think that one of the things he will be best remembered for is something he did not say. There are various versions of it, but they all amount to the same thing:

> If the bee disappeared off the surface of the globe, then man would have only four years of life left. No more bees, no more pollination, no more plants, no more animals, no more man.

This statement could be dissected and disproved in numerous ways. For a start, there are over 20,000 species of bees, so what is 'the bee'? Plus most of our crops are not pollinated by bees (or by any other insects), but are wind-pollinated grasses such as wheat and rice, or may be propagated using tubers, as in the case of potatoes and yams. Pollinators are important for agriculture, as we saw in Chapter 7, but we'd not starve without them.

But, regardless of the scientific veracity, what is particularly annoying about this is that *Einstein never said it!* That deserves emphasis because, although many people have made it clear on social media, and have pointed it out to journalists and the public at large, I frequently see it stated that Einstein issued such a warning. And documentary film makers and journalists are still using it to support their work, even on cinema posters and magazine covers. As far as anyone is aware Einstein had no interest in bees whatsoever, and the source of the quote is unclear, though it seems to be an amalgam of statements by other people (Quote Investigator 2013, Breeze 2014).

The infamous Einstein quote is not the only mis-fact that is bandied around concerning pollinators and pollination: pollinator conservation seems to be full of them, and some scientists are themselves guilty of making inaccurate or overblown statements, especially about honey bees (see Ollerton *et al.* 2012). Among the wider public, as Wilson *et al.* (2017) note, 'interest exceeds understanding in public support of bee conservation'. It's not helped by what I've termed the 'honey bee bullshit machine' – a near-continuous stream of misinformation and non-facts that has emerged over the past couple of decades. Here are some examples that I've seen, all of which are factually incorrect either in part or completely, as the earlier chapters of this book will have demonstrated:

> Healthy honey bee populations are vital to food and crop production, and the natural environment. In the UK, honey bees are responsible for 80% of pollination, and a third of the food we eat is pollinated by bees.
>
> Records show that in the 1950s there were 50 species of native bee in the UK, today there are just 25.
>
> The honey bee is the only bee that is looked after by man.
>
> There are no feral honey bee colonies left as they simply cannot survive a varroa infestation or diseases that they would be exposed to without the beekeeper's assistance.

One of the flip sides of having expert knowledge on a topic is the frustration of seeing inaccurate claims made or information presented as fact, in the media and online. No doubt experts in other disciplines get equally frustrated, and I think that as scientists and academics we have a role in society of providing evidence and advice. This is not always a comfortable experience. When I've pointed out some of the misinformation about honey bees on social media, for example, occasionally people have demanded to know what my 'hidden agenda' is and whether I might be funded by a big agrochemical firm. My 'agenda' is always accuracy and evidence-based action. And I've never received such funding. But there's a bigger question here about the role of environmental scientists in wider society, and the extent to which we should simply do the research and the teaching, provide data and evidence as needed, and otherwise keep quiet. In other words, should scientists like myself who study biodiversity, or those whose work is on pollution or wastes management or climate change, also be campaigners and polemicists? There is a difficult line to walk between scientist as campaigner and scientist as neutral presenter of facts, but scientists are citizens too, often concerned about the wider implications of what they work on. I think that polemicist/activist is quite an admirable position for a scientist to take in many ways, as long as the rhetoric is backed up by sound science (Dicks 2013). It's also brave, given that perceptions of scientists can change the likelihood of their research being funded or even published – reviews and reviewers are rarely as objective as we would like to believe.

In my blog and on Twitter and Facebook I've made no secret of the fact that I take certain positions on subjects such as the impact of poorly conceived development on nature reserves, the fallacies of political spin, and future developments in biodiversity conservation. Those are positions that are determined as much by my personal motivations as an 'environmentalist' (a term I don't like but which is widely understood and will do for now) as they are by my professional role as a university scientist who undertakes research and teaching. I am not a neutral observer, though I try to be an objective one.

I'm certainly not alone in this regard, and there are many scientists who agree, some of whom now have very high profiles. For instance, Dave Goulson's campaigning against the use of pesticides has brought him much support but also abuse online. As Sponsler *et al.* (2019) note, 'the relationship between pesticides and pollinators is a complex *socioecological* system' [my emphasis] in which government policies, scientific evidence and the opinions of the public, farmers, NGOs and other involved stakeholders such as campaign groups all vie for attention (Figure 13.3). The three important domains of concern from an environmental perspective (pesticide use, exposure to pollinators, and subsequent effects on ecosystems) are actually part of a larger and more complex network of social and environmental cause and effect.

In Germany, Joern Fischer and colleagues have used their Ideas for Sustainability blog (ideas4sustainability.wordpress.com) to develop some notions about the kind of science that we should be doing 'for it to be of use in the sense of creating a better, more sustainable world'. I'd add that what is important is not just the science that we do, but how we present that science (the passion and the storytelling) to a range of audiences, and also the personal (political) positions we take

FIGURE 13.3 A poster by Birmingham-based satirical artist 'Foka Wolf' captures much of the public concern over the production and use of pesticides by large agrochemical firms such as the fabricated global company Megacorp – slogan 'Profit Before People'.

on the issues that the science illuminates. Doubtless this has always been true, but the domination of the internet means that suddenly everyone knows 'the facts', whether or not those 'facts' are grounded in truth. Interest in pollinator conservation from the start of the millennium has risen in parallel with the use of the internet as a system for disseminating knowledge, and so 'saving the bees' has become a meme which is accepted uncritically by most people. Overall this is probably a good thing, but it grates when the claims that are being promoted are all too often untrue or only partial truths.

There are wider, global alliances of scientists and citizens who are concerned about the future. One of the most high-profile is the Alliance of World Scientists (scientistswarning.forestry.oregonstate.edu) who have to date produced several *World Scientists' Warnings to Humanity*, between them signed by over 25,000 scientists from more than 180 countries, across all disciplines. In the last year Extinction Rebellion has also become a very high-profile nexus of activism, writing and civil disobedience. I've been happy to support both, signing letters and providing advice when asked. A close friend and his wife and daughter were all arrested in London during the Easter 2019 demonstrations and I have nothing but admiration for them. I like to think that I would have been involved if I'd been in the country, but I was in Tenerife running a field course for my students and then doing some follow-up data collection. In fact, I think education and evidence are as important as campaigning: we do what we do. But all of this is not enough if the ears of governments are closed to the issues around declines in pollinators in particular and biodiversity more broadly. Given this, it's time to consider where the political and non-governmental action on conserving pollinators is at the moment, and what the future holds for such activity.

'Do wasps and nettles sting each other?' The role of NGOs and citizen science

This question was posed by a member of the audience at a talk I gave at an environmental event in Northampton a few years ago. I've done many such talks over the years, to gardening and natural history groups, Transition Towns, the Round Table, the Women's Institute, and so forth. I like doing them because the audiences are always attentive and the questions often get to the heart of an issue. Talking with wider society is an important thing for scientists to do. The question of whether wasps and nettles sting each other was a bit tangential to the subject of my talk on the importance of pollinators. Caught on the hop, I fudged an answer. Thinking about it later I concluded that they don't: wasps are too well armoured to be provoked by nettles into stinging them if they were to land on a leaf. But the question does, I think, serve as a useful metaphor that highlights the potential for conflict between NGOs, well-meaning members of the public and citizen scientists on the one hand, and the actual needs of scientists, policy makers and pollinators on the other. I will illustrate this by means of two examples, the first from the USA, the second from the UK.

The monarch butterfly (*Danaus plexippus*) is an iconic migrating species that in eastern and central North America travels from Mexico to Canada and back, over the course of a few generations. West of the Rocky Mountains it mainly travels back to southern California to spend the winter. This behaviour, and their vast overwintering assemblages, have become the focus of intense efforts to understand monarch ecology and biology (Agrawal 2017). It is especially urgent because monarch populations have declined substantially in recent decades, particularly in the west, and there's a clear need to find out why this has happened. Climate change has been implicated, as have pesticides and other aspects of intensive agricultural management. Citizen scientists have played an important role in understanding the patterns of this decline, and continue to do so via the Monarch Watch initiative (www.monarchwatch.org). Monarch caterpillar host plants are mainly milkweeds (*Asclepias* spp. – Apocynaceae), and especially common milkweed, *A. syriaca*. However, this is an important agricultural weed that has been eradicated from huge tracts of farmland, and so Monarch Watch and conservation NGOs have encouraged the public and businesses with land holdings to plant milkweeds. Unfortunately, we know that not all *Asclepias* species are suitable for monarch caterpillars, and one in particular is a cause for concern. Tropical milkweed (*A. curassavica*) is one of the showiest in the genus and very popular with gardeners. However, it's not native to North America and, unlike native species, produces leaves and flowers all year round if conditions are suitable. This means that it harbours huge populations of a protozoan parasite of monarchs (*Ophryocystis elektroscirrha*). When eaten by caterpillars in large numbers, this parasite causes reduced adult body size, shorter lifespan, fewer successful matings, and a decline in flight ability that can affect migration. When grown in the north the evergreen tropical milkweed can also confuse monarchs into breeding when they ought to be migrating. Although NGOs such as the Xerces Society are trying to educate the public about this problem (e.g. Wheeler 2018), tropical milkweed is still being recommended by some plant nurseries as a food source for monarchs.

The irony of well-intentioned citizens possibly doing more harm than good by their activities is one which crops up repeatedly in pollinator conservation – for example, keeping honey bees at such high densities that they out-compete wild pollinators. There's a double irony in the case of the monarchs, as they are not especially effective at pollinating milkweed flowers despite being one of the poster species of North American 'Save the Pollinators' campaigns (Agrawal 2017).

Encouraging the public to 'plant for pollinators' has been one of the central planks of pollinator conservation strategies by governments and NGOs in many countries (see Chapter 9). Another has been to get them involved in regularly recording the pollinators on their patch, be it a garden, park or local wild habitat. One example is the Friends of the Earth 'Great British Bee Count' (GBBC), a citizen science project designed to augment the monitoring work being done by specialist groups such as the Bees, Ants and Wasps Recording Society (BWARS – see Chapter 10). However, many scientists have mixed feelings about the GBBC and similar schemes. On the one hand it's great to engage the public in campaigns that raise awareness of the importance of pollinators, and to get them out looking at bees and other insects. But the reality is that the hundreds of thousands of records submitted to the GBBC have very limited scientific value, despite what Friends of the Earth might claim. That's because it is very, very difficult to identify bees to even broad groups without some training, and (apart from a few distinctive bees) impossible to determine them to species level unless you are a specialist. I've been studying pollinators for thirty years and there are whole groups within our *c.*270 native species that I have great problems identifying, deferring to the expertise of real specialists such as the members of BWARS.

BWARS itself made public its concerns at the quality of the data being submitted to the GBBC, and the fact that records cannot be checked because no photograph was taken and (worse) there are no collected insect specimens to compare. I am all in favour of getting the public engaged with pollinators and biodiversity more broadly, but let's also be realistic about what can be achieved by these kinds of campaigns. A much better approach is illustrated by the Bees in the Burbs project run by Kit Prendergast in Australia (www.facebook.com/groups/Beesintheburbs), in which members of the public are urged to upload photographs of native bees so that they can be given a provisional identification. Images are a good starting point for identification, though for many groups, without a physical specimen it's only possible to determine them to genus or even family. There are other ways in which the public can contribute to our understanding of pollinators and pollination, which I'll discuss in the final chapter.

Where are we at the moment? The importance of IPBES

The overarching themes of this book have been about understanding the biodiversity of pollinators and pollination; the science behind their study; why it's important; how it all contributes to human well-being (including both economic and intangible benefits); and how policy informed by science can help in the conservation of pollinators and the ecosystems that they support. These are all issues that have a global perspective beyond the bounds of my home country of the United Kingdom, or even my continent of Europe, because species, ecosystems and the threats to them are not constrained by borders. The International Pollinator Initiative *Plan of Action 2018–2030* (www.cbd.int/sbstta/sbstta-22-sbi-2/sbstta-22-ipi-draft.pdf) lists in its annex almost forty documents, plans, codes, protocols, tools, assessments, reports and initiatives, and this is far from a comprehensive list. Has any of this made a positive difference? Are pollinator populations and habitats in better shape now than they were, say, twenty years ago?

The answer is that it depends where in the world we are looking. In the UK, bee and flower-visiting wasp extinctions seem to have halted and the decline in bee pollinators across the country appears to have levelled out and may even be starting to reverse (see Figures 10.2 and 10.3). Some of this good news is doubtless due to habitat restoration and local rewilding initiatives (Chapter 12) as well as a raised awareness of good urban management and gardening practices (Chapters 8 and 9). One of the other things that has happened is that attention has focused on the role of pesticides in pollinator decline (see Chapter 10), leading to partial bans on neonicotinoid pesticides in the EU and elsewhere. As of late 2020, the status of the moratorium post-Brexit is unclear, as trade negotiations are taking priority.

In other parts of the world the situation is less clear. We do know, however, that habitat destruction is starting to increase again across large parts of South America and Africa, and this is certainly causing concern among environmentalists for all sorts of reasons, not just because of its effects on pollinators. Climate change is another huge concern, of course (Chapter 6), as drought, fire and floods affect large swathes of habitat in Australia, India, the USA and elsewhere.

In the UK, at least, another important development has been the implementation of a Pollinator Monitoring Scheme (PoMS – www.ceh.ac.uk/our-science/projects/pollinator-monitoring). This brings together government agencies, NGOs, universities and citizen scientists to engage in a standardised framework to begin to understand trends in pollinator diversity and abundance across a range of taxa, not just bees. In contrast to some of the other citizen science recording schemes, the PoMS has been well structured to ensure the scientific value of the data. There are similar initiatives at various stages of development in different parts of the world; for example I am currently involved with SURPASS (Safeguarding Pollination Services in a Changing World), which has as one of its aims the implementation of pollinator monitoring systems in some Latin American countries (bee-surpass.org).

National and regional initiatives are clearly important, but so is global action led by the United Nations, which has had an interest in pollinators since at least the 1990s, as we have seen. One of the most significant events in world conservation in recent years has been the establishment of IPBES – the Intergovernmental Science-Policy Platform on Biodiversity and Ecosystem Services (www.ipbes.net). IPBES was established in 2012 and in many ways is a parallel entity to the IPCC (Intergovernmental Panel on Climate Change), bringing together scientists, policy makers and stakeholders, with a mission 'to strengthen the science–policy interface for biodiversity and ecosystem services for the conservation and sustainable use of biodiversity, long-term human well-being and sustainable development'.

Tellingly, the first major output from IPBES was an *Assessment Report on Pollinators, Pollination and Food Production*, published in 2016. I acted as an expert reviewer for the document and got to know it very well indeed, despite its size – it runs to over 550 pages. The report represents a huge body of evidence gathering by natural and social scientists, as well as indigenous knowledge holders, that will act as a benchmark for future global action on pollinators. It's worth quoting some of the headline figures from the report (Box 13.2).

Box 13.2 Headline figures from the IPBES *Assessment Report* (2016)

- **20,000** – number of species of wild bees; there are also some species of butterflies, moths, wasps, beetles, birds, bats and other vertebrates that contribute to pollination [note that this predates my assessment of pollinator diversity – see Chapter 2]
- **75%** – percentage of the world's food crops that depend at least in part on pollination

- **US$235–577 billion** – annual value of global crops directly affected by pollinators
- **300%** – increase in volume of agricultural production dependent on animal pollination in the past 50 years
- **almost 90%** – percentage of wild flowering plants that depend to some extent on animal pollination [based on my published estimate – see Chapter 1]
- **1.6 million tonnes** – annual honey production from the western honey bee
- **16.5%** – percentage of vertebrate pollinators threatened with extinction globally
- **40%** – percentage of invertebrate pollinator species (particularly bees and butterflies) facing extinction

Pollination ecology and pollinator conservation is a fast-moving field, and there have already been significant scientific and policy developments since the IPBES assessment was finalised. To coincide with the release of the report, some of the authors published important articles in the two most prestigious scientific journals. The first was 'Safeguarding pollinators and their values to human well-being' (Potts *et al.* 2016, *Nature*), followed by 'Ten policies for pollinators' (Dicks *et al.* 2016, *Science*). In the latter paper the authors set out a series of recommendations for politicians. Here's their list, with some annotations [in square brackets]:

1. Raise pesticide regulatory standards [to include our most important pollinators – wild bees and other insects]
2. Promote integrated pest management (IPM) [rather than automatically feeding the profits of agrochemical companies]
3. Include indirect and sublethal effects in GM crop risk assessments
4. Regulate movement of managed pollinators [lots of evidence that poor husbandry is a major cause of colony collapse disorder, for example]
5. Develop incentives, such as insurance schemes, to help farmers benefit from ecosystem services instead of agrochemicals
6. Recognise pollination as an agricultural input in extension services
7. Support diversified farming systems [does Brexit provide an opportunity to do this in the UK?]

8 Conserve and restore 'green infrastructure' (a network of habitats that pollinators can move between) in agricultural and urban landscapes [already lots being done on this in urban areas but much less in rural areas – see Chapter 8]

9 Develop long-term monitoring of pollinators and pollination [started for the UK and some other countries – see Chapter 10]

10 Fund participatory research on improving yields in organic, diversified and ecologically intensified farming

Overall it was a sensible set of recommendations, and I would have added just two: first, to develop education and awareness programmes on the importance of natural capital and ecosystem services, aimed at farmers, civil servants, politicians, planners, business and industry, developers, and others; and second, to build consideration of natural capital into local planning systems so that the loss of habitats (including trees, ponds, etc.) is properly accounted for, something which is happening regionally as an outcome of the Nene Valley NIA project (see above and Chapter 7).

Getting politicians to take notice of these recommendations in an age where scientific experts are derided as no different to 'soothsayers and astrologers' will be a challenge, though. In the USA, between 2000 and 2017, individual states brought in 110 new laws and policies relating to pollinator conservation, including aspects of beekeeping, use of pesticides, public awareness, scientific research and conservation of habitat (Hall and Steiner 2019). As the authors note, these laws 'narrate an evolution of bureaucratic thinking on insects'. However, they concluded that this 'bureaucratic thinking' had not addressed almost half of ten policy targets that had been proposed by scientists in the USA as important for pollinator conservation.

Final thoughts

The science underlying the ecology, biodiversity and natural history of ecological processes and the ecosystem services they provide to society is hugely complex. That's true even for a well-defined aspect of ecology such as animal pollination of crops and wild plants. As someone who has studied pollination ecology for over thirty years I know how little we truly understand – yet this is supposed to be one of the more 'simple' ecosystem services, according to a UK Parliamentary Office of Science and Technology (POST) briefing document from 2007. Translating that scientific complexity and uncertainty into concepts and language that can be understood by governments and wider society is a challenging task. However, I'd argue that in order to conserve plant–pollinator interactions we don't need to know the finer details and dynamics of the species/communities/ecosystems involved (as interesting as they are). What we require is as much natural and semi-natural habitat within a landscape as is possible, appropriately managed (or left alone), and with as few anthropogenic stressors (such as pesticides and other pollutants) as possible. And we've known that for many years, long before the start of the current campaigns to 'save the pollinators'. Yet governments and agri-business consistently fail to deliver this basic requirement, and our natural environment is at risk of becoming ever less diverse and less hospitable to biodiversity. In the UK, species-rich grasslands are still declining long after it was pointed out that over 90% had disappeared: planting of new wild-flower meadows is all very well but it is not equivalent to saving old ones with their complex communities of fungi, bacteria and invertebrates. Loss of biodiversity is a pattern that is being repeated all over the world, despite the best intentions of NGOs and the strategies and initiatives of governments and businesses.

In this chapter I've laid out a broad view of where the political and societal interest in pollinators came from, how it has manifested itself in wider society, and what this means now. But as we have seen recently in Brazil and the USA, a change of government can result in huge policy shifts that reverse progress made in environmental protection. Some of this impacts directly on pollinators: President Donald Trump's pledge to build a wall along the US–Mexican border to prevent illegal migration is being challenged by the National Butterfly Center in southern Texas because of its effects on both resident and (with a direct connection back to *The Forgotten Pollinators*, where much of this began) migratory species (Zurcher 2019). The fates of pollinators have always been tightly bound to government policies, and as we saw in Chapter 10 this has been true for at least a hundred years. It is up to us as citizens, as consumers, and as members of NGOs to ensure that pollinator conservation is always on the political agenda. Indeed, local actions and schemes are just as important as national or international initiatives. Local government can play a role here, as when Northamptonshire Councillor Jim Hakewill became Mayor of Kettering in 2012 and chose bee conservation as the theme of his time in office. Likewise, a number of the UK's Local Nature Partnerships (LNPs) have focused some of their attention on raising awareness of the issues around pollinator decline and encouraging citizens to take action. Northamptonshire LNP (on which I sit) has 'Power to the Pollinators' as one of its priority themes, which involves engaging with county residents via talks, advising on management and planting in public spaces, and so on. As we'll see in the final chapter, everyone can make a contribution in so many different ways.

Chapter 14

Studying pollinators and pollination

Four main aims motivated me to write this book. The first was to share some of the fascination I have with the interactions between plants and their pollinators, the outcome of those relationships for both groups of organisms, and their importance to both nature and society. It's a fascination that has enriched my life enormously: not only has it been at the centre of my career to date, but it's given me opportunities to travel, to go to places that many only see through television and cinema, to share my knowledge, to interact with brilliant colleagues, and to discover things that no other biologist has documented. A second aim was to dispel some of the myths and misunderstandings around pollinators and their role in plant reproduction. These are long-dispelled ideas that still appear in some textbooks, but there are also subtleties and complexities of the interaction that are often not widely appreciated. Myths include the idea that wind pollination was somehow the 'primitive' state in the evolution of flowering plants (Chapter 2), that honey bees pollinate most crops and wild plants, or indeed that bees in general are the only significant group of pollinators (Chapters 2 and 7). Misunderstandings include the idea that pollinators and flowers somehow 'cooperate', as opposed to mutually exploiting one another, or that plants with highly specialised flower structure, such as orchids, only have a single pollinator per species (Chapter 4).

The information contained in this book that supports these first two aims has come from research by professional scientists, myself included, but also from the work of highly knowledgeable non-professional naturalists and members of the public. All of these groups of individuals have made a contribution to our understanding of the ecology, evolution and conservation of pollinators and their interactions with flowers. But here's the thing: there is an enormous amount that we still do not understand about plant–pollinator interactions, and my third aim for the book was therefore to demonstrate just how much there is still to discover, and to stimulate further study. Hence, this chapter is written for enthusiastic naturalists of all abilities and backgrounds who may not have realised that they can make important contributions to this most fascinating area of ecological enquiry. However, I hope that it's also useful for scientists who are already working on pollinators and flowers, for whom some of the sources listed below may be new. Likewise, the chapter may benefit scientists from other disciplines thinking of moving into the field, and students at all levels who need some initial pointers. The ideas and resources presented below are not comprehensive, nor are they meant to be: the intention is to provide a taster of the richness of information that is available (often freely) and how this can inform and aid the study of pollinators and pollination. Although I have tried to be as international as possible, inevitably there's going to be an emphasis on the resources I know best – from the UK and other parts of the Anglophone world. However, on my blog I've included some examples of websites from elsewhere, and in other languages.

As for the fourth aim, I'll discuss it at the end of this chapter.

Identifying plants and pollinators

The first step to understanding more about plant–pollinator interactions is to know what it is that you are looking at. For all of us, the process of identification starts at a high taxonomic level: is it a bee or a fly or a butterfly? Then, with time and experience, it becomes possible to identify at ever finer levels, to distinguish bumblebees, honey bees and solitary bees, for instance. Identifying to species level is relatively easy for some groups, such as butterflies, but much harder for bees and hoverflies, for example. Even ecologists doing research on pollinators struggle without specialist help, and (if exact identifications are not crucial to the question being addressed) may lump pollinators into 'functional groups' or 'guilds' of species, such as 'large bees', 'small bees', 'small flies', 'honey bees' and 'others' (see Figure 6.3).

The availability and quality of identification guides to different groups of pollinators varies enormously across the world. Europe is well catered for, with some great books on bees, hoverflies, butterflies and moths. I have listed a small selection of these in Box 14.1, along with a couple of useful guides from North America and India. Many other countries and regions have similar guides, as they do for birds of course, which is useful if you are studying hummingbirds or sunbirds.

Plants are even better served with field guides than most of the pollinator groups, and new guides to regions not previously covered are appearing every year. However, mobile phone apps for plant identification have recently started to be developed, such as PlantSnap and PlantNet. Although I don't have much experience with these apps, I've seen others use them, and they can be very accurate for common species, at least in Britain. That accuracy will no doubt increase in the future. The era of *Star Trek* technology in which a species can be 'scanned' visually, or its DNA sequenced in the field, may not be so far away.

There are plenty of free online resources for identifying plants and pollinators too. For British bees I can recommend Steven Falk's Flickr site (www.flickr.com/photos/63075200@N07/collections) and the BWARS web pages (www.bwars.com). Several butterfly groups have online identification guides too, such as Butterfly Conservation's *Identify a Butterfly* (butterfly-conservation.org/butterflies/

Box 14.1 Insect pollinators: some identification guides

- Steven Falk & Richard Lewington, *Field Guide to the Bees of Great Britain and Ireland* (Bloomsbury, 2015; also published in Dutch in 2020, with some additional species)
- Stuart Ball & Roger Morris, *Britain's Hoverflies: A Field Guide* (Princeton University Press; 2nd edition, 2015)
- Alan Stubbs & Steven Falk, *British Hoverflies: An Illustrated Identification Guide.* British Entomological & Natural History Society; 2nd edition, 2002)
- Mark P. van Veen, *Hoverflies of Northwest Europe: Identification Keys to the Syrphidae* (KNNV; 2nd edition, 2010)
- Luis Martín, Alberto Castiel & Elisa Sandoval, *Guía de Campo de los Polinizadores de España* (Ediciones Mundi-Prensa, 2015)
- David Newland, Robert Still, Andy Swash & David Tomlinson, *Britain's Butterflies* (Princeton University Press, 2nd edition, 2020)
- David Newland, Robert Still & Andy Swash, *Britain's Day-Flying Moths* (Princeton University Press, 2nd edition, 2019)
- Paul Waring, Martin Townsend & Richard Lewington, *Field Guide to the Moths of Great Britain and Ireland* (Bloomsbury, 3rd edition, 2017)
- Jeffrey H. Skevington & Michelle M. Locke, *Field Guide to the Flower Flies of Northeastern North America* (Princeton University Press, 2019)
- B. Mitra, *Flower Visiting Flies (Diptera, Insecta) of Kolkata and Surroundings* (Zoological Survey of India, 2008)

identify-a-butterfly). There are also specialist groups on Facebook for Hymenoptera, African butterflies, plants, Australian bees, and so forth where enthusiasts and professionals provide help to identify species from photographs that you upload. When you do this it's important to provide as much detail as possible, such as date, locality, weather, time of day, behaviour, flowers that were being foraged, caterpillar host plants, nesting sites and so forth. When you use these sites, you are not only getting help with identifying species, you are also providing information that helps us to understand the distribution and ecology of pollinators and the plants they use, when during the year they are active, and how all of this changes over time.

Identifying by using all the senses that you have

I often tell my students that in order to begin to study pollinators and their interactions with flowers, you need two sets of very sophisticated equipment: your senses and a brain. That's a basic requirement, and with these you can do a lot. However, if you has limited hearing or vision, there's still much that you can learn in the field. For example, with my eyes closed I can distinguish between honey bees, bumblebees and solitary bees in the genus *Anthophora*, simply by the sounds they make (a kind of aural 'jizz' if you like – see below). Bumblebees produce a deeper buzz than honey bees, and *Anthophora* bees are very fast and frenetic in flight. That's based on experience but no real effort: I'm sure that anyone could do much better if they practised. I know of a couple of research projects that are trying to produce automated identification systems based on the sounds insects make, with varying degrees of success. It may never be possible to identify insects to species level using sound, but categorisation to higher taxonomic levels might be feasible.

Some bees have distinctive smells too, and of course flowers are the epitome of organisms that can be distinguished by their scent, shape and texture, so limited eyesight is not a hurdle in this respect.

The requirement of a brain, in this instance, is for you to approach every encounter with consideration, curiosity, and a desire to understand. That's a great starting point for anyone, and with experience, reading, and interacting with other enthusiasts, comes greater knowledge

Once you become familiar with the pollinators in your own region you will begin to realise that there are some species that, with a little experience, can be identified at a glance, sometimes on the wing. Birders talk of the 'jizz' of a bird that enables them to identify a species in flight from a distance, and the same is possible for many of the more distinctive butterflies and day-flying moths.[4] There are also some bees that fall into this category, but they are fewer in number.

Identifying pollinators from photographs is an easier proposition, as I noted in the previous chapter, but only if the image is good and the species is distinctive. A few years ago I thought I had a new species of bee for our garden, and indeed for Northamptonshire as a whole: the little flower bee (*Anthophora bimaculata*). One solitary bee authority confirmed it from my photographs and I duly posted the story on my blog, only for another authority to question it and suggest that it was in fact a different species (*A. quadrimaculata*), which has since been confirmed in the garden. This disagreement by experts illustrates the point that specimens are needed to determine most bee species. Collecting specimens is a topic that divides opinion, and many naturalists and environmentalists are against it: 'We are supposed to be conserving these species, not killing them', goes the argument. I appreciate these concerns, but collecting a modest number of individuals for the purposes of identification does more good than harm unless the species is ultra-rare. However, you need to know what you are doing, and I'd recommend talking with specialist entomologists before trying to build up a collection of specimens.

Recording and submitting observations

Once you have started identifying pollinators and their plants, what should you do with that information? Well, a starting point is to keep a garden list, or a list of species that you have encountered in a park or nature reserve, or on your allotment or local patch. When you are confident of your identifications, and can back them up with images, then you can submit them to natural history sites like iRecord (www.brc.ac.uk/irecord) or the African Butterfly Database (www.abdb-africa.org). In the UK, local Biological Records Centres and the County Recorders for bees, flies, beetles, butterflies, plants etc. would also be interested in your records. You can find contact details by googling appropriate terms, or getting in touch with your local Wildlife Trust. In Ireland there's the National Biodiversity Data Centre (www.biodiversityireland.ie) and the *All-Ireland Pollinator Plan* (www.pollinators.ie), which has information on how to get involved in a range of citizen science initiatives such as bumblebee and butterfly monitoring schemes. Many other countries have similar websites and schemes for recording observations. As I've noted in some of the previous chapters, use of this data by scientists shows that amateur naturalists can make a significant contribution to scientific understanding of the distribution of pollinators and plants, and therefore to nature conservation. Knowing where species are present, or indeed where they are absent, can be very helpful if we wish to conserve them, particularly rare species in fragmented habitats.

The advent of digital photography and high-quality camera phones means that images can be taken and shared much more quickly than was previously possible, giving rise to recording apps such as iRecord Butterflies (butterfly-conservation.org/our-work/recording-and-monitoring/irecord-butterflies). Photographs can

FIGURE 14.1 Painted lady butterfly (*Vanessa cardui*) perching, and possibly feeding, on yarrow (*Achillea millefolium*). Note the wear on the wings – this individual is probably a long-distance migrant. Photo by Mark Hindmarch.

contain much more than just an image of the subject, however. Take Figure 14.1, for instance. This photograph of a rather tatty-looking painted lady butterfly (*Vanessa cardui*) is date-stamped, so we know when in the season it was photographed. Some smartphones allow one to add a location to images, so we might also have information about where in the world it was taken (I use a free app called GPS Status to record latitude, longitude and altitude for images when I'm collecting data in the field). The butterfly is on yarrow (*Achillea millefolium*), so we suspect that's one of its nectar plants. There are other plants present, so we know something of the ecological context of the habitat in which the butterfly was living. Finally, the wings are extremely worn and faded: this is an old butterfly that has probably migrated here from the continent. The wing damage might even indicate that it's been attacked by a bird at some point. So much information in just one image. In our global study of Apocynaceae pollination systems that I've discussed in previous chapters we used the images from the Atlas of African Lepidoptera project, one of a number of citizen science projects covering a wide range of organisms (vmus.adu.org.za/vm_projects.php), in just this way, and it gave us a huge amount of information on butterflies as flower visitors to members of this family in Africa. The citizen scientists who uploaded these photographs made an important contribution to our work.

Understanding pollination: some simple garden experiments

Observing the species in your own patch of the world and seeing how they change over time can add important information to our knowledge of the natural world. Understanding of the declines of large, once-common moths in Britain would not have been possible without the regular light trapping undertaken by moth enthusiasts, for instance (see Chapter 10). However, some may wish to take their studies further by actually conducting experiments in the garden, to start to really appreciate the ecology of these animals and how pollination functions as an ecological process. There's a venerable history of such research – most of Charles Darwin's work on plants was conducted in his own gardens and greenhouses. More recently, other scientists have also used their own patch for data collection, myself included (see Chapter 9). It also provides an opportunity for kids to learn more about nature, so it's good to get them involved too. I've set out a few ideas in the following paragraphs. But really, the only limit to what you can discover is your own imagination: spend time sitting in your garden or local park, closely observing and thinking, and let the questions come to you.

Observing and thinking are particularly effective when it comes to unveiling the precise nature of the interactions between a flower and its pollinator(s). As we saw in Chapter 3, flowers can have some interesting and incredibly intricate adaptations to being pollinated by a single pollinator, or by a range of different species. Watching the insects, birds and other animals that interact with the flowers gives an insight into how those flowers have evolved. For example, watching bumblebees chewing holes in the base of *Fuchsia* flowers, and stealing the nectar without picking up or depositing any pollen, suddenly makes sense when you realise that many of these flowers have evolved to be pollinated by long-tongued hummingbirds that can easily access the nectar from the front. For some short-tongued bees going in from the back is easier, though others land on the reproductive parts and crawl in from the front. Every type of flower has a similar story to tell, and unravelling those stories can be fascinating.

The relative effectiveness of different types of 'bee hotels' and other artificial nest sites for pollinators is still largely unknown. Does it matter if these are made of wood, stone or plastic? Is bamboo better than holes drilled in wood? Do the same holes get used by the same species each year, or does it vary? To what extent does the diameter or the depth of the holes make a difference? There are

FIGURE 14.2 A nest box for cavity-nesting bees that incorporates a viewing port. These should only be opened occasionally, to reduce disturbance to the mother bees. In this model the whole nesting unit can be slid out to check the nests, and for cleaning.

lots of questions that can be asked, providing endless opportunities for testing different styles and configurations of bee hotels – and what works in one garden, or area of a garden, may not work in another, so the effects of different habitats can also be studied. Some models of bee hotel have clear plastic viewing ports that you can uncover to check the nests, which adds an extra dimension to any study of bee nesting biology (Figure 14.2). Again, this is something that could really appeal to children.

Similar sorts of questions can be asked about ground-nesting bees and, if you have space, you could experiment with cutting lawns to different heights or assessing the bees' soil preferences. Some species seem to go for clay, others for sandy soil, on either flat or sloping sites, in the shade or in the sun. This is still an area with a lot of unanswered questions, and there is scope for the general public to make a real contribution. A citizen science project run by Steph Maher and Tom Ings at Anglia Ruskin University asked people to send in records of large mining bee aggregations. Three hundred and ninety-four records were submitted, including details of the type of bee, locality, size of the aggregation, topography, shade, and so forth. This produced a huge amount of data for Steph's PhD, helping scientists to understand more about how female mining bees make decisions on where to site their nests. As Steph and her colleagues pointed out in a publication stemming from the work, ground-nesting bee aggregations 'are both ephemeral and cryptic structures in the landscape', and such a significant data set could not have been generated without the aid of the public (Maher *et al.* 2019).

There's also still a lot we don't know about why some garden plants are more popular than others for pollinators. Is this related to how abundant different flowers are? Are some nectars or pollens toxic? How often do pollinators switch between different types of flowers, or do they show flower constancy and mainly move between the same types of flowers? How long do mother bees stay outside their nest during foraging trips? How frequently do parasitic bees and wasps visit these nests? Using a voice recorder or the video facility on your mobile phone can help you to track fast-moving individuals and to study their behaviour by timing how long they remain on flowers, when they move, and so forth. Voice recorders are especially useful for following territorial bee species such as the wool carder bee (*Anthidium manicatum*) that move rapidly to chase off competitors.

The tree bumblebee nest in our garden that I mentioned in Chapter 11 was a source of fascination for me during the early summer of 2012. Binoculars in hand and cup of coffee by my side, I spent time observing the bees entering and leaving the eaves of the house. The results it generated were not unexpected: in the morning, more bees leave the nest than return, whereas later in the day it's about equal as bees come and go in their foraging trips, and activity peaks during the afternoon (Figure 14.3). However, it gave a fascinating insight into the behaviour of a bee that was still relatively poorly studied at that time. I also noticed that the only flight orientations I saw occurred in the late afternoon and early evening. These are brief trips from the nest that the newly emerged worker bees make,

FIGURE 14.3 The average number of tree bumblebees (*Bombus hypnorum*) entering and leaving a garden nest per 5-minute observation period, at different times of the day. Data collected in our garden in Northampton, June and July 2012.

flying backwards and forwards in front of the entrance before returning, in order to learn the location of the nest. It was a very small sample size (just four bees), but I wonder if that's a general pattern for this species.

For a lot of edible garden plant varieties, we know very little about how important pollinators are for ensuring fruit and seed set. By covering flower buds with small net bags such as the voile ones that are used for wedding gifts, and comparing bagged and unbagged flowers, you can understand how important those pollinators are for ensuring good crop yields (see also 'pollination bags' in Wikipedia). I did this a few years ago on a large blue passion flower (*Passiflora caerulea*), the vivid orange fruits of which are edible (Figure 14.4). I learned a few things from this experiment, including that the plant does not set fruit without pollinators – bagged flowers produced no fruit, as you can see in the smaller graph in Figure 14.4. So I conclude that pollinators are necessary for fruit set. I suspect that the species is self-compatible, as this was the only individual in the vicinity, but that would need to be tested by performing hand pollinations. Fruits differ greatly in size, by a factor of more than two, from about 23 mm to 57 mm in length, and from 19 mm to 34 mm in width. The size of the fruit, not surprisingly, correlates with the number of seeds it contains, which can range from just two to more than 190. This tells me that the amount of pollen being delivered to each flower over its lifetime also varies – because, of course, one pollen grain fertilises one ovule to produce one seed. I suspect that large bumblebees are better pollinators of blue passion flower than smaller honey bees, solitary bees and hoverflies, because they more frequently contact the stigmas while foraging for nectar (see Figures 9.3b and 14.4). But again, this needs to be more formally tested by counting the number of contacts made by each group of insects, or counting pollen grains deposited after one visit to a newly opened flower. Finally, I learned that my sample size was

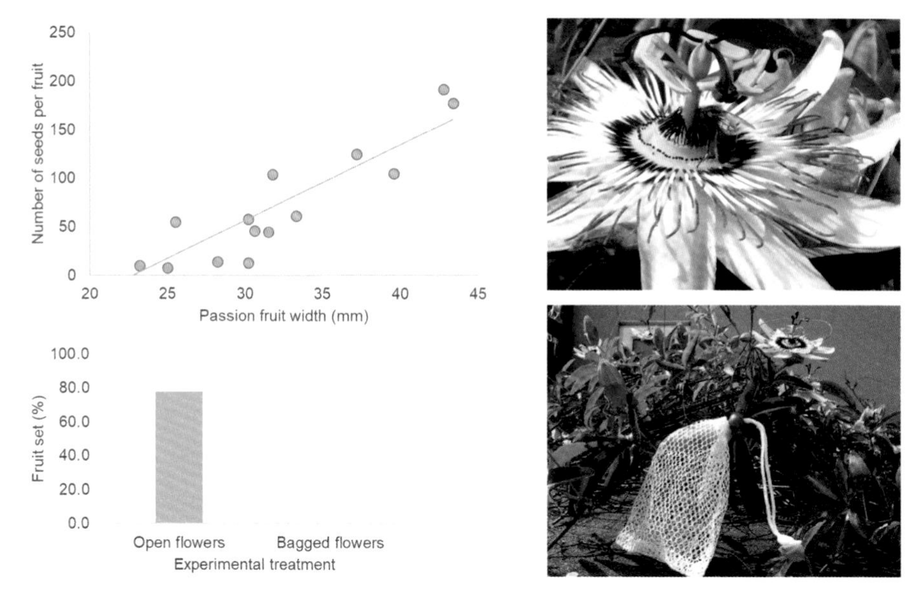

FIGURE 14.4 Overview of my passion fruit experiment. One of the pollinator-excluded flowers is shown in the lower right image (yes, that is the type of bag that's used for washing machine tablets). The top right image is a honey bee – notice how it's not large enough to contact the anthers or stigma, and contrast it with the bumblebee in Figure 9.3b. There is a positive correlation between fruit size and number of seeds in each fruit (top graph). Bagging flowers results in no fruit set at all (bottom graph). Data collected in our Northampton garden.

too small – I started by bagging more buds, but some of these were lost when the flowers went mouldy in the rain and my wife Karin got a bit too enthusiastic with her secateurs. That's common, though – I know from previous experience that you need to bag at least twice as many buds as you think you will need.

I chose blue passion flower for this experiment because the flowers are large and easy to manipulate, and because I could find no information about the pollination biology of the species in the UK. But that's true for many garden plants, and so there's a lot still to be learned and lots of opportunities for people to find out more about the pollination biology of the plants they grow in their gardens by performing simple experiments such as this. Two books on the techniques used by pollination ecologists have been published that provide more ideas and information: *Techniques for Pollination Biologists* (Kearns and Inouye 1993), and *Practical Pollination Biology* (Dafni *et al.* 2005).

Sharing your findings

For most people, collecting information about plants and pollinators and sharing it on iRecord is sufficient. But some may want to share their findings more widely. Social media is the most immediate way of doing so, and every day my Twitter

Box 14.2 Pollinator and pollination blogs

- Emily Scott's Adventuresinbeeland's Blog: adventuresinbeeland.com
- The University of KwaZulu-Natal's Pollination Research Lab: pollinationresearch.wordpress.com/blog
- Michael Whitehead's blog on plants, pollination, evolution, ecology, natural history, and beer-making using wild yeasts from nectar: michaelwhitehead.net
- Bad Beekeeping Blog: badbeekeepingblog.com
- The Bartomeus Lab blog on ecology, global change and pollinators: bartomeuslab.com/blog
- Natural Newstead – Observations of flora, fauna and landscape in central Victoria, Australia (very good for bird pollinators): geoffpark.wordpress.com
- Under the Banyan (strong focus on fig trees): underthebanyan.blog
- Ray Cannon's nature notes: rcannon992.com
- Ecological Interactions Lab – Diego Vázquez's Group in Argentina: interactio.org
- Murtagh's Meadow – Ramblings of an Irish ecologist and gardener: murtaghsmeadow.wordpress.com
- Bees in a French Garden: beesinafrenchgarden.wordpress.com
- Ecology is Not a Dirty Word: ecologyisnotadirtyword.com
- Philip Strange Science and Nature Writing: philipstrange.wordpress.com

feed is full of tweets from people who have photographed pollinators in their garden, place of work or local nature reserve; using appropriate hashtags such #bees, #butterflies or #pollinators ensures that they are seen by a wider audience. A more permanent, but labour-intensive, way of sharing findings can be to set up a Facebook page devoted to your area, or to begin a blog. I started blogging in 2012 (jeffollerton.co.uk) as a way of sharing some of the writing that I was doing anyway without any kind of outlet for it. There are some really interesting blogs out there which focus partly or solely on pollinators, and the examples listed in Box 14.2 will give you an idea of the range of what is available in the blogosphere.

Another way to share your observations is through the more traditional forum of writing articles for magazines, and there are any number of local natural history journals (printed and electronic) that are eager to publish such work. For some, this interest in documenting the natural world leads eventually to writing a book – and I've listed some examples in a later section.

Some thoughts on keeping honey bees

When thinking about pollinators, and studying or conserving them, many people's first thought is of honey bees – and they wonder about keeping a hive or two in their garden or allotment. In 2012, in an article in the Royal Horticultural Society's *Plantsman* magazine entitled 'The importance of native pollinators', I wrote:

> By all means, if you wish to help honey bees then keep some hives. But that will not help our more important native pollinators: it is the equivalent of trying to conserve songbirds by opening a chicken farm.

In retrospect I think I'd modify that statement, because there are certainly rare breeds of chickens (and varieties of honey bees) that deserve to be conserved despite being anthropogenic in origin: human-modified biodiversity is biodiversity nonetheless. But the core of the statement holds true: keeping honey bees is not, on its own, going to do much for wild pollinators, which are often much more threatened than their managed relatives (see Chapter 10). However, beekeeping is a fascinating pastime in its own right and can teach one a lot about bee biology. Although it's relatively easy to get hold of the necessary equipment and stock to become a novice beekeeper, I'd urge anyone who wishes to do so to get in touch with one of their local beekeeping groups – the British Beekeeping Association website (www.bbka.org.uk) can help you find one.

Getting expert advice is important, because keeping bees is not easy; it takes years to perfect the craft, and even experienced beekeepers make mistakes and lose colonies. Nor is it perfectly safe. A couple of years ago I was asked by a beekeeping society to do a talk for them on the diversity of pollinators, and I spoke to a group of almost fifty interested beekeepers. After my talk came the club notices – and one item of news was especially poignant. A member's husband (who was not himself a beekeeper, simply an observer of his wife's hobby) had recently died and it was confirmed that this had been caused by anaphylactic shock from a bee sting. It was a sad and sobering moment for the society members and for me. But it brought home the fact that beekeeping is not a hobby (or even an occupation) to be taken up lightly. Indeed, some beekeepers I have spoken to have strong misgivings about the way that beekeeping is being promoted by companies, the media and celebrities as 'the next thing' for people to try, especially in urban areas. Aside from the (admittedly rare) chance of being seriously hurt or killed by bees, beekeeping is a technically demanding hobby that requires study and skill if the colony is not to fail in its first year. If you're serious about beekeeping, don't let any of this put you off, but talk to experienced beekeepers and go to an apiary open day such as those run by Bee Urban in London (Figure 14.5).

FIGURE 14.5 An open day at the Bee Urban apiary in Kennington, London. Attendees get to take part in a hive inspection, learn more about keeping bees, and taste some excellent honey-based products.

Learning more about plant–pollinator interactions

So many books have been published about pollinators in recent years that it's hard to know where to start. Many of these focus on bees, or even just honey bees, mixed with personal stories – what I like to term 'auto-bee-ography' – and such books can provide a good and accessible introduction to the lives of pollinators and flowers. Examples include Brigit Strawbridge Howard's *Dancing with Bees* (2019), Helen Jukes' *A Honeybee Heart Has Five Openings* (2018), and Thomas Seeley's *The Lives of Bees* (2019). Dave Goulson has written a series of auto-bee-ographical accounts of his work with bumblebees and other pollinators, his attempts to restore a French grassland, and the value of wildlife gardening: *A Sting in the Tale* (2013), *A Buzz in the Meadow* (2014), *Bee Quest* (2017b) and *The Garden Jungle* (2019). His volume on *Bumblebees: Behaviour, Ecology, and Conservation* (2003b), while more technical, is packed with information, as is Ted Benton's *Bumblebees* (2006). Apart from the Falk and Lewington book on bees, mentioned above, the other recent comprehensive work on this topic is the two-volume *Handbook of the Bees of the British Isles* by George Else and Mike Edwards (2018). For a more global perspective on this group of insects, *Bees: A Natural History* by Chris O'Toole (2013) is a great introduction, and packed with ideas for further reading and garden projects. *The Solitary Bees* by Bryan Danforth and colleagues (2019) provides a comprehensive account of the non-social species and their diverse habits. For younger readers there's also a wealth of books available, and as an introduction to bees I can recommend *The Bee Book* by Fergus Chadwick *et al.* (2016).

Moving away from the ubiquitous bees, Erica McAlister's *The Secret Life of Flies* (2017) includes a chapter about dipteran pollinators. *Wily Violets and Underground Orchids* (1989) and *The Rose's Kiss* (1999), both by Peter Bernhardt, give interesting insights into pollination from the flower's perspective. Michael Proctor, Peter Yeo and Andrew Lack's *The Natural History of Pollination* (1996), which I cited in Chapter 13, is getting a little dated now but is still packed full of good, basic knowledge on the topic. Pat Willmer's more recent *Pollination and Floral Ecology* (2011) is incredibly comprehensive, although it contains a few errors related to aspects of the botany and the interaction networks.

For those wishing to go deeper and access the scientific literature, keyword searches in Google Scholar are the best approach. Some of the articles and book chapters that are found will be freely available as PDFs, but others may be behind a publisher's paywall. Contacting the author via email will usually result in them sending you a copy, however. The *Journal of Pollination Ecology* (www.pollinationecology.org) is free to access and has an associated *Pollination Magazine* (jpollecol.blogspot.com) in which scientists present their findings to a wider audience.

Web sites about pollinators and pollination vary in their quality, and it pays to read what they say with a critical eye and to learn which are the reputable sources. Some are promoted by agrochemical companies and so, not surprisingly, take a particular view of how pesticides are affecting pollinator populations. Some examples of websites about bees and other pollinators, many of which are not in English, can be found on my blog, which I regularly update.

One of the very best ways to discover more about pollinators and pollination is to watch some of the many videos of lectures by experts that are available on YouTube. New ones are added regularly and a search using appropriate key words will open up a world of knowledge.

Podcasts have become very popular in recent years and one of my favourites is the *In Defense of Plants* series (www.indefenseofplants.com); search for my name and you'll find an interview I did about pollination in 'my' plant family, the Apocynaceae – and there's lots of other good material on that site.

Finally, look out for lectures and courses on identification skills, pollinator ecology and flower biology delivered by natural history societies and NGOs. In the UK this means organisations such as the Wildlife Trusts and the Field Studies Council, but similar groups exist in many parts of the world, and most have websites.

The fourth aim

As I was completing the first draft of this book in the summer of 2019 I had the unexpected pleasure of reading *Enlightenment Now* by Steven Pinker. I referred to this book in Chapter 13, and Pinker's thesis is broadly correct: since the Enlightenment of the eighteenth century, humans have made enormous progress, globally and in most continents and countries, in a range of areas including health, education, personal liberty and knowledge. But this has come at a cost to the Earth that has left a legacy of environmental pollution, destruction and climate change for ourselves and our descendants. In some places that pollution and destruction has halted and is beginning to be reversed, but in many regions it is as bad as it ever was, if not worse. Overall, the planet is in a poorer state than it was, say, fifty years ago, and climate change is not going to make it any better over the next fifty. Action by individuals, businesses and governments is urgently needed to change behaviours and policies.

I hope that this book has given you an insight into the world of pollinators and the plants that they pollinate, and has kindled a desire to learn more. Because my fourth aim in writing this book is to help ensure that these species, and their relationships, are not lost. Pollinators and their interactions with plants can act as a rallying point for galvanising the kind of societal change that is required, because pollination is woven through our society, our agriculture, and the ecology on which we depend. The world would be a poorer place, both biologically and culturally, if plants had not evolved relationships with the animals that pollinate them. We have a duty to ensure their future – and developing our own relationships with, and understanding of, these organisms will go a long way towards achieving this. The question is not if, but how, we choose to do that.

Notes

1. Sarma *et al.* (2007) refer to the snail as *Lamellaxis gracile*, and the plant as *Volvulopsis nummularium*. These are now both outdated synonyms.
2. In its earlier stages this work was carried out in collaboration with her husband, the late Denis Owen. Denis was the second supervisor of my PhD and first supervisor to quite a number of entomologists who work on pollinating insects, including Dave Goulson and Tim Shreeve. His death just after retiring was a loss to insect ecology, but he left a large and sometimes idiosyncratic legacy of research papers and books.
3. I first read that in King's novel *From a Buick 8*, but a later search on Google suggests that it was originally an Archie Bunker line.
4. The derivation of the word 'jizz' (also spelled 'gis' and 'giss') is unknown. I, like many others, had always assumed that it stood for 'general impression of size and shape'. But this has recently been debunked by Greenwood and Greenwood (2018).

References

Agrawal, A. (2017) *Monarchs and Milkweed: A Migrating Butterfly, a Poisonous Plant, and Their Remarkable Story of Coevolution*. Princeton University Press, Princeton, NJ.

Aguilar, R., Martén-Rodríguez, S., Avila-Sakar, G. *et al.* (2015) A global review of pollination syndromes: a response to Ollerton *et al.* 2015. *Journal of Pollination Ecology* 17: 126–128.

Aizen, M.A. and Harder, L.D. (2009) The global stock of domesticated honeybees is growing slower than agricultural demand for pollination. *Current Biology* 19: 915–918.

Aizen, M.A., Garibaldi, L.A., Cunningham, S.A. and Klein, A.M. (2008) Long-term global trends in crop yield and production reveal no current pollination shortage but increasing pollinator dependency. *Current Biology* 18: 1572–1575.

Alarcón, R., Waser, N.M. and Ollerton, J. (2008) Year-to-year variation in the topology of a plant–pollinator interaction network. *Oikos* 117: 1796–1807.

Alayón, D.I.O. (2013) The Macaronesian bird-flowered element as a model system to study the evolution of ornithophilous floral traits. *Vieraea* 41: 73–89.

Alexander, K.N.A. (2002) The invertebrates of living and decaying timber in Britain and Ireland: a provisional annotated checklist. English Nature Research Reports no. 467. English Nature, Peterborough.

Alexander, K.N.A., Bower, L. and Green, G.H. (2014) A remarkable saproxylic insect fauna from a traditional orchard in Worcestershire – but are the species resident or transitory? *British Journal of Entomology and Natural History* 27: 221–229.

Algert, S.J., Baameur, A., Diekmann, L.O., Gray, L. and Ortiz, D. (2016) Vegetable output, cost savings, and nutritional value of low-income families' home gardens in San Jose, CA. *Journal of Hunger and Environmental Nutrition* 11: 328–336.

Amorim, F.W., Ballarin, C.S., Mariano, G. *et al.* (2020) Good heavens what animal can pollinate it? A fungus-like holoparasitic plant potentially pollinated by opossums. *Ecology* 101: e03001. doi: 10.1002/ecy.3001.

Anderson, L.D. and Atkins, E.L. (1968) Pesticide usage in relation to beekeeping. *Annual Review of Entomology* 13: 213–238.

Anon. (1919) The bee industry. *British Bee Journal* 25 September 1919: 419.

Aono, Y. and Kazui, K. (2008), Phenological data series of cherry tree flowering in Kyoto, Japan, and its application to reconstruction of springtime temperatures since the 9th century. *International Journal of Climatology* 28: 905–914.

Aono, Y. and Saito, S. (2010) Clarifying springtime temperature reconstructions of the medieval period by gap-filling the cherry blossom phenological data series at Kyoto, Japan. *International Journal of Biometeorology* 54: 211–219.

Arbetman, M., Meeus, I. and Morales, C.L., Aizen, M.A. and Smagghe, G. (2013) Alien parasite hitchhikes to Patagonia on invasive bumblebee. *Biological Invasions* 15: 489–494.

Archer, C.R., Pirk, C.W.W., Carvalheiro, L.G. and Nicolson, S.W. (2014) Economic and ecological implications of geographic bias in pollinator ecology in the light of pollinator declines. *Oikos* 123: 401–406.

Armbruster, W.S. (2017) The specialization continuum in pollination systems: diversity of concepts and implications for ecology, evolution and conservation. *Functional Ecology* 31: 88–100.

Armbruster, W.S. and Baldwin, B.G. (1998) Switch from specialized to generalized pollination. *Nature* 394: 632.

Armbruster, W.S., Lee, J. and Baldwin, B.G. (2009) Macroevolutionary patterns of defense and pollination in *Dalechampia* vines: adaptation, exaptation, and evolutionary novelty. *Proceedings of the National Academy of Sciences of the USA* 106: 18085–18090. doi: 10.1073/pnas.0907051106.

Atkins, P.J. (2003) Is it urban? The relationship between food production and urban space in Britain, 1800–1950. In Hietala, M. and Vahtikari, T. (eds), *The Landscape of Food*. Finnish Literature Society, Helsinki, pp. 133–144.

Bacha, E.L. (1992) *150 Years of Coffee*. Salamandra Consultoria Editorial, Rio de Janeiro, Brazil.

Bailes, E., Ollerton, J., Pattrick, J. and Glover, B.J. (2015) How can an understanding of plant–pollinator interactions contribute to global food security? *Current Opinion in Plant Biology* 26: 72–79.

Bailey, L. and Perry, J.N. (1982) The diminished incidence of *Acarapis woodi* (Rennie) (Acari: Tarsonemidae), in Britain. *Bulletin of Entomological Research* 72: 655–662.

Baldock, K.C.R. (2020) Opportunities and threats for pollinator conservation in global towns and cities. *Current Opinion in Insect Science* 38: 63–71. doi: 10.1016/j.cois.2020.01.006.

Baldock, K.C.R., Goddard, M.A., Hicks, D.M. *et al.* (2015) Where is the UK's pollinator biodiversity? The importance of urban areas for flower-visiting insects. *Proceedings of the Royal Society B* 282: 20142849. doi: 10.1098/rspb.2014.2849.

Baldock, K.C.R., Goddard, M.A., Hicks, D.M. *et al.* (2019) A systems approach reveals urban pollinator hotspots and conservation opportunities. *Nature Ecology & Evolution* 3: 363–373.

Balfour, N.J., Fensome, K.A., Samuelson, E.E. and Ratnieks, F.L. (2015) Following the dance: ground survey of flowers and flower-visiting insects in a summer foraging hotspot identified via honey bee waggle dance decoding. *Agriculture, Ecosystems and Environment* 213: 265–271.

Balfour, N., Ollerton, J., Castellanos, M.C. and Ratnieks, F.L.W. (2018) British phenological records indicate high diversity and extinction rates among late-summer-flying pollinators. *Biological Conservation* 222: 278–283.

Ballantyne, G., Baldock, K.C.R. and Willmer, P.G. (2015) Constructing more informative plant–pollinator networks: visitation and pollen deposition networks in a heathland plant community. *Proceedings of the Royal Society B* 282: 20151130. doi: 10.1098/rspb.2015.1130.

Bartomeus, I., Ascher, J.S., Gibbs, J. *et al.* (2013) Historical changes in northeastern US bee pollinators related to shared ecological traits. *Proceedings of the National Academy of Sciences of the USA* 110: 4656–4660. doi: 10.1073/pnas.1218503110.

Bascompte, J. and Jordano, P. (2014) *Mutualistic Networks*. Princeton University Press, Princeton, NJ.

Bate, J. (2003) *John Clare: a Biography*. Farrar, Straus & Giroux, New York, NY.

Baude, M., Kunin, W., Boatman, N. *et al.* (2016) Historical nectar assessment reveals the fall and rise of floral resources in Britain. *Nature* 530: 85–88.

Beattie, A.J. (2006) The evolution of ant pollination systems. *Botanische Jahrbücher* 127: 43–55.

Beattie, A.J., Turnbull, C., Knox, R.B. and Williams, E.G. (1984) Ant inhibition of pollen function: a possible reason why ant pollination is rare. *American Journal of Botany* 71: 421–426.

Benedek, P. (1972) Possible indirect effect of weed control on population changes of wild bees pollinating lucerne. *Acta Phytopathologica Academia Scientifica Hungaria* 7: 267–278.

Benton, T. (2006) *Bumblebees*. New Naturalist 98. Collins, London.

Benton, T. (2008) *Bombus ruderarius* (Müller, 1776): current knowledge of its autecology and reasons for decline. http://hymettus.org.uk/downloads/B.%20ruderarius%20report08.pdf (accessed August 2020).

Bentos, T.V., Mesquita, R.C.G. and Williamson, G.B. (2008) Reproductive phenology of Central Amazon pioneer trees. *Tropical Conservation Science* 1: 186–203.

Bernhardt, P. (1989) *Wily Violets and Underground Orchids: Revelations of a Botanist*. University of Chicago Press, Chicago.

Bernhardt, P. (1999) *The Rose's Kiss: A Natural History of Flowers*. University of Chicago Press, Chicago.

Biella, P., Akter, A., Ollerton, J. *et al.* (2019) Experimental loss of generalist plants reveals alterations in plant–pollinator interactions and a constrained flexibility of foraging. *Scientific Reports* 9: 1–13.

Biella, P., Akter, A., Ollerton, J., Nielsen, A. and Klecka, J. (2020a) An empirical attack tolerance test alters the structure and species richness of plant–pollinator networks. *Functional Ecology* (in press).

Biella, P., Ćetković, A., Gogala, A. *et al.* (2020b) North-westward range expansion of the bumblebee *Bombus haematurus* into Central Europe is associated with warmer winters and niche conservatism. *Insect Science*. doi: 10.1111/1744-7917.12800.

Biesmeijer, J.C., Roberts, S.P.M., Reemer, M. *et al.* (2006) Parallel declines in pollinators and insect-pollinated plants in Britain and the Netherlands. *Science* 313: 351–354.

Bischoff, I., Eckelt, E. and Kuhlmann, M. (2004) On the biology of the ivy-bee *Colletes hederae* Schmidt & Westrich, 1993 (Hymenoptera, Apidae). *Bonner Zoologische Beiträge* 53: 27–35.

Blitzer, E.J., Gibbs, J., Park, M.G. and Danforth, B.N. (2016) Pollination services for apple are dependent on diverse wild bee communities. *Agriculture, Ecosystems and Environment* 221: 1–7.

Breeze, T. (2014) Common misconceptions: why Einstein isn't an authority on bees. https://caerreading.wordpress.com/2014/06/11/common-misconceptions-why-einstein-isnt-an-authority-on-bees (accessed August 2020).

Breeze, T.D., Bailey, A.P., Balcombe, K.G. and Potts, S.G. (2011) Pollination services in the UK: how important are honeybees? *Agriculture, Ecosystems and Environment* 142: 137–143.

Brewer, D. (2018) *Birds New to Science: Fifty Years of Avian Discoveries*. Christopher Helm, London.

Briggs, H. (2015) Bees 'get a buzz' from pesticides. BBC News. www.bbc.co.uk/news/science-environment-32399907 (accessed August 2020).

Brittain, C., Kremen, C., Garber, A. and Klein, A.-M. (2014) Pollination and plant resources change the nutritional quality of almonds for human health. *PLoS One* 9: e90082. doi: 10.1371/journal.pone.0090082.

Brosi, B.J., Niezgoda, K. and Briggs, H.M. (2017) Experimental species removals impact the architecture of pollination networks. *Biology Letters* 13: 20170243. doi: 10.1098/rsbl.2017.0243.

Broughton, R.K., Hebda, G., Maziarz, M. *et al.* (2015) Nest-site competition between bumblebees (Bombidae), social wasps (Vespidae) and cavity-nesting birds in Britain and the Western Palearctic. *Bird Study* 62: 427–437.

Brown, J. and Cunningham, S.A. (2019) Global-scale drivers of crop visitor diversity and the historical development of agriculture. *Proceedings of the Royal Society B* 286: 20192096. doi: 10.1098/rspb.2019.2096.

Buglife (2019) Species management sheet: sea aster mining bee. https://cdn.buglife.org.uk/2019/07/Sea-aster-mining-bee-management-guidance-sheet.pdf (accessed August 2020).

Bumblebee Conservation Trust (2020) Short-haired bumblebee reintroduction project. www.bumblebeeconservation.org/short-haired-bumblebee-reintroduction-project (accessed August 2020).

Burkill, I.H. (1897) Fertilization of some spring flowers on the Yorkshire coast. *Journal of Botany* 35: 92–189.

Burkle, L.A., Marlin, J.C. and Knight, T.M. (2013) Plant–pollinator interactions over 120 years: loss of species, co-occurrence and function. *Science* 339: 1611–1615.

BWARS (2019) Look out for a new bee-fly. https://www.bwars.com/content/look-out-new-bee-fly (accessed August 2020).

Camargo, J.M.F. and Roubik, D.W. (1991) Systematics and bionomics of the apoid obligate necrophages: the *Trigona hypogea* group. *Biological Journal of the Linnean Society* 44: 13–39.

Cambridge Conservation Initiative (2018) *The Pollination Deficit: Towards Supply Chain Resilience in the Face of Pollinator Decline*. CCI, Cambridge. www.cisl.cam.ac.uk/resources/publication-pdfs/the-pollination-deficit.pdf (accessed August 2020).

Cameron, S.A., Lozier, J.D., Strange, J.P. *et al.* (2011) Patterns of widespread decline in North American bumble bees. *Proceedings of the National Academy of Sciences of the USA* 108: 662–667. doi: 10.1073/pnas.1014743108.

Camps-Calvet, M., Langemeyer, J., Calvet-Mir, L. and Gomez-Baggethun, E. (2016) Ecosystem services provided by urban gardens in Barcelona, Spain: insights for policy and planning. *Environmental Science and Policy* 62: 14–23.

Cantor, M., Pires, M.M., Marquitti, F.M.D. *et al.* (2017) Nestedness across biological scales. *PLoS One* 12: e0171691. doi: 10.1371/journal.pone.0171691.

Capinera, J.L. (2017) Biology and food habits of the invasive snail *Allopeas gracile* (Gastropoda: Subulinidae). *Florida Entomologist* 100: 116–123.

Cardinal, S. and Danforth, B.N. (2013) Bees diversified in the age of eudicots. *Proceedings of the Royal Society B* 280: 20122686. doi: 10.1098/rspb.2012.2686.

Cardoso, P. (2012) Habitats Directive species lists: urgent need of revision. *Insect Conservation and Diversity* 5: 169–174. doi: 10.1111/j.1752-4598.2011.00140.x.

Carson, R. (1962) *Silent Spring*. Houghton Mifflin, USA.

Carthew, S.M. and Goldingay, R.L. (1997) Non-flying mammals as pollinators. *Trends in Ecology and Evolution* 12: 104–108.

Carvalheiro, L.G., Kunin, W.E., Keil, P. *et al.* (2013) Species richness declines and biotic homogenisation have slowed down for NW-European pollinators and plants. *Ecology Letters* 16: 870–878. doi: 10.1111/ele.12121.

Carvell, C., Bourke, A.F.G., Osborne, J.L. and Heard, M.S. (2015) Effects of an agri-environment scheme on bumblebee reproduction at local and landscape scales. *Basic and Applied Ecology* 16: 519–530.

Chadwick, F., Fitzmaurice, B., Alton, S. and Earl, J. (2016) *The Bee Book*. Dorling Kindersley, London.

Chaplin-Kramer, R., Dombeck, E., Gerber, J. *et al.* (2014) Global malnutrition overlaps with pollinator-dependent micronutrient production. *Proceedings of the Royal Society B* 281: 20141799. doi: 10.1098/rspb.2014.1799.

Christenhusz, M.J.M. and Byng, J.W. (2016) The number of known plants species in the world and its annual increase. *Phytotaxa* 261: 201–217.

Clapham, A.R., Tutin, T.G. and Moore, D.M. (1990) *Flora of the British Isles*. Cambridge University Press, Cambridge.

Classen, A., Peters, M.K., Ferger, S.W. *et al.* (2014) Complementary ecosystem services provided by pest predators and pollinators increase quantity and quality of coffee yields. *Proceedings of the Royal Society B* 281: 20133148. doi: 10.1098/rspb.2013.3148.

Compton, S.G., Thornton, I.W.B., New, T.R. and Underhill, L. (1988) The colonization of the Krakatau islands by fig wasps and other chalcids (Hymenoptera, Chalcidoidea). *Philosophical Transactions of the Royal Society B* 322: 459–470. doi: 10.1098/rstb.1988.0138.

Coombs, G., Dold, A.P. and Peter, C.I. (2011) Generalized fly-pollination in *Ceropegia ampliata* (Apocynaceae–Asclepiadoideae): the role of trapping hairs in pollen export and receipt. *Plant Systematics and Evolution* 296: 137–148.

Cox, P.A. (1983) Extinction of the Hawaiian avifauna resulted in a change of pollinators for the ieie, *Freycinetia arborea*. *Oikos* 41: 195–199.

Cox, P.A. and Knox, R.B. (1988) Pollination postulates and two-dimensional pollination in hydrophilous Monocotyledons. *Annals of the Missouri Botanical Garden* 75: 811–818.

Cozien, R.J., van der Niet, T., Johnson, S.D. and Steenhuisen, S.-L. (2019) Saurian surprise: lizards pollinate South Africa's enigmatic hidden flower. *Ecology* 100: e02670. doi: 10.1002/ecy.2670.

Cranmer, L., McCollin, D. and Ollerton, J. (2012) Landscape structure influences pollinator movements and directly affects plant reproductive success. *Oikos* 121: 562–568.

Crepet, W.L. (1984) Advanced (constant) insect pollination mechanisms: pattern of evolution and implications vis-à-vis angiosperm diversity. *Annals of the Missouri Botanical Garden* 71: 607–630.

Cross, I. and Notton, D.G. (2017) Small-headed resin bee, *Heriades rubicola*, new to Britain (Hymenoptera: Megachilidae). *British Journal of Entomology and Natural History* 30: 1–8.

Crowther, L.P., Hein, P.-L. and Bourke, A.F.G. (2014) Habitat and forage associations of a naturally colonising insect pollinator, the tree bumblebee *Bombus hypnorum*. *PLoS One* 9: e107568. doi: 10.1371/journal.pone.0107568.

Crowther, L.P., Wright, D.J., Richardson, D.S., Carvell, C. and Bourke, A.F.G. (2019) Spatial ecology of a range-expanding bumble bee pollinator. *Ecology and Evolution* 9: 986–997. doi: 10.1002/ece3.4722.

Dafni, A., Kevan, P. and Husband, B.C. (2005) *Practical Pollination Biology*. Enviroquest Ltd, Canada.

Dafni, A., Kevan, P., Gross, C. and Goka, K. (2010) *Bombus terrestris*, pollinator, invasive and pest: an assessment of problems associated with its widespread introductions for commercial purposes. *Applied Entomology and Zoology* 45: 101–113. doi: 10.1303/aez.2010.101.

Dainese, M., Martin, E.A., Aizen, M.A. *et al.* (2019) A global synthesis reveals biodiversity-mediated benefits for crop production. *Science Advances* 5: eaax0121. doi: 10.1126/sciadv.aax0121.

Dalsgaard, B. (2011) Nectar-feeding and pollination by the Cuban green woodpecker (*Xiphidiopicus percussus*) in the West Indies. *Ornitologia Neotropical* 22: 447–451.

Dalsgaard, B., Magård, E., Fjeldså, J. *et al.* (2011) Specialization in plant-hummingbird networks is associated with species richness, contemporary precipitation and Quaternary climate-change velocity. *PLoS One* 6(10): e25891. doi: 10.1371/journal.pone.0025891.

Dalsgaard, B., Trøjelsgaard, K., Martín González, A.M. *et al.* (2013) Historical climate-change influences modularity of pollination networks. *Ecography* 36: 1331–1340. doi: 10.1111/j.1600-0587.2013.00201.x.

Dams, M. and Dams, L. (1977) Spanish rock art depicting honey gathering during the Mesolithic. *Nature* 268: 228–230.

Danforth, B.N. (2002) Evolution of sociality in a primitively eusocial lineage of bees. *Proceedings of the National Academy of Sciences of the USA* 99: 286–290. doi: 10.1073/pnas.012387999.

Danforth, B.N., Minckley, R.L., Neff, J.L. and Fawcett, F. (2019) *The Solitary Bees: Biology, Evolution, Conservation.* Princeton University Press, Princeton, NJ.

Darwin, C. (1859) *On the Origin of Species by Means of Natural Selection, or the Preservation of Favoured Races in the Struggle for Life.* John Murray, London.

Darwin, C. (1862) *On the Various Contrivances by which British and Foreign Orchids are Fertilised by Insects, and on the Good Effects of Intercrossing.* John Murray, London.

Darwin, C. (1877) *The Different Forms of Flowers on Plants of the Same Species.* John Murray, London.

Darwin, C. (1887) *The Life and Letters of Charles Darwin, Including an Autobiographical Chapter*, ed. F. Darwin. Vol. 1. John Murray, London.

Darwin, C. (1988) *Charles Darwin's Beagle Diary*, ed. R.D. Keynes. Cambridge University Press, Cambridge.

Davies, Z.G., Fuller, R.A., Loram, A. *et al.* (2009) A national scale inventory of resource provision for biodiversity within domestic gardens. *Biological Conservation* 142: 761–771.

de Brito, V.L.G., Rech, A.R., Ollerton, J. and Sazima, M. (2017) Nectar production, reproductive success and the evolution of generalised pollination within a specialised pollen-rewarding plant family: a case study using *Miconia theizans. Plant Systematics and Evolution* 303: 709–718.

Defra (2014) *The National Pollinator Strategy: for Bees and Other Pollinators in England.* Defra, York. www.gov.uk/government/publications/national-pollinator-strategy-for-bees-and-other-pollinators-in-england (accessed August 2020).

Defra (2020) *Agriculture in the United Kingdom 2019.* www.gov.uk/government/statistics/agriculture-in-the-united-kingdom-2019 (accessed August 2020).

Dellinger, A.S., Scheer, L.M., Artuso, S. *et al.* (2019) Bimodal pollination systems in Andean Melastomataceae involving birds, bats, and rodents. *The American Naturalist* 194: 104–116.

Devoto, M., Bailey, S. and Memmott, J. (2011) The 'night shift': nocturnal pollen-transport networks in a boreal pine forest. *Ecological Entomology* 36: 25–35.

Dias, B., Raw, A. and Imperatriz-Fonseca, V. (1999) *International Pollinators Initiative: the São Paulo Declaration on Pollinators.* Ministry of the Environment, Brasília.

Diller, C., Castañeda-Zárate, M. and Johnson, S.D. (2019) Generalist birds outperform specialist sunbirds as pollinators of an African *Aloe. Biology Letters* 15: 20190349. doi: 10.1098/rsbl.2019.0349.

Dicks, L.V. (2013) Bees, lies and evidence-based policy. *Nature* 7437: 283.

Dicks, L.V., Viana, B., Bommarco, R. *et al.* (2016) Ten policies for pollinators. *Science* 354: 975–976.

Domingos, M.A., Nadia, T.L. and Machado, I.C. (2017) Complex flowers and rare pollinators: Does ant pollination in *Ditassa* show a stable system in Asclepiadoideae (Apocynaceae)? *Arthropod–Plant Interactions* 11: 339–349.

Dormann, C.F., Gruber, B. and Fruend, J. (2008) Introducing the bipartite package: analysing ecological networks. *R News* 8 (2): 8–11.

Dover, J.W. and Fry, G.L.A. (2001) Experimental simulation of some visual and physical components of a hedge and the effects on butterfly behaviour in an agricultural landscape. *Entomologia Experimentalis et Applicata* 100: 221–233.

Duan, Y.-W., Ren, H., Li, T. *et al.* (2019) A century of pollination success revealed by herbarium specimens of seed pods. *New Phytologist* 224: 1512–1517.

Duffy, K.J., Patrick, K.L. & Johnson, S.D. (2020) Outcrossing rates in a rare 'ornithophilous' aloe are correlated with bee visitation. *Plant Systematics and Evolution* 306: 23.

Dupont, Y.L. and Olesen, J.M. (2004) Fugleblomster på de Kanariske Øer. *Naturens Verden* 87: 2–11.

Dyg, P.M., Christensen, S. and Peterson, C.J. (2019) Community gardens and wellbeing amongst vulnerable populations: a thematic review. *Health Promotion International* 2019: 1–14. doi: 10.1093/heapro/daz067.

Eggs, B. and Sanders, D. (2013) Herbivory in spiders: the importance of pollen for orb-weavers. *PLoS One* 8 (11): e82637. doi: 10.1371/journal.pone.0082637.

Eilers, E.J., Kremen, C., Greenleaf, S.S., Garber, A.K. and Klein, A.-M. (2011) Contribution of pollinator-mediated crops to nutrients in the human food supply. *PLoS One* 6: e21363. doi: 10.1371/journal.pone.0021363.

Eisikowitch, D. (1980) The role of dark flowers in the pollination of certain Umbelliferae. *Journal of Natural History* 14: 737–742.

Ellis, A.M., Myers, S.S. and Ricketts, T.H. (2015) Do pollinators contribute to nutritional health? *PLoS One* 10: e114805. doi: 10.1371/journal.pone.0114805.

Else, G.R. and Edwards, M. (2018) *Handbook of the Bees of the British Isles*. Ray Society, London.

Erenler, H. (2013) The diversity of pollinators in the gardens of large English country houses. PhD thesis, University of Northampton.

Erenler, H.E., Gillman, M.P. and Ollerton, J. (2020) Impact of extreme events on pollinator assemblages. *Current Opinion in Insect Science* 38: 34–39.

European Commission (2020) EU National Apiculture Programmes https://ec.europa.eu/agriculture/sites/agriculture/files/honey/programmes/programmes_en.pdf (accessed August 2020).

Faegri, K. and van der Pijl, L. (1966) *The Principles of Pollination Ecology*. Pergamon Press, Oxford.

Falk, S. and Lewington, R. (2015) *Field Guide to the Bees of Great Britain and Ireland*. Bloomsbury, London.

Falk, S.J. and Earwaker, R. (2019) Dusky-horned nomad bee (*Nomada bifasciata*), new to Britain (Hymenoptera: Apidae). *British Journal of Entomology and Natural History* 32: 170–175.

Farley, P. and Roberts, M.S. (2011) *Edgelands: Journeys into England's True Wilderness.* Jonathan Cape, London.

Fernández de Castro, A.G., Moreno-Saiz, J.C. and Fuertes-Aguilar, J. (2017) Ornithophily for the nonspecialist: differential pollination efficiency of the Macaronesian island paleoendemic *Navaea phoenicea* (Malvaceae) by generalist passerines. *American Journal of Botany* 104: 1556–1568. doi: 10.3732/ajb.1700204.

Feurdean, A., Ruprecht, E., Molnár, Z., Hutchinson, S.M. and Hickler, T. (2018) Biodiversity-rich European grasslands: ancient, forgotten ecosystems. *Biological Conservation* 228: 224–232. doi: 10.1016/j.biocon.2018.09.022.

Finch, J.T.D., Power, S.A., Welbergen, J.A. and Cook, J.M. (2018) Two's company, three's a crowd: co-occurring pollinators and parasite species in *Breynia oblongifolia* (Phyllanthaceae). *BMC Evolutionary Biology* 18: 93. doi: 10.1186/s12862-018-1314-y.

Fitch, G., Wilson, C.J., Glaum, P. *et al.* (2019) Does urbanization favour exotic bee species? Implications for the conservation of native bees in cities. *Biology Letters* 15: 20190574. doi: 10.1098/rsbl.2019.0574.

Fleming, T.H. and Holland, J.N. (1998) The evolution of obligate pollination mutualisms: senita cactus and senita moth. *Oecologia* 114: 368–375.

Fleming, T.H., Shaley, C.T., Holland, J.N., Nason, J.D. and Hamrick, J.L. (2001) Sonoran desert columnar cacti and the evolution of generalized pollination systems. *Ecological Monographs* 71: 511–530.

Fleming, T.H., Geiselman, C. and Kress, W.J. (2009) The evolution of bat pollination: a phylogenetic perspective. *Annals of Botany* 104: 1017–1043. doi: 10.1093/aob/mcp197.

Forbes, S.J. and Northfield, T.D. (2017) Increased pollinator habitat enhances cacao fruit set and predator conservation. *Ecological Applications* 27: 887–899.

Forrest, J.R.K. (2015) Plant–pollinator interactions and phenological change: what can we learn about climate impacts from experiments and observations? *Oikos* 124: 4–13.

Fox, K., Vitt, P., Anderson, K. *et al.* (2013) Pollination of a threatened orchid by an introduced hawk moth species in the tallgrass prairie of North America. *Biological Conservation* 167: 316–324.

Fox, R. (2013) The decline of moths in Great Britain: a review of possible causes. *Insect Conservation and Diversity* 6: 5–19.

Friends of the Earth (2017) Is ragwort poisonous? A ragwort mythbuster. https://friendsoftheearth.uk/nature/ragwort-poisonous-ragwort-mythbuster (accessed August 2020).

Gallai, N., Salles, J.-M., Settele, J. and Vaissière, B.E. (2009) Economic valuation of the vulnerability of world agriculture confronted with pollinator decline. *Ecological Economics* 68: 810–821.

Garbuzov, M., Fensome, K.A. and Ratnieks, F.L. (2015) Public approval plus more wildlife: twin benefits of reduced mowing of amenity grass in a suburban public park in Saltdean, UK. *Insect Conservation and Diversity* 8: 107–119.

Garibaldi, L.A., Carvalheiro, L.G., Vaissière, B.E. *et al.* (2016) Mutually beneficial pollinator diversity and crop yield outcomes in small and large farms. *Science* 351: 388–391.

Garratt, M.P.D., Truslove, C.L., Coston, D.J. *et al.* (2014) Pollination deficits in UK apple orchards. *Journal of Pollination Ecology* 12: 9–14.

Gess, S.K. (1996) *The Pollen Wasps: Ecology and Natural History of the Masarinae.* Harvard University Press, Cambridge, MA.

Ghazoul, J. (2015) Qualifying pollinator decline evidence. *Science* 348: 981–982.

Gilburn, A.S., Bunnefeld, N., Wilson, J.M. *et al.* (2015) Are neonicotinoid insecticides driving declines of widespread butterflies? *PeerJ* 3: e1402. doi: 10.7717/peerj.1402.

Godfray, H.C.J., Blacquière, T., Field, L.M. *et al.* (2015) A restatement of recent advances in the natural science evidence base concerning neonicotinoid insecticides and insect pollinators. *Proceedings of the Royal Society B* 282: 20151821. doi: 10.1098/rspb.2015.1821.

Godínez-Álvarez, H. (2004) Pollination and seed dispersal by lizards: a review. *Revista Chilena de Historia Natural* 77: 569–577.

Goldblatt, P. and Manning, J.C. (2000) The long-proboscid fly pollination system in Southern Africa. *Annals of the Missouri Botanical Garden* 87: 146–170.

Goldsmith, K.M. and Goldsmith, T.H. (1982) Sense of smell in the black-chinned hummingbird. *Condor* 84: 237–238.

Gonzalez, V.H., Cruz, P., Folks, N. *et al.* (2018) Attractiveness of the dark central floret in wild carrots: do umbel size and height matter? *Journal of Pollination Ecology* 23: 98–101.

Gonzalez-Megias, A., Menéndez, R., Roy, D., Brereton, T. and Thomas, C.D. (2008) Changes in the composition of British butterfly assemblages over two decades. *Global Change Biology* 14: 1464–1474.

Gori, D.F. (1989) Floral color change in *Lupinus argenteus* (Fabaceae): why should plants advertise the location of unrewarding flowers to pollinators? *Evolution* 43: 870–881.

Gorostiague, P., Sajama, J. and Ortega-Baes, P. (2018) Will climate change cause spatial mismatch between plants and their pollinators? A test using Andean cactus species. *Biological Conservation* 226: 247–255.

Goulson, D. (2003a) Effects of introduced bees on native ecosystems. *Annual Review of Ecology, Evolution, and Systematics* 34: 1–26.

Goulson, D. (2003b) *Bumblebees: Behaviour, Ecology, and Conservation.* Oxford University Press, Oxford.

Goulson, D. (2009) Evaluating the role of ecological isolation in maintaining the species boundary in *Silene dioica* and *S. latifolia. Plant Ecology* 205: 201–211.

Goulson, D. (2013) *A Sting in the Tale.* Jonathan Cape, London.

Goulson, D. (2014) *A Buzz in the Meadow.* Jonathan Cape, London.

Goulson, D. (2017a) Are robotic bees the future? www.sussex.ac.uk/lifesci/goulsonlab/blog/robotic-bees (accessed August 2020).

Goulson, D. (2017b) *Bee Quest*. Jonathan Cape, London.

Goulson, D. (2019) *The Garden Jungle: or Gardening to Save the Planet*. Jonathan Cape, London.

Goulson, D. and Jerrim, K. (1997) Maintenance of the species boundary between *Silene dioica* and *S. latifolia* (red and white campion). *Oikos* 78: 254–266.

Goulson, D. and Williams, P.H. (2001) *Bombus hypnorum* (Hymenoptera: Apidae), a new British bumblebee? *British Journal of Entomology and Natural History* 14: 129–131.

Goulson, D., Peat, J., Stout, J.C. *et al.* (2002) Can alloethism in workers of the bumblebee *Bombus terrestris* be explained in terms of foraging efficiency? *Animal Behaviour* 64: 123–130.

Goulson, D., McGuire, K., Munro, E.E. *et al.* (2009) Functional significance of the dark central floret of *Daucus carota* (Apiaceae) L.; is it an insect mimic? *Plant Species Biology* 24: 77–82.

Goulson, D., Nicholls, E., Botías, C. and Rotheray, E.L. (2015) Bee declines driven by combined stress from parasites, pesticides, and lack of flowers. *Science* 347: 1255957.

Goulson, D., O'Connor, S. and Park, K.J. (2018) Causes of colony mortality in bumblebees. *Animal Conservation* 21: 45–53.

Green, C.G. (2017) The status of global pollinators according to IUCN Red Lists. Unpublished MSc Thesis, University College Dublin.

Greenberg, J. (2017) No, coffee is not the second-most traded commodity after oil. PolitiFact. www.politifact.com/factchecks/2017/may/08/starbucks/no-coffee-not-second-most-traded-commodity-after-o/ (accessed August 2020).

Greenleaf, S.S., Williams, N.M., Winfree, R. and Kremen, C. (2007) Bee foraging ranges and their relationship to body size. *Oecologia* 153: 589–596.

Greenwood, J.J.D. and Greenwood, J.G. (2018) The origin of the birdwatching term 'jizz'. *British Birds* 111: 292–294.

Grixti, J.C. (2009) Decline of bumble bees (*Bombus*) in the North American Midwest. *Biological Conservation* 142: 75–84.

Groom, S.V.C. and Rehan, S.M. (2018) Climate-mediated behavioural variability in facultatively social bees. *Biological Journal of the Linnean Society* 125: 165–170.

Haffner, J. (2015) The dangers of eco-gentrification: what's the best way to make a city greener? *The Guardian* 6 May 2015. www.theguardian.com/cities/2015/may/06/dangers-ecogentrification-best-way-make-city-greener (accessed August 2020).

Hall, D.M. and Steiner, R. (2019) Insect pollinator conservation policy innovations at subnational levels: lessons for lawmakers. *Environmental Science and Policy* 93: 118–128.

Hall, D.M., Camilo, G.D., Tonietto, R.K. *et al.* (2017) The city as a refuge for insect pollinators. *Conservation Biology* 31: 24–29. doi: 10.1111/cobi.12840.

Hallsworth, A. and Wong, A. (2015) Urban gardening realities: the example case study of Portsmouth, England. *International Journal on Food System Dynamics* 6: 1–11.

Hansen, D.M., Beer, K. and Müller, C.B. (2006) Mauritian coloured nectar no longer a mystery: a visual signal for lizard pollinators. *Biology Letters* 2: 165–168. doi: 10.1098/rsbl.2006.0458.

Harder, L.D. and Barrett, S.C.H. (1995) Mating cost of large floral displays in hermaphrodite plants. *Nature* 373: 512–515.

Harding, P.T. and Plant, R.A. (1978) A second record of *Cerambyx cerdo* L. (Coleoptera: Cerambycidae) from sub-fossil remains in Britain. *Entomologist's Gazette* 29: 150–152.

Hawkes, W. (2016) Flight of the bumblebee – part I. BiOME Ecology. https://biomeecology.com/nature/insects/2016/06/flight-bumblebee-part-i (accessed August 2020).

Heiduk, A., Brake, I., Tolasch, T., Frank, J. *et al.* (2010) Scent chemistry and pollinator attraction in the deceptive trap flowers of *Ceropegia dolichophylla*. *South African Journal of Botany* 76: 762–769.

Heiduk, A., Kong, H., Brake, I. *et al.* (2015) Deceptive *Ceropegia dolichophylla* fools its kleptoparasitic fly pollinators with exceptional floral scent. *Frontiers in Ecology and Evolution* 3. doi: 10.3389/fevo.2015.00066.

Heiduk, A., Brake, I., von Tschirnhaus, M. *et al.* (2016) *Ceropegia sandersonii* mimics attacked honeybees to attract kleptoparasitic flies for pollination. *Current Biology* 26: 2787–2793.

Heiduk, A., Brake, I., von Tschirnhaus, M. *et al.* (2017) Floral scent and pollinators of *Ceropegia* trap flowers. *Flora* 232: 169–182.

Heinrich, B. (2004) *Bumblebee Economics*, 2nd edition. Harvard University Press, Cambridge, MA.

Henry, M., Cerrutti, N., Aupinel, P. *et al.* (2015) Reconciling laboratory and field assessments of neonicotinoid toxicity to honeybees. *Proceedings of the Royal Society B* 282: 20152110. doi: 10.1098/rspb.2015.2110.

Herbertsson, L., Lindström, S.A.M., Rundlöf, M., Bommarco, R. and Smith, H.G. (2016) Competition between managed honeybees and wild bumblebees depends on landscape context. *Basic and Applied Ecology* 17: 609–616.

Heringer, H., Palmeira, L.R.M., Alves, A.C.F. *et al.* (2005) Estudo da capacidade olfatória em três representantes da subfamília Trochilinae: *Eupetomena macroura* (Gould, 1853), *Thalurania furcata eriphile* (Lesson, 1832) e *Amazilia lactea* (Lesson, 1832). Proceedings VII Congress of the Brazilian Society of Ecologia 20th to 25th November 2005. http://seb-ecologia.org.br/revistas/indexar/anais/viiceb/resumos/318a.pdf (accessed August 2020).

Herrera, C.M. (1996) Floral traits and plant adaptation to insect pollinators: a devil's advocate approach. In Lloyd, D.G. and Barrett, S.C.H. (eds), *Floral Biology*. Springer, Boston, MA, pp. 65–87.

Herrera, C.M. (2019) Complex long-term dynamics of pollinator abundance in undisturbed Mediterranean montane habitats over two decades. *Ecological Monographs* 89: e01338. doi: 10.1002/ecm.1338.

Herrera, C.M. (2020) Gradual replacement of wild bees by honeybees in flowers of the Mediterranean Basin over the last 50 years. *Proceedings of the Royal Society B* 287: 20192657. doi: 10.1098/rspb.2019.2657.

Heymann, E.W. (2011) Florivory, nectarivory, and pollination: a review of primate–flower interactions. *Ecotropica* 17: 41–52.

Hillebrand, H. (2004) On the generality of the latitudinal diversity gradient. *The American Naturalist* 163: 192–211.

Hingston, A. (2007) The potential impact of the large earth bumblebee *Bombus terrestris* (Apidae) on the Australian mainland: lessons from Tasmania. *Victorian Naturalist* 124: 110–116.

Hingston, A.B. and McQuillan, P.B. (2000) Are pollination syndromes useful predictors of floral visitors in Tasmania? *Austral Ecology* 25: 600–609.

Hinkelman, J. (2020) Earliest behavioral mimicry and possible food begging in a Mesozoic alienopterid pollinator. *Biologia* 75: 83–92.

Hinkelman, J. and Vršanská, L.A. (2020) Myanmar amber cockroach with protruding feces contains pollen and a rich microcenosis. *The Science of Nature* 107: 13.

Hogendoorn, K., Gross, C., Sedgley, M. and Keller, M. (2006) Increased tomato yield through pollination by native Australian *Amegilla chlorocyanea* (Hymenoptera: Anthophoridae). *Journal of Economic Entomology* 99: 828–833.

Hohmann, H. (1993) *Bienen, Wespen und Ameisen der Kanarischen Inseln*. Übersee-Museum, Bremen.

Holland, J.N. and Fleming, T.H. (1999) Mutualistic interactions between *Upiga virescens* (Pyralidae), a pollinating seed-consumer, and *Lophocereus schottii* (Cactaceae). *Ecology* 80: 2074–2084.

Holzschuh, A., Dudenhöffer, J.-H. and Tscharntke, T. (2012) Landscapes with wild bee habitats enhance pollination, fruit set and yield of sweet cherry. *Biological Conservation* 153: 101–107.

Howard, B.S. (2019) *Dancing with Bees*. Chelsea Green Publishing, London.

Høye, T., Post, E., Schmidt, N., Trøjelsgaard, K. and Forchhammer, M.C. (2013) Shorter flowering seasons and declining abundance of flower visitors in a warmer Arctic. *Nature Climate Change* 3: 759–763.

Huang, Z.H., Song, Y.P. and Huang, S.Q. (2017) Evidence for passerine bird pollination in *Rhododendron* species. *AoB Plants* 9, plx062. doi: 10.1093/aobpla/plx062.

Hughes, W.O.H., Oldroyd, B.P., Beekman, M. and Ratnieks, F.L.W. (2008) Ancestral monogamy shows kin selection is key to the evolution of eusociality. *Science* 320: 1213–1216.

Hung, K.-L.J., Kingston, J.M., Lee, A., Holway, D.A. and Kohn, J.R. (2019) Non-native honey bees disproportionately dominate the most abundant floral resources in a biodiversity hotspot. *Proceedings of the Royal Society B* 286: 20182901. doi: 10.1098/rspb.2018.2901.

Inoue, M.N., Yokoyama, J. and Washitani, I. (2008) Displacement of Japanese native bumblebees by the recently introduced *Bombus terrestris* (L.) (Hymenoptera: Apidae). *Journal of Insect Conservation* 12: 135–146.

International Coffee Organization (2020) Historical Data on the Global Coffee Trade. www.ico.org/new_historical.asp (accessed August 2020).

IPBES (2016) *The Assessment Report of the Intergovernmental Science-Policy Platform on Biodiversity and Ecosystem Services on Pollinators, Pollination and Food Production*. Potts, S.G., Imperatriz-Fonseca, V.L. and Ngo, H.T. (eds). Secretariat of the Intergovernmental Science-Policy Platform on Biodiversity and Ecosystem Services, Bonn, Germany. doi: 10.5281/zenodo.3402856.

Jacobs, J.H., Clark, S.J., Denholm, I. *et al.* (2010) Pollinator effectiveness and fruit set in common ivy, *Hedera helix* (Araliaceae). *Arthropod–Plant Interactions* 4: 19–28.

Jakobsson, S., Bernes, C., Bullock, J.M., Verheyen, K. and Lindborg, R. (2018) How does roadside vegetation management affect the diversity of vascular plants and invertebrates? A systematic review. *Environmental Evidence* 7: 17. doi: 10.1186/s13750-018-0129-z.

Janzen, D.H. (1974) The deflowering of Central America. *Natural History* 83: 48–53.

Johansen, C.A. (1977) Pesticides and pollinators. *Annual Review of Entomology* 22: 177–192.

Johnson, A.L., Fetters, A.M. and Ashman, T.L. (2017) Considering the unintentional consequences of pollinator gardens for urban native plants: is the road to extinction paved with good intentions? *New Phytologist* 215: 1298–1305. doi: 10.1111/nph.14656.

Johnson, S.D. (2004) An overview of plant–pollinator relationships in southern Africa. *International Journal of Tropical Insect Science* 24: 45–54.

Johnson, S.D. (2006) Pollinator-driven speciation in plants. In Harder, L.D. and Barrett, S.C.H. (eds), *Ecology and Evolution of Flowers*. Oxford University Press, Oxford, pp. 295–310.

Johnson, S.D. (2010) The pollination niche and its role in the diversification and maintenance of the southern African flora. *Philosophical Transactions of the Royal Society B* 365: 499–516. doi: 10.1098/rstb.2009.0243.

Johnson, S.D. and Steiner, K.E. (1997) Long-tongued fly pollination and evolution of floral spur length in the *Disa draconis* complex (Orchidaceae). *Evolution* 51: 45–53.

Johnson, S.D. and Steiner, K.E. (2000) Generalization versus specialization in plant pollination systems. *Trends in Ecology and Evolution* 15: 140–143.

Johnson, S.D. and Wester, P. (2017) Stefan Vogel's analysis of floral syndromes in the South African flora: an appraisal based on 60 years of pollination studies. *Flora* 232: 200–206.

Johnson, S.D., Moré, M., Amorim, F.W. *et al.* (2017) The long and the short of it: a global analysis of hawkmoth pollination niches and interaction networks. *Functional Ecology* 31: 101–115. doi: 10.1111/1365-2435.12753.

Jones, C.M. and Brown, M.J. (2014) Parasites and genetic diversity in an invasive bumblebee. *Journal of Animal Ecology* 83: 1428–1440.

Jordano, P. (1987) Patterns of mutualistic interactions in pollination and seed dispersal: connectance, dependence asymmetries, and coevolution. *The American Naturalist* 129: 657–677.

Jukes, H. (2018) *A Honeybee Heart Has Five Openings*. Scribner, London.

Junker, R.R. and Parachnowitsch, A.L. (2015) Working towards a holistic view on flower traits: how floral scents mediate plant–animal interactions in concert with other floral characters. *Journal of the Indian Institute of Science* 95: 43–67.

Kaiser-Bunbury, C., Mougal, J., Whittington, A. *et al.* (2017) Ecosystem restoration strengthens pollination network resilience and function. *Nature* 542: 223–227.

Kato, M., Takimura, A. and Kawakita, A. (2003) An obligate pollination mutualism and reciprocal diversification in the tree genus *Glochidion* (Euphorbiaceae). *Proceedings of the National Academy of Sciences of the USA* 100: 5264–5267.

Kawakita, A. and Kato, M. (2004) Obligate pollination mutualism in *Breynia* (Phyllanthaceae): further documentation of pollination mutualism involving *Epicephala* moths (Gracillariidae). *American Journal of Botany* 91: 1319–1325.

Kay, K.M. and Sargent, R.D. (2009) The role of animal pollination in plant speciation: integrating ecology, geography, and genetics. *Annual Review of Ecology, Evolution and Systematics* 40: 637–656.

Kearns, C.A. and Inouye, D.W. (1993) *Techniques for Pollination Biologists*. University Press of Colorado, Niwot, CO.

Kearns, C.A., Inouye, D.W. and Waser, N.M. (1998) Endangered mutualisms: the conservation of plant–pollinator interactions. *Annual Review of Ecology and Systematics* 29: 83–112. doi: 10.1146/annurev.ecolsys.29.1.83.

Kevan, P.G. (1970) High arctic insect–flower visitor relations: the inter-relationships of arthropods and flowers at Lake Hazen, Ellesmere Island, Northwest Territories. PhD thesis, University of Alberta, Canada.

Kevan, P.G. (1974) Pollination, pesticides, and environmental quality. *BioScience* 24: 198–199.

Kevan, P.G. (1975) Pollination and environmental conservation. *Environmental Conservation* 2: 293–298.

Kevan, P.G. (1977) Blueberry crops in Nova Scotia and New Brunswick – pesticides and crop reductions. *Canadian Journal of Agricultural Economics* 25: 61–64.

Kinlock, N.L., Prowant, L., Herstoff, E.M. *et al.* (2017) Explaining global variation in the latitudinal diversity gradient: meta-analysis confirms known patterns and uncovers new ones. *Global Ecology and Biogeography* 27: 125–141.

Kirby, W. (1802) *Monographia Apum Angliae, or, An Attempt to Divide into Their Natural Genera and Families, Such Species of the Linnean Genus Apis as Have Been Discovered in England*. J. Raw, Ipswich.

Kirby-Lambert, C. (2016) *Nomada alboguttata* Herrich-Schäffer, 1839 new to the British Isles and *Nomada zonata* Panzer, 1798 first record for mainland Britain. *BWARS Newsletter* Autumn 2016: 29–31.

Kite, G.C., Hetterscheid, W.L.A., Lewis, M.J. *et al.* (1998) Inflorescence odours and pollinators of *Arum* and *Amorphophallus* (Araceae). In Owens, S.J. and Rudall, P.J. (eds), *Reproductive Biology in Systematics, Conservation and Economic Botany.* Royal Botanic Gardens, Kew, pp. 295–315.

Klein, A.-M., Steffan-Dewenter, I. and Tscharntke, T. (2003) Fruit set of highland coffee increases with the diversity of pollinating bees. *Proceedings of the Royal Society B* 270: 955–961. doi: 10.1098/rspb.2002.2306.

Klein, A.-M., Vaissière, B.E., Cane, J.H. *et al.* (2007) Importance of pollinators in changing landscapes for world crops. *Proceedings of the Royal Society of London B* 274: 303–313.

Klein, A.-M., Freitas, B.M., Bomfim, I.G.A. *et al.* (2020) *Insect Pollination of Crops in Brazil: A Guide for Farmers, Gardeners, Politicians and Conservationists.* Available for free download from: www.nature.uni-freiburg.de/ressourcen/publikationen-pdfs/cpb-book-icpr-ff3-160-ebook-sklein.pdf (accessed August 2020).

Knudsen, J.T., Tollsten, L., Groth, I., Bergstrom, G. and Raguso, R.A. (2004) Trends in floral scent chemistry in pollination syndromes: floral scent composition in hummingbird-pollinated taxa. *Botanical Journal of the Linnean Society* 146: 191–199. doi: 10.1111/j.1095-8339.2004.00329.x.

Kosior, A., Celary, W., Olejniczak, P. *et al.* (2007) The decline of the bumble bees and cuckoo bees (Hymenoptera: Apidae: Bombini) of western and central Europe. *Oryx* 41: 79–88.

Kratochwil, A. (2016) Review of the Icelandic bee fauna (Hymenoptera: Apoidea: Anthophila). *Integrative Systematics: Stuttgart Contributions to Natural History* 9: 217–227.

Labandeira, C.C. (2010) The pollination of Mid Mesozoic seed plants and the early history of long-proboscid insects. *Annals of the Missouri Botanical Garden* 97: 469–513.

Labandeira, C.C., Kvaček, J. and Mostovski, M. (2007) Pollination drops, pollen, and insect pollination of Mesozoic gymnosperms. *Taxon* 56: 663–695.

Labandeira, C.C., Yang, Q., Santiago-Blay, J.A. *et al.* (2016) The evolutionary convergence of mid-Mesozoic lacewings and Cenozoic butterflies. *Proceedings of the Royal Society B* 283: 20152893. doi: 10.1098/rspb.2015.2893.

Lack, A.J. and Diaz, A. (1991) The pollination of *Arum maculatum* L.: a historical review and new observations. *Watsonia* 18: 333–342.

Lamborn, E. and Ollerton, J. (2000) Experimental assessment of the functional morphology of inflorescences of *Daucus carota* (Apiaceae): testing the 'fly catcher effect'. *Functional Ecology* 14: 445–454.

Land Trust (2020) Canvey Wick. https://thelandtrust.org.uk/space/canvey-wick/?doing_wp_cron=1584545812.9599471092224121093750 (accessed August 2020).

Larson, B.M.H., Kevan, P.G. and Inouye, D.W. (2001) Flies and flowers: taxonomic diversity of anthophiles and pollinators. *The Canadian Entomologist* 133: 439–465.

Lawton, J.H., Brotherton, P.N.M., Brown, V.K. *et al.* (2010) *Making Space for Nature: a Review of England's Wildlife Sites and Ecological Network*. Defra, London.

Leong, M., Dunn, R.R. and Trautwein, M.D. (2018) Biodiversity and socioeconomics in the city: a review of the luxury effect. *Biology Letters* 14: 20180082. doi: 10.1098/rsbl.2018.0082.

Lord, J.M. (1991) Pollination and seed dispersal in *Freycinetia baueriana*, a dioecious liane that has lost its bat pollinator. *New Zealand Journal of Botany* 29: 83–86.

Loughlin, N.J.D., Gosling, W.D., Mothes, P. and Montoya, E. (2018) Ecological consequences of post-Columbian indigenous depopulation in the Andean–Amazonian corridor. *Nature Ecology and Evolution* 2: 1233–1236.

Lyver, P., Perez, E., Carneiro da Cunha, M. and Roué, M. (eds) (2015) *Indigenous and Local Knowledge about Pollination and Pollinators associated with Food Production: Outcomes from the Global Dialogue Workshop (Panama 1–5 December 2014)*. Unesco, Paris.

Machado, C.A., Robbins, N., Gilbert, M.T.P. and Herre, E.A. (2005) Critical review of host specificity and its coevolutionary implications in the fig/fig-wasp mutualism. *Proceedings of the National Academy of Sciences of the USA* 102 (Suppl. 1): 6558–6565. doi: 10.1073/pnas.0501840102.

Magallón, S., Gómez-Acevedo, S., Sánchez-Reyes, L.L. and Hernández-Hernández, T. (2015) A metacalibrated time-tree documents the early rise of flowering plant phylogenetic diversity. *New Phytologist* 207: 437–453. doi: 10.1111/nph.13264.

Maguiña, R. and Muchhala, N. (2017) Do artificial nectar feeders affect bat–plant interactions in an Ecuadorian cloud forest? *Biotropica* 49: 586–592.

Maher, S., Manco, F. and Ings, T.C. (2019) Using citizen science to examine the nesting ecology of ground-nesting bees. *Ecosphere* 10: e02911. doi: 10.1002/ecs2.2911.

Mallinger, R.E., Gaines-Day, H.R. and Gratton, C. (2017) Do managed bees have negative effects on wild bees? A systematic review of the literature. *PLoS One* 12: e0189268. doi: 10.1371/journal.pone.0189268.

Manley, R., Boots, M. and Wilfert, L. (2017) Condition-dependent virulence of slow bee paralysis virus in *Bombus terrestris*: are the impacts of honeybee viruses in wild pollinators underestimated? *Oecologia* 184: 305–315. doi: 10.1007/s00442-017-3851-2.

Marlin, J.C. and LaBerge, W.E. (2001) The native bee fauna of Carlinville, Illinois, revisited after 75 years: a case for persistence. *Conservation Ecology* 5: 9. www.consecol.org/vol5/iss1/art9.

Martin, E.C. and McGregor, S.E. (1973) Changing trends in insect pollination of commercial crops. Annual Review of Entomology 18: 207–226.

Martin, G.R. (2020) *Bird Senses: How and What Birds See, Hear, Smell, Taste, and Feel*. Pelagic, Exeter.

Martín González, A.M., Dalsgaard, B., Ollerton, J. *et al.* (2009) Effects of climate on pollination networks in the West Indies. *Journal of Tropical Ecology* 25: 493–506.

Maruyama, P.K., Vizentin-Bugoni, J., Sonne, J. *et al.* (2016) The integration of alien plants in plant-hummingbird pollination networks across the Americas:

the importance of species traits and insularity. *Diversity and Distributions* 22: 672–681.

Matsumura, C., Yokoyama, J. and Washitani, I. (2004) Invasion status and potential ecological impacts of an invasive alien bumblebee, *Bombus terrestris* L. (Hymenoptera: Apidae) naturalized in Southern Hokkaido, Japan. *Global Environmental Research* 8: 51–66.

Mayer, C., Soka, G. and Picker, M. (2006) The importance of monkey beetle (Scarabaeidae: Hopliini) pollination for Aizoaceae and Asteraceae in grazed and ungrazed areas at Paulshoek, Succulent Karoo, South Africa. *Journal of Insect Conservation* 10: 323.

Mayr, G. (2007) New specimens of the early Oligocene Old World hummingbird *Eurotrochilus inexpectatus*. *Journal of Ornithology* 148: 105–111.

Mayr, G. and Micklich, N. (2010) New specimens of the avian taxa *Eurotrochilus* (Trochilidae) and *Palaeotodus* (Todidae) from the early Oligocene of Germany. *Paläontologische Zeitschrift* 84: 387–395.

Mayr, G. and Wilde, V. (2014) Eocene fossil is earliest evidence of flower-visiting by birds. *Biology Letters* 10: 20140223. doi: 10.1098/rsbl.2014.0223.

McAlister, E. (2017) *The Secret Life of Flies*. Natural History Museum, London.

Memmott, J., Waser, N.M. and Price, M.V. (2004) Tolerance of pollination networks to species extinctions. *Proceedings of the Royal Society B* 271: 2605–2611. doi: 10.1098/rspb.2004.2909.

Mertens, J.E.J., Tropek, R., Dzekashu, F.F. *et al.* (2017) Communities of flower visitors of *Uvariopsis dioica* (Annonaceae) in lowland forests of Mt. Cameroon, with notes on its potential pollinators. *African Journal of Ecology* 56: 146–152.

Micheneau, C., Fournel, J., Warren, B.H. *et al.* (2010) Orthoptera, a new order of pollinator. *Annals of Botany* 105: 355–364.

Michener, C.D. (2007) *The Bees of the World*. Johns Hopkins University Press, Baltimore, MD.

Mikkola, K. (1984) Migration of wasp and bumble bee queens across the Gulf of Finland (Hymenoptera: Vespidae and Apidae). *Notulae Entomologicae* 643: 125–128.

Millard, J.W., Freeman, R. and Newbold, T. (2020) Text-analysis reveals taxonomic and geographic disparities in animal pollination literature. *Ecography* 43: 44–59. doi: 10.1111/ecog.04532.

Minckley, R.L., Roulston, T.H. and Williams, N.M. (2013) Resource assurance predicts specialist and generalist bee activity in drought. *Proceedings of the Royal Society B* 280: 20122703. doi: 10.1098/rspb.2012.2703.

Moles, A. and Ollerton, J. (2014) Is the notion that species interactions are stronger and more specialized in the tropics a zombie idea? (guest post). Dynamic Ecology. https://dynamicecology.wordpress.com/2014/05/13/is-the-notion-that-species-interactions-are-stronger-and-more-specialized-in-the-tropics-a-zombie-idea-guest-post (accessed August 2020).

Moles, A. and Ollerton, J. (2016) Is the notion that species interactions are stronger and more specialized in the tropics a zombie idea? *Biotropica* 48: 141–145.

Moore, R.T. (1947) Habits of male hummingbirds near their nests. *The Wilson Bulletin* 59: 21–25.

Morales, C.L., Arbetman, M.P., Cameron, S.A. and Aizen, M.A. (2013) Rapid ecological replacement of a native bumble bee by invasive species. *Frontiers in Ecology and the Environment* 11: 529–534.

Morelle, R. (2014) Tree bumblebee: 'record sightings' for invasive bee. BBC News. www.bbc.co.uk/news/science-environment-27701591 (accessed August 2020).

Muchhala, N. (2006) Nectar bat stows huge tongue in its rib cage. *Nature* 444: 701–702.

Nabhan, G.P. and Buchmann, S. (1996) *The Forgotten Pollinators*. Island Press, Washington, DC.

Nagamitsu, T. and Inoue, T. (1997) Cockroach pollination and breeding system of *Uvaria elmeri* (Annonaceae) in a lowland mixed-dipterocarp forest in Sarawak. *American Journal of Botany* 84: 208–213.

Nahas, L., Gonzaga, M.O. and Del-Claro, K. (2017) Wandering and web spiders feeding on the nectar from extrafloral nectaries in neotropical savanna. *Journal of Zoology* 301: 125–132.

National Biodiversity Data Centre (2015) *All-Ireland Pollinator Plan 2015–2020*. National Biodiversity Data Centre Series No. 3. NBDC, Waterford.

National Research Council (2007) *Status of Pollinators in North America*. National Academies Press, Washington, DC.

NatWest (2019) Where next for rapeseed prices? https://natwestbusinesshub.com/articles/where-next-for-rapeseed-prices (accessed August 2020).

Navarro-Pérez, M.L., López, J., Fernández-Mazuecos, M. *et al.* (2013) The role of birds and insects in pollination shifts of *Scrophularia* (Scrophulariaceae). *Molecular Phylogenetics and Evolution* 69: 239–254.

Navarro-Pérez, M.L., López, J., Rodríguez-Riaño, T. *et al.* (2017) Confirmed mixed bird–insect pollination system of *Scrophularia trifoliata* L., a Tyrrhenian species with corolla spots. *Plant Biology* 19: 460–468.

Nedelcheva, A. (2011) Observations on the wall flora of Kyustendil (Bulgaria). *EurAsian Journal of BioSciences* 5: 80–90.

New, T.R. and Thornton, I.W.B. (1992) Butterflies of Anak Krakatau, Indonesia: faunal development in early succession. *Journal of the Lepidopterists' Society* 46: 83–86.

Ngo, H.T., Mojica, A.C. and Packer, L. (2011) Coffee plant–pollinator interactions: a review. *Canadian Journal of Zoology* 89: 647–660.

Nicholls, E., Ely, A., Birkin, L., Basu, P. and Goulson, D. (2020) The contribution of small-scale food production in urban areas to the sustainable development goals: a review and case study. *Sustainability Science*. doi: 10.1007/s11625-020-00792-z.

Nieto, A., Roberts, S.P.M., Kemp, J. *et al.* (2014) *European Red List of Bees*. European Union, Luxembourg.

Notton, D.G. (2016) Grass-carrying wasp, *Isodontia mexicana* (De Saussure), genus and species new to Britain (Hymenoptera: Sphecidae). *British Journal of Entomology and Natural History* 29: 241–245.

Notton, D.G. (2018) The spider wasp, *Agenioideus apicalis* (Hymenoptera: Pompilidae) new to Britain, and a second British record of *Agenioideus sericeus*. *British Journal of Entomology and Natural History* 31: 17–25.

Notton, D.G. and Norman, H. (2017) Hawk's-beard nomad bee, *Nomada facilis*, new to Britain (Hymenoptera: Apidae). *British Journal of Entomology and Natural History* 30: 201–214.

Nowakowski, M. and Pywell, R.F. (2016) *Habitat Creation and Management for Pollinators*. Centre for Ecology and Hydrology, Wallingford. www.ceh.ac.uk/book-habitat-creation-and-management-pollinators (accessed August 2020).

Nunes, C.E., Peñaflor, M.F., Bento, J.M., Salvador, M.J. and Sazima, M. (2016) The dilemma of being a fragrant flower: the major floral volatile attracts pollinators and florivores in the euglossine-pollinated orchid Dichaea pendula. *Oecologia* 182: 933–946.

Oelschlägel, B., Nuss, M., von Tschirnhaus, M. *et al.* (2015) The betrayed thief – the extraordinary strategy of *Aristolochia rotunda* to deceive its pollinators. *New Phytologist* 206: 342–351. doi: 10.1111/nph.13210.

Ojeda, I., Santos-Guerra, A., Jaén-Molina, R. *et al.* (2012) The origin of bird pollination in Macaronesian *Lotus* (Loteae, Leguminosae). *Molecular Phylogenetics and Evolution* 62: 306–318.

Olesen, J.M. (1985) The Macaronesian bird–flower element and its relation to bird and bee opportunists. *Botanical Journal of the Linnean Society* 91: 395–414.

Olesen, J.M. and Valido, A. (2003) Lizards as pollinators and seed dispersers: an island phenomenon. *Trends in Ecology and Evolution* 18: 177–181.

Olesen, J.M., Eskildsen, L.I. and Venkatasamy, S. (2002) Invasion of pollination networks on oceanic islands: importance of invader complexes and endemic super generalists. *Diversity and Distributions* 8: 181–192.

Olesen, J.M., Alarcón, M., Ehlers, B.K., Aldasoro, J.J. and Roquet, C. (2012) Pollination, biogeography and phylogeny of oceanic island bellflowers (Campanulaceae). *Perspectives in Plant Ecology, Evolution and Systematics* 14: 169–182.

Ollerton, J. (1993) Ecology of flowering and fruiting in *Lotus corniculatus* L. PhD thesis, Oxford Brookes University.

Ollerton, J. (1998) Sunbird surprise for syndromes. *Nature* 394: 726–727. doi: 10.1038/29409.

Ollerton, J. (1999) The evolution of pollinator–plant relationships within the arthropods. In Melic, A., DeHaro, J.J., Mendez, M. and Ribera, I. (eds), *Evolution and Phylogeny of the Arthropoda*. Entomological Society of Aragon, Zaragoza, pp. 741–758.

Ollerton, J. (2016) How to deal with bumblebees in your roof. https://jeffollerton.wordpress.com/2016/07/01/how-to-deal-with-bumblebees-in-your-roof (accessed August 2020).

Ollerton, J. (2017) Pollinator diversity: distribution, ecological function, and conservation. *Annual Review of Ecology, Evolution and Systematics* 48: 353–376.

Ollerton, J. & Cranmer, L. (2002) Latitudinal trends in plant–pollinator interactions: are tropical plants more specialised? *Oikos* 98: 340–350.

Ollerton, J. and Diaz, A. (1999) Evidence for stabilising selection acting on flowering time in *Arum maculatum* (Araceae): the influence of phylogeny on adaptation. *Oecologia* 119: 340–348.

Ollerton, J. and Liede, S. (1997) Pollination systems in the Asclepiadaceae: a survey and preliminary analysis. *Biological Journal of the Linnean Society* 62: 593–610.

Ollerton, J. and Raguso, R.A. (2006) The sweet stench of decay. *New Phytologist* 172: 382–385. doi: 10.1111/j.1469-8137.2006.01903.x.

Ollerton, J., Johnson, S.D., Cranmer, L. and Kellie, S. (2003) The pollination ecology of an assemblage of grassland asclepiads in South Africa. *Annals of Botany* 92: 807–834. doi: 10.1093/aob/mcg206.

Ollerton, J., Johnson, S.D. and Hingston, A.B. (2006) Geographical variation in diversity and specificity of pollination systems. In Waser, N.M. and Ollerton, J. (eds), *Plant–Pollinator Interactions: From Specialization to Generalization*. University of Chicago Press, Chicago, IL, pp. 283–308.

Ollerton, J., Stott, A., Allnutt, E., Shove, S., Taylor, C. and Lamborn, E. (2007a) Pollination niche overlap between a parasitic plant and its host. *Oecologia* 151: 473–485.

Ollerton, J., Grace, J. and Smith, K. (2007b) Pollinator behaviour and adaptive floral colour change in *Anthophora alluadii* (Hymenoptera: Apidae) and *Erysimum scoparium* (Brassicaceae) on Tenerife. *Entomologia Generalis* 29: 253–268.

Ollerton, J., Killick, A., Lamborn, E., Watts, S. and Whiston, M. (2007c) Multiple meanings and modes: on the many ways to be a generalist flower. *Taxon* 56: 717–728.

Ollerton, J., Cranmer, L., Stelzer, R., Sullivan, S. and Chittka, L. (2009a) Bird pollination of Canary Island endemic plants. *Naturwissenschaften* 96: 221–232.

Ollerton, J., Alarcón, R., Waser, N.M. *et al.* (2009b) A global test of the pollination syndrome hypothesis. *Annals of Botany* 103: 1471–1480. doi: 10.1093/aob/mcp031.

Ollerton, J., Masinde, S., Meve, U., Picker, M. and Whittington, A. (2009c) Fly pollination in *Ceropegia* (Apocynaceae: Asclepiadoideae): biogeographic and phylogenetic perspectives. *Annals of Botany* 103: 1501–1514. doi: 10.1093/aob/mcp072.

Ollerton, J., Tarrant, S. and Winfree, R. (2011) How many flowering plants are pollinated by animals? *Oikos* 120: 321–326. doi: 10.1111/j.1600-0706.2010.18644.x.

Ollerton, J., Price, V., Armbruster, W.S. *et al.* (2012) Overplaying the role of honey bees as pollinators: a comment on Aebi and Neumann (2011). *Trends in Ecology and Evolution* 27: 141–142.

Ollerton, J., Erenler, H., Edwards, M. and Crockett, R. (2014) Extinctions of aculeate pollinators in Britain and the role of large-scale agricultural changes. *Science* 346: 1360–1362.

Ollerton, J., Rech, A.R., Waser, N.M. and Price, M.V. (2015) Using the literature to test pollination syndromes: some methodological cautions. *Journal of Pollination Ecology* 16: 119–125.

Ollerton, J., Rouquette, J.R. and Breeze, T.D. (2016) Insect pollinators boost the market price of culturally important crops: holly, mistletoe and the spirit of Christmas. *Journal of Pollination Ecology* 19: 93–97.

Ollerton, J., Dötterl, S., Ghorpadé, K. *et al.* (2017) Diversity of Diptera families that pollinate *Ceropegia* (Apocynaceae) trap flowers: an update in light of new data and phylogenetic analyses. *Flora* 234: 233–244.

Ollerton, J., Liede-Schumann, S., Endress, M.E. *et al.* (2019a) The diversity and evolution of pollination systems in large plant clades: Apocynaceae as a case study. *Annals of Botany* 123: 311–325. doi: 10.1093/aob/mcy127.

Ollerton, J., Koju, N.P., Maharjan, S.R. and Bashyal, B. (2019b) Interactions between birds and flowers of *Rhododendron* spp., and their implications for mountain communities in Nepal. *Plants, People, Planet*. doi: 10.1002/ppp3.10091.

Orford, K.A., Vaughan, I.P. and Memmott, J. (2015) The forgotten flies: the importance of non-syrphid Diptera as pollinators. *Proceedings of the Royal Society B* 282: 20142934. doi: 10.1098/rspb.2014.2934.

Ortega-Olivencia, A., Rodríguez-Riaño, T., Valtueña, F.J., López, J. and Devesa, J.A. (2005) First confirmation of a native bird-pollinated plant in Europe. *Oikos* 110: 578–590.

Ortega-Olivencia, A., Rodríguez-Riaño, T., Pérez-Bote, J.L. *et al.* (2012) Insects, birds and lizards as pollinators of the largest-flowered *Scrophularia* of Europe and Macaronesia. *Annals of Botany* 109: 153–167. doi: 10.1093/aob/mcr255.

O'Toole, C. (2013) *Bees: A Natural History*. Firefly Press, Cardiff.

Owen, J. (2010) *Wildlife of a Garden: A Thirty-Year Study*. Royal Horticultural Society, Peterborough.

Parliamentary Office of Science and Technology (2007) Ecosystem Services. POSTnote 281 www.parliament.uk/documents/post/postpn281.pdf (accessed August 2020).

Parmesan, C., Ryrholm, N., Stefanescu, C. *et al.* (1999) Poleward shifts in geographical ranges of butterfly species associated with regional warming. *Nature* 399: 579–583.

Parrish, J.A.D. and Bazzaz, F.A. (1978) Pollination niche separation in a winter annual community. *Oecologia* 35: 133–140.

Pauw, A. (2007) Collapse of a pollination web in small conservation areas. *Ecology* 88: 1759–1769.

Pellmyr, O., Thompson, J.N., Brown, J. and Harrison, R.G. (1996) Evolution of pollination and mutualism in the yucca moth lineage. *The American Naturalist* 148: 827–847.

Peñalver, E., Labandeira, C.C., Barrón, E. *et al.* (2012) Thrips pollination of Mesozoic gymnosperms. *Proceedings of the National Academy of Sciences of the USA* 109: 8623–8628. doi: 10.1073/pnas.1120499109.

Peñalver, E., Arillo, A., Pérez-de la Fuente, R. *et al.* (2015) Long-proboscid flies as pollinators of Cretaceous gymnosperms. *Current Biology* 14: 1917–1923. doi: 10.1016/j.cub.2015.05.062.

Pérez-Barrales, R., Arroyo, J. and Armbruster, W.S. (2007) Differences in pollinator faunas may generate geographic differences in floral morphology and integration in *Narcissus papyraceus* (Amaryllidaceae). *Oikos* 116: 1904–1918.

Pérez-Barrales, R., Pino, R., Albaladejo, R.G. and Arroyo, J. (2009), Geographic variation of flower traits in *Narcissus papyraceus* (Amaryllidaceae): do pollinators matter? *Journal of Biogeography* 36: 1411–1422.

Perino, A., Pereira, H.M., Navarro, L.M. *et al.* (2019) Rewilding complex ecosystems. *Science* 364: eaav5570. doi: 10.1126/science.aav5570.

Peris, D., Pérez-de la Fuente, R., Peñalver, E. *et al.* (2017) False blister beetles and the expansion of gymnosperm-insect pollination modes before angiosperm dominance. *Current Biology* 27: 897–904. doi: 10.1016/j.cub.2017.02.009.

Peterken, G. (2019) Defining 'natural woodland'. *British Wildlife* 30: 157–159.

Phillips, B.B., Shaw, R.F., Holland, M.J. *et al.* (2018a) Drought reduces floral resources for pollinators. *Global Change Biology* 24: 3226–3235. doi: 10.1111/gcb.14130.

Phillips, B.B., Williams, A., Osborne, J.L. and Shaw, R.F. (2018b) Shared traits make flies and bees effective pollinators of oilseed rape (*Brassica napus* L.). *Basic and Applied Ecology* 32: 66–76.

Phillips, R.D., Peakall, R., van der Niet, T. and Johnson, S.D. (2020) Niche perspectives on plant–pollinator interactions. *Trends in Plant Science*. doi: 10.1016/j.tplants.2020.03.009.

Pinker, S. (2018) *Enlightenment Now: the Case for Reason, Science, Humanism, and Progress*. Penguin, London.

Pitts, J.P., Wasbauer, M.S. and von Dohlen, C.D. (2005) Preliminary morphological analysis of relationships between the spider wasp subfamilies (Hymenoptera: Pompilidae): revisiting an old problem. *Zoologica Scripta* 35: 63–84.

Polte, S. and Reinhold, K. (2013) The function of the wild carrot's dark central floret: attract, guide or deter? *Plant Species Biology* 28: 81–86.

Potts, S.G., Roberts, S.P.M., Dean, R. *et al.* (2010a) Declines of managed honey bees and beekeepers in Europe. *Journal of Apicultural Research* 49: 15–22.

Potts, S.G., Biesmeijer, J.C., Kremen, C. *et al.* (2010b) Global pollinator declines: trends, impacts and drivers. *Trends in Ecology and Evolution* 25: 345–353.

Potts, S.G., Imperatriz-Fonseca, V., Ngo, H.T. *et al.* (2016) Safeguarding pollinators and their values to human well-being. *Nature* 540: 220–229.

Potts, S.G., Neumann, P., Vaissiere, B. and Vereecken, N.J. (2018) Robotic bees for crop pollination: why drones cannot replace biodiversity. *Science of the Total Environment* 642: 665–667.

Powell, J.A. (1992) Interrelationships of yuccas and yucca moths. *Trends in Ecology and Evolution* 7: 10–15.

Powney, G.D., Harrower, C., Outhwaite, C. and Isaac, N.J.B. (2019a) UK Biodiversity Indicators 2019. D1c Status of pollinating insects. Technical background document. JNCC/Centre for Ecology and Hydrology, UK.

Powney, G.D., Carvell, C., Edwards, M. *et al.* (2019b) Widespread losses of pollinating insects in Britain. *Nature Communications* 10: 1018. doi: 10.1038/s41467-019-08974-9.

Prendergast, K. (2017) The false allure of robotic facsimiles of pollinating animals. *Australian Wildlife* 2: 17–22.

Prevéy, J.S., Rixen, R., Rüger, N. *et al.* (2019) Warming shortens flowering seasons of tundra plant communities. *Nature Ecology and Evolution* 3: 45–52. doi: 10.1038/s41559-018-0745-6.

Proctor, M. and Yeo, P. (1973) *The Pollination of Flowers*. New Naturalist 54. Collins, London.

Proctor, M., Yeo, P. and Lack, A. (1996) *The Natural History of Pollination*. New Naturalist 83. Collins, London.

Prŷs-Jones, O.E., Kristjánsson, K. and Ólafsson, E. (2016) Hitchhiking with the Vikings? The anthropogenic bumblebee fauna of Iceland – past and present. *Journal of Natural History*. 50: 45–46.

Pyke, G.H. (2016) Floral nectar: pollinator attraction or manipulation? *Trends in Ecology and Evolution* 31: 339–341.

Qiu, Y., Chen, B.J.W., Song, Y. *et al.* (2016) Composition, distribution and habitat effects of vascular plants on the vertical surfaces of an ancient city wall. *Urban Ecosystems* 19: 939–948.

Quote Investigator (2013) If the bee disappeared off the face of the earth, man would only have four years left to live. https://quoteinvestigator.com/2013/08/27/einstein-bees (accessed August 2020).

Rader, R., Bartomeus, I., Garibaldi, L.A. *et al.* (2016) Non-bee insects are important contributors to global crop pollination. *Proceedings of the National Academy of Sciences of the USA* 113: 146–151.

Rahman, Md.L. (2010) The potential of restored landfill sites for biodiversity conservation in the UK. PhD thesis, University of Northampton.

Rahman, Md.L., Tarrant, S., McCollin, D. and Ollerton, J. (2013) Plant community composition and attributes reveal conservation implications for newly created grassland on capped landfill sites. *Journal for Nature Conservation* 21: 198–205.

Ramirez, N. and Brito, Y. (1992) Pollination biology in a palm swamp community in the Venezuelan central plains. *Botanical Journal of the Linnean Society* 110: 277–302.

Ramírez, S.R., Gravendeel, B., Singer, R.B., Marshall, C.R. and Pierce, N.E. (2007) Dating the origin of the Orchidaceae from a fossil orchid with its pollinator. *Nature* 448: 1042–1045.

Rasmont, P. (1989) Espèces de bourdons en expansion en Belgique (Hymenoptera, Apidae). *Notes Fauniques de Gembloux* 18: 57–64.

Rasmont, P., Franzén, M., Lecocq, T. *et al.* (2015) *Climatic Risk and Distribution Atlas of European Bumblebees. BioRisk* 10 (special issue), 246 pp.

Ravoet, J., De Smet, L., Meeus, I. *et al.* (2014) Widespread occurrence of honey bee pathogens in solitary bees. *Journal of Invertebrate Pathology* 122: 55–58.

Rech, A.R., Dalsgaard, B., Sandel, B. *et al.* (2016) The macroecology of animal versus wind pollination: ecological factors are more important than historical climate stability. *Plant Ecology and Diversity* 9: 253–262.

Regan, E.C., Santini, L., Ingwall-King, L. *et al.* (2015) Global trends in the status of bird and mammal pollinators. *Conservation Letters* 8: 397–403. doi: 10.1111/conl.12162.

Reilly, J.R., Artz, D.R., Biddinger, D. *et al.* (2020) Crop production in the USA is frequently limited by a lack of pollinators. *Proceedings of the Royal Society B* 287: 20200922. doi: 10.1098/rspb.2020.0922.

Remsen, J.V., Stiles, F.G. and Scott, P.E. (1986) Frequency of arthropods in stomachs of tropical hummingbirds. *The Auk* 103: 436–441.

Ren, D. (1998) Flower-associated Brachycera flies as fossil evidence for Jurassic angiosperm origins. *Science* 280: 85–88.

Ren, D., Labandeira, C.C., Santiago-Blay, J.A. *et al.* (2009) A probable pollination mode before angiosperms: Eurasian, long-proboscid scorpionflies. *Science* 326: 840–847.

Richards, A.J. (1997) *Plant Breeding Systems*. Chapman & Hall, London.

Rivers, M.C., Beech, E., Bazos, I. *et al.* (2019) *European Red List of Trees*. IUCN, Cambridge and Brussels.

Robertson, C. (1928) *Flowers and Insects: Lists of Visitors of Four Hundred and Fifty-three Flowers*. Privately published, Carlinville, IL. www.biodiversitylibrary.org/item/43820 (accessed August 2020).

Rollings, R. and Goulson, D. (2019) Quantifying the attractiveness of garden flowers for pollinators. *Journal of Insect Conservation* 23: 803–817. doi: 10.1007/s10841-019-00177-3.

Rosas-Guerrero, V., Aguilar, R., Martén-Rodríguez, S. *et al.* (2014) A quantitative review of pollination syndromes: do floral traits predict effective pollinators? *Ecology Letters* 17: 388–400. doi: 10.1111/ele.12224.

Roubik, D.W. (1978) Competitive interactions between neotropical pollinators and Africanized honey bees. *Science* 201: 1030–1032.

Rouquette, J. (2016) *Natural Capital and Ecosystem Services in the Nene Valley: Mapping and Valuation*. Natural Capital Solutions, Northampton. www.naturalcapitalsolutions.co.uk/previous-projects/case-study-2 (accessed August 2020).

Rudall, P.J., Alves, M. and Sajo, Md.G. (2016) Inside-out flowers of *Lacandonia brasiliana* (Triuridaceae) provide new insights into fundamental aspects of floral patterning. *PeerJ* 4: e1653. doi: 10.7717/peerj.1653.

Ruggiero, M., Ascher, J. *et al.* (2020) ITIS bees: ITIS world bee checklist (version Sep 2009). In Roskov, Y., Ower, G., Orrell, T. *et al.* (eds), *Species 2000 & ITIS Catalogue of Life*, 2020-06-04 Beta. Species 2000: Naturalis, Leiden. www.catalogueoflife.org/col (accessed August 2010).

Russo, L. (2016) Positive and negative impacts of non-native bee species around the world. *Insects* 7 (4): 69. doi: 10.3390/insects7040069.

Sakai, K. and Nagai, S. (1998) *The Cetoniine Beetles of the World*. Mushi-Sha, Tokyo.

Salisbury, A., Armitage, J., Bostock, H. *et al.* (2015a) Enhancing gardens as habitats for flower-visiting aerial insects (pollinators): should we plant native or exotic species? *Journal of Applied Ecology* 52: 1156–1164. doi: 10.1111/1365-2664.12499.

Salisbury, A., Malumphy, C. and Barclay, M. (2015b) Recent records of Capricorn beetles *Cerambyx* spp. in England and Wales (Cerambycidae). *The Coleopterist* 24: 40–44.

Samways, M.J. (2015) Future-proofing insect diversity. *Current Opinion in Insect Science* 12: 71–78. doi: 10.1016/j.cois.2015.09.008.

Sandom, C.J., Ejrnæs, R., Hansen, M.D.D. and Svenning, J.-C. (2014) High herbivore density associated with vegetation diversity in interglacial ecosystems. *Proceedings of the National Academy of Sciences of the USA* 111: 4162–4167. doi: 10.1073/pnas.1311014111.

Santos, A.A., Leijs, R., Picanço, M.C., Glatz, R. and Hogendoorn, K. (2020) Modelling the climate suitability of green carpenter bee (*Xylocopa aerata*) and its nesting hosts under current and future scenarios to guide conservation efforts. *Austral Ecology* 45: 271–282.

Sarma, K., Tandon, R., Shivanna, K.R. and Ram, M.H.Y. (2007) Snail-pollination in *Volvulopsis nummularium*. *Current Science* 93: 826–831.

Saunders, M.E. Janes, J.K. and O'Hanlon, J.C. (2020) Moving on from the insect apocalypse narrative: engaging with evidence-based insect conservation. *BioScience* 70: 80–89.

Schleuning, M., Fründ, J., Klein, A.-M. *et al.* (2012) Specialization of mutualistic interaction networks decreases toward tropical latitudes. *Current Biology* 22: 1925–1931. doi: 10.1016/j.cub.2012.08.015.

Schmid-Hempel, R., Eckhardt, M., Goulson, D. *et al.* (2014) The invasion of southern South America by imported bumblebees and associated parasites. *Journal of Animal Ecology* 83: 823–837. doi: 10.1111/1365-2656.12185.

Schmidt, K. and Westrich, P. (1993) *Colletes hederae* n. sp., eine bisher unerkannte, auf Efeu (*Hedera*) spezialisierte Bienenart (Hymenoptera: Apoidea). *Entomologische Zeitschrift* 103: 89–112.

Scott-Brown, A.S., Arnold, S.E.J., Kite, G.C. *et al.* (2019) Mechanisms in mutualisms: a chemically mediated thrips pollination strategy in common elder. *Planta* 250: 367–379.

Seeley, T.D. (2019) *The Lives of Bees*. Princeton University Press, Princeton, NJ.

Senapathi, D., Carvalheiro, L.G., Biesmeijer, J.C. *et al.* (2015) The impact of over 80 years of land cover changes on bee and wasp pollinator communities in England. *Proceedings of the Royal Society B* 282: 20150294. doi: 10.1098/rspb.2015.0294.

Settele, J., Bishop, J. and Potts, S.G. (2016) Climate change impacts on pollination. *Nature Plants* 2: 16092.

Shanahan, M. (2016) *Ladders to Heaven*. Unbound, UK.

Shuttleworth, A. and Johnson, S.D. (2008) Bimodal pollination by wasps and beetles in the African milkweed *Xysmalobium undulatum*. *Biotropica* 40: 568–574.

Shuttleworth, A. and Johnson, S.D. (2009a) New records of insect pollinators for South African asclepiads (Apocynaceae: Asclepiadoideae). *South African Journal of Botany* 75: 689–698.

Shuttleworth, A. and Johnson, S.D. (2009b) A key role for floral scent in a wasp-pollination system in *Eucomis* (Hyacinthaceae), *Annals of Botany* 103: 715–725.

Singh, A. (2011) Observations on the vascular wall flora of Banaras Hindu University Campus, India. *Bulletin of Environment, Pharmacology and Life Sciences* 1: 33–39.

Sirohi, M.H., Jackson, J., Edwards, M. and Ollerton, J. (2015) Diversity and abundance of solitary and primitively eusocial bees in an urban centre: a case study from Northampton (England). *Journal of Insect Conservation* 19: 487–500.

Siviter, H., Horner, J., Brown, M.J.F. and Leadbeater, E. (2020) Sulfoxaflor exposure reduces egg laying in bumblebees *Bombus terrestris*. *Journal of Applied Ecology* 57: 160–169. doi: 10.1111/1365-2664.13519.

Smith, M.R., Singh, G.M., Mozaffarian, D. and Myers, S.S. (2015) Effects of decreases of animal pollinators on human nutrition and global health: a modelling analysis. *Lancet* 386: 1964–1972. doi: 10.1016/S0140-6736(15)61085-6.

Soga, M., Gaston, K.J. and Yamaura, Y. (2017) Gardening is beneficial for health: a meta-analysis. *Preventive Medicine Reports* 5: 92–99.

Sonne, J., Kyvsgaard, P., Maruyama, P.K. *et al.* (2016) Spatial effects of artificial feeders on hummingbird abundance, floral visitation and pollen deposition. *Journal of Ornithology* 157: 573–581.

Soroye, P., Newbold, T. and Kerr, J. (2020) Climate change contributes to widespread declines among bumble bees across continents. *Science* 367: 685–688. [See also the commentary by Bridle and van Rensburg, pp. 626–627 of the same issue.]

Sponsler, D.B., Grozinger, C.M., Hitaj, C. *et al.* (2019) Pesticides and pollinators: a socioecological synthesis. *Science of the Total Environment* 662: 1012–1027. doi: 10.1016/j.scitotenv.2019.01.016.

Stanley, D., Garratt, M., Wickens, J. *et al.* (2015a) Neonicotinoid pesticide exposure impairs crop pollination services provided by bumblebees. *Nature* 528: 548–550.

Stanley, D., Smith, K. and Raine, N. (2015b) Bumblebee learning and memory is impaired by chronic exposure to a neonicotinoid pesticide. *Scientific Reports* 5: 16508. doi: 10.1038/srep16508.

Steiner, K.E. (1998) Beetle pollination of peacock moraeas (Iridaceae) in South Africa. *Plant Systematics and Evolution* 209: 47–65.

Stone, G., Willmer, P. and Nee, S. (1996) Daily partitioning of pollinators in an African *Acacia* community. *Proceedings of the Royal Society of London B* 263: 1389–1393. doi: 10.1098/rspb.1996.0203.

Stone, G.N., Willmer, P. and Rowe, J.A. (1998) Partitioning of pollinators during flowering in an African *Acacia* community. *Ecology* 79: 2808–2827.

Tarrant, S. (2009) The potential of restored landfill sites to support pollinating insects. PhD thesis, University of Northampton.

Tarrant, S., Ollerton, J., Rahman, Md.L., Tarrant, J. and McCollin, D. (2013) Grassland restoration on landfill sites in the East Midlands, UK: an evaluation of floral resources and pollinating insects. *Restoration Ecology* 21: 560–568.

Thebo, A.L., Drechsel, P. and Lambin, E.F. (2014) Global assessment of urban and peri-urban agriculture: irrigated and rainfed croplands. *Environmental Research Letters* 9: 114002. doi: 10.1088/1748-9326/9/11/114002.

Thomas, J.A. (1995) The ecology and conservation of *Maculinea arion* and other European species of large blue butterfly. In Pullin, A.S. (ed.), *Ecology and Conservation of Butterflies*. Springer, Dordrecht, pp. 180–197.

Thomsen, P.F. and Willerslev, E. (2015) Environmental DNA: an emerging tool in conservation for monitoring past and present biodiversity. *Biological Conservation* 183: 4–18. doi: 10.1016/j.biocon.2014.11.019.

Tiusanen, M., Hebert, P.D.N., Schmidt, N.M. and Roslin, T. (2016) One fly to rule them all: muscid flies are the key pollinators in the Arctic. *Proceedings of the Royal Society B* 283: 20161271. doi: 10.1098/rspb.2016.1271.

Toledo-Hernández, M., Wanger, T.C. and Tscharntke, T. (2017) Neglected pollinators: can enhanced pollination services improve cocoa yields? A review. *Agriculture, Ecosystems and Environment* 247: 137–148.

Tremblay, R.L. (1992) Trends in the pollination ecology of the Orchidaceae: evolution and systematics. *Canadian Journal of Botany* 70: 642–650.

Valido, A., Dupont, Y.L. and Hansen, D.M. (2002) Native birds and insects, and introduced honeybees visiting *Echium wildpretii* (Boraginaceae) in the Canary Islands. *Acta Oecologica* 23: 413–419.

Valido, A., Dupont, Y.L. and Olesen, J.M. (2004) Bird–flower interactions in the Macaronesian islands. *Journal of Biogeography* 31: 1945–1953.

Valido, A., Rodríguez-Rodríguez, M.C. and Jordano, P. (2019) Honeybees disrupt the structure and functionality of plant–pollinator networks. *Scientific Reports* 9: 4711. doi: 10.1038/s41598-019-41271-5.

Vamosi, J.C. and Vamosi, S.M. (2010) Key innovations within a geographical context in flowering plants: towards resolving Darwin's abominable mystery. *Ecology Letters* 13: 1270–1279.

van der Niet, T. and Johnson, S.D. (2012) Phylogenetic evidence for pollinator-driven diversification of angiosperms. *Trends in Ecology and Evolution* 27: 353–361.

van der Niet, T., Peakall, R. and Johnson, S.D. (2014) Pollinator-driven ecological speciation in plants: new evidence and future perspectives. *Annals of Botany* 113: 199–212. doi: 10.1093/aob/mct290.

van der Wal, R., Anderson, H., Robinson, A. *et al.* (2015) Mapping species distributions: a comparison of skilled naturalist and lay citizen science recording. *Ambio* 44 (Suppl. 4): 584–600. doi: 10.1007/s13280-015-0709-x.

van Klink, R., Bowler, D.E., Gongalsky, K.B. *et al.* (2020) Meta-analysis reveals declines in terrestrial but increases in freshwater insect abundances. *Science* 368: 417–420.

van Tussenbroek, B.I., Villami, N., Márquez-Guzmán, J. *et al.* (2016) Experimental evidence of pollination in marine flowers by invertebrate fauna. *Nature Communications* 7: 12980. doi: 10.1038/ncomms12980.

Vanbergen, A.J., Heard, M.S., Breeze, T., Potts, S.G. and Hanley, N. (2014) Status and value of pollinators and pollination services. A report for the Department for Environment, Food and Rural Affairs (Defra UK).

Varatharajan, R., Maisnam, S., Shimray, C.V. and Rachana, R.R. (2016) Pollination potential of thrips (Insecta: Thysanoptera): an overview. *Zoo's Print* 31: 6–12.

Vereecken, N.J., Streinzer, M., Ayasse, M. *et al.* (2011), Integrating past and present studies on *Ophrys* pollination – a comment on Bradshaw *et al. Botanical Journal of the Linnean Society*, 165: 329–335. doi: 10.1111/j.1095-8339.2011.01112.x.

Vizentin-Bugoni, J., Maruyama, P.K.M., Souza, C.S. *et al.* (2018) Plant–pollinator networks in the tropics: a review. In Dáttilo, W. and Rico-Gray, V. (eds), *Ecological Networks in the Tropics.* Springer, Cham, pp. 73–91.

Vlasáková, B., Pinc, J., Jůna, F. and Varadínová, Z.K. (2019) Pollination efficiency of cockroaches and other floral visitors of *Clusia blattophila. Plant Biology* 21: 753–761.

Vogel, S. (1954) Blütenbiologische Typen als Elemente der Sippengliederung. *Botanische Studien (Jena)* 1: 1–338.

Vogel, S. (1961) Die Bestäubung der Kesselfallen-Blüten von *Ceropegia. Beiträge zur Biologie der Pflanzen* 36: 159–237.

Vogel, S. (1990) *The Role of Scent Glands in Pollination: On the Structure and Function of Osmophores.* Routledge, London.

Vogel, S., Westerkamp, C., Thiel, B. and Gessner, K. (1984) Ornithophilie auf den Canarischen Inseln. *Plant Systematics and Evolution* 146: 225–248.

Walsh, S.K., Pender, R.J., Junker, R.R. *et al.* (2019) Pollination biology reveals challenges to restoring populations of *Brighamia insignis* (Campanulaceae), a critically endangered plant species from Hawai'i. *Flora* 259: 151448. doi: 10.1016/j.flora.2019.151448.

Wardhaugh, C.W. (2015) How many species of arthropods visit flowers? *Arthropod–Plant Interactions* 9: 547–565. doi: 10.1007/s11829-015-9398-4.

Warren, M. (1999) Poleward shifts in geographical ranges of butterfly species associated with regional warming. *Nature* 399: 579–583.

Waser, N.M. and Price, M.V. (1981) Pollinator choice and stabilizing selection for flower color in *Delphinium nelsonii. Evolution* 35: 376–390.

Waser, N.M. and Price, M.V. (1990) Pollination efficiency and effectiveness of bumble bees and hummingbirds visiting *Delphinium nelsonii. Collectanea Botanica (Barcelona)* 19: 9–20.

Waser, N.M. and Price, M.V. (1991) Outcrossing distance effects in *Delphinium nelsonii*: pollen loads, pollen tubes, and seed set. *Ecology* 72: 171–179.

Waser, N.M. and Ollerton, J. (eds) (2006) *Plant–Pollinator Interactions: from Specialization to Generalization*. University of Chicago Press, Chicago, IL.
Waser, N.M., Chittka, L., Price, M.V., Williams, N.M. and Ollerton, J. (1996) Generalization in pollination systems, and why it matters. *Ecology* 77: 1043–1060.
Waser, N.M., Ollerton, J. and Erhardt, A. (2011) Typology in pollination biology: lessons from an historical critique. *Journal of Pollination Ecology* 3: 1–7.
Waser, N.M., Ollerton, J. and Price, M.V. (2015) Response to Aguilar *et al.*'s (2015) critique of Ollerton *et al.* (2009). *Journal of Pollination Ecology* 17: 1–2.
Watts, S., Huamán Ovalle, D., Moreno Herrera, M. and Ollerton, J. (2012) Pollinator effectiveness of native and non-native flower visitors to an apparently generalist Andean shrub, *Duranta mandonii* (Verbenaceae). *Plant Species Biology* 27: 147–158.
Watts, S., Dormann, C.F., Martín González, A.M. and Ollerton, J. (2016) The influence of floral traits on specialisation and modularity of plant–pollinator networks in a biodiversity hotspot in the Peruvian Andes. *Annals of Botany* 118: 415–429. doi: 10.1093/aob/mcw114.
Weiss, M.R. and Lamont, B.B. (1997) Floral colour change and insect pollination: a dynamic relationship. *Israel Journal of Plant Sciences* 45: 185–199.
Weissmann, J.A., Picanço, A., Borges, P.A.V. and Schaefer, H. (2017) Bees of the Azores: an annotated checklist (Apidae, Hymenoptera). *ZooKeys* 642: 63–95. doi: 10.3897/zookeys.642.10773.
Wenzel, A., Grass, I., Belavadi, V.V. and Tscharntke, T. (2020) How urbanization is driving pollinator diversity and pollination – a systematic review. *Biological Conservation* 241: 108321. doi: 10.1016/j.biocon.2019.108321.
Wester, P. (2015) The forgotten pollinators: first field evidence for nectar-feeding by primarily insectivorous elephant-shrews. *Journal of Pollination Ecology* 16: 108–111.
Wester, P. (2019) First observations of nectar-drinking lizards on the African mainland. *Plant Ecology and Evolution*, 152: 78–83. doi: 10.5091/plecevo.2019.1513.
Westerfelt, P., Widenfalk, O., Lindelöw, Å., Gustafsson, L. and Weslien, J. (2015) Nesting of solitary wasps and bees in natural and artificial holes in dead wood in young boreal forest stands. *Insect Conservation and Diversity* 8: 493–504.
Wheeler, J. (2018) Tropical milkweed – a no-grow. Xerces Society blog. https://xerces.org/blog/tropical-milkweed-a-no-grow (accessed August 2020).
Wilfert, L., Long, G., Leggett, H.C. *et al.* (2016) Deformed wing virus is a recent global epidemic in honeybees driven by *Varroa* mites. *Science* 351: 594–597.
Williams, P.H. and Osborne, J.L. (2009) Bumblebee vulnerability and conservation world-wide. *Apidologie* 40: 367–387.
Willis, J.C. and Burkill, I.H. (1895) Flowers and insects in Great Britain. Part I. *Annals of Botany* 9: 227–273.
Willis, J.C. and Burkill, I.H. (1903a) Flowers and insects in Great Britain. Part II. *Annals of Botany* 17: 313–349.
Willis, J.C. and Burkill, I.H. (1903b) Flowers and insects in Great Britain. Part III. *Annals of Botany* 17: 539–570.

Willis, J.C. and Burkill, I.H. (1908) Flowers and insects in Great Britain. Part IV. *Annals of Botany* 22: 603–649.

Willmer, P. (2011) *Pollination and Floral Ecology*. Princeton University Press, Princeton, NJ.

Wilson, G.F. (1946) Recent developments in garden pest control. *Journal of the Royal Horticultural Society* 69: 334–343.

Wilson, J.S., Forister, M.L. and Carril, O.M. (2017) Interest exceeds understanding in public support of bee conservation. *Frontiers in Ecology and the Environment* 15: 460–466.

Wilson, P., Castellanos, M.C., Wolfe, A.D. and Thomson, J.D. (2006) Shifts between bee and bird pollination in penstemons. In Waser, N.M. and Ollerton, J. (eds), *Plant–Pollinator Interactions: From Specialization to Generalization*. University of Chicago Press, Chicago, IL, pp. 47–68.

Wood, C.J., Pretty, J. and Griffin, M. (2016) A case–control study of the health and well-being benefits of allotment gardening. *Journal of Public Health* 38: e336–e344. doi: 10.1093/pubmed/fdv146.

Woodcock, B., Isaac, N., Bullock, J. *et al.* (2016) Impacts of neonicotinoid use on long-term population changes in wild bees in England. *Nature Communications* 7: 12459. doi: 10.1038/ncomms12459.

Wright, G., Softley, S. and Earnshaw, H. (2015) Low doses of neonicotinoid pesticides in food rewards impair short-term olfactory memory in foraging-age honeybees. *Scientific Reports* 5: 15322. doi: 10.1038/srep15322.

Wulf, A. (2015) *The Invention of Nature: the Adventures of Alexander von Humboldt, the Lost Hero of Science*. John Murray, London.

Wyns, D. (2014) Hybrid carrot seed pollination. Bee Informed. https://beeinformed.org/2014/09/05/hybrid-carrot-seed-pollination-2 (accessed August 2020).

Xie, Z., Williams, P.H. and Tang, Y. (2008) The effect of grazing on bumblebees in the high rangelands of the eastern Tibetan Plateau of Sichuan. *Journal of Insect Conservation* 12: 695–703.

Xiong, W., Ollerton, J., Liede-Schumann, S. *et al.* (in press) Specialized cockroach pollination in the rare and endangered plant *Vincetoxicum hainanense* (Apocynaceae, Asclepiadoideae) in China. *American Journal of Botany*.

Yañez, O., Piot, N., Dalmon, A. *et al.* (2020) Bee viruses: routes of infection in Hymenoptera. *Frontiers in Microbiology* 11: 943. doi: 10.3389/fmicb.2020.00943.

Yang, L.-Y., Machado, C.A., Dang, X.-D. *et al.* (2015) The incidence and pattern of co-pollinator diversification in dioecious and monoecious figs. *Evolution* 69: 294–304. doi: 10.1111/evo.12584.

Yang, X. and Miyako, E. (2020) Soap bubble pollination. *iScience* 23: 101188. doi: 10.1016/j.isci.2020.101188.

Yoder, J.B., Smith, C.I. and Pellmyr, O. (2010) How to become a yucca moth: minimal trait evolution needed to establish the obligate pollination mutualism. *Biological Journal of the Linnean Society* 100: 847–855. doi: 10.1111/j.1095-8312.2010.01478.x.

Zomlefer, W.B. and Giannasi, D.E. (2005) Floristic survey of Castillo de San Marcos National Monument, St. Augustine, Florida. *Castanea* 70: 222–236.

Zurcher, A. (2019) The butterflies that could stop Trump's wall. BBC News. www.bbc.co.uk/news/world-us-canada-47736573 (accessed August 2020).

Index